This book is timely and compelling in addressing alignments of governance and climate change mitigation in the urban environment in China. [. . .] The analysis as a whole, with arenas appreciated and in depth city assessments addressing contexts, dynamics and challenges in sectors so pertinent to climate change mitigation, neatly threads together theory and practice in making sense of urban environmental governance. In doing so, it is to be commended as adding considerably to a vital and ever expanding body of literature and knowledge.

Ian Thynne, Australian National University and University of Hong Kong

The two authors make a serious scholarly and inspiring effort to advance the theoretical frontier and enrich knowledge by examining this quickly emerging topic of environmental protection in China from a governance perspective. . . . their study has effectively achieved the objective of advancing a proper understanding of the governance dynamics and challenges of urban climate mitigation in Chinese cities, making an important contribution to the theoretical inquiry of environmental governance. This book will inspire further theoretical and empirical research efforts on climate change governance in general and in China in particular.

Carlos Wing-Hung Lo, The Hong Kong Polytechnic University

Climate Change Governance in Chinese Cities

In the last thirty years, China has experienced rapid economic development and urbanisation, which has resulted in high levels of environmental degradation and has put considerable pressure on the country's infrastructure and natural resources. As China commits to considerably lower the carbon intensity of its economy, this volume analyses and explains the governance of climate change mitigation responses in major Chinese cities.

The book focuses specifically on two highly carbon intensive sectors, buildings and transport, in Guangzhou, Shenzhen and Hong Kong to explore how collaborative municipal networks function in practice in Chinese cities. The authors find that effective coordination relies on the political will of local administrative elites, the political significance attached to climate change issues, the legitimate authority granted to the coordinating agency, and human and financial capitals. Collaboration is hampered by limited span of network engagement, inadequate authority of the primary network participants, insufficient input and output legitimacy of the sectoral innovations, and missing linkages across functionally segregated sectors. The book concludes that the enhanced collaboration and coordination between networks that has emerged in the process of low carbon transitions is transforming the Chinese environmental state into a more pluralistic, inclusive and legitimate one.

This book will be of interest to researchers and practitioners across disciplines including Chinese studies, environmental politics and policy, urban studies, and planning and geography.

Qianqing Mai is a Researcher in the Department of Public Policy, City University of Hong Kong.

Maria Francesch-Huidobro is Assistant Professor, Department of Public Policy, City University of Hong Kong, and Honorary Assistant Professor of the Kadoorie Institute, The University of Hong Kong.

Routledge studies in Asia and the environment

The role of Asia will be crucial in tackling the world's environmental problems. The primary aim of this series is to publish original, high quality, research level work by scholars in both the East and the West on all aspects of Asia and the environment. The series aims to cover all aspects of environmental issues, including how they relate to economic development, sustainability, technology, society and government policies, and to include all regions of Asia.

1 **Climate Change Governance in Chinese Cities**
Qianqing Mai and Maria Francesch-Huidobro

Upcoming titles:

2 **Deliberating Environmental Policy in India**
Participation and the Role of Advocacy
Sunayana Ganguly

3 **Sustainability in Contemporary Rural Japan**
Challenges and Opportunities
Edited by Stephanie Assmann

4 **Borneo and the Environment**
Challenges and Change
Peter Eaton and Mark Cleary

Climate Change Governance in Chinese Cities

Qianqing MAI
and
Maria FRANCESCH-HUIDOBRO

Routledge
Taylor & Francis Group

LONDON AND NEW YORK

First published 2015 by Routledge

2 Park Square, Milton Park, Abingdon, Oxfordshire OX14 4RN
711 Third Avenue, New York, NY 10017

Routledge is an imprint of the Taylor & Francis Group, an informa business

First issued in paperback 2017

British Library Cataloguing in Publication Data
A catalogue record for this book is available from the British Library

Library of Congress Cataloging-in-Publication Data
Mai, Qianqinq.
Climate change governance in Chinese cities / Qianqinq Mai and Maria
    Francesch-Huidobro.
        pages cm. — (Routledge studies in Asia and the environment)
    Includes bibliographical references and index.
    1. Climatic changes—Government policy—China.   2. Climatic
changes—China.   3. Climate change mitigation—China.   I. Francesch-
Huidobro, Maria.   II. Title.
    QC903.2.C5M35 2015
    363.738'74560951—dc23
    2014022353

ISBN: 978-1-138-78542-7 (hbk)
ISBN: 978-0-8153-6706-2 (pbk)

Typeset in Times New Roman
by Apex CoVantage, LLC

To my family and friends
(Qianqing Mai)

To my students
(Maria Francesch-Huidobro)

Contents

PART III
Challenges of climate change governance in China   221

Figures

Tables

Foreword

The study of environmental governance in China is both discouraging and inspiring. Discouragement comes from the growing pain of observing that the dominant ideology in Chinese governments at both the Central and local levels manifests the persisting conflicts between environmental protection and economic growth under the national strategy of pursuing sustainable development. Take a glance at the landscape of environmental governance in today's China after decades of the nationwide effort in institution and capacity building for green evolution. Politically, the bureaucratic status of environmental protection authorities continues to be weak vis-à-vis the commissions and governmental agencies in charge of planning and economic development. Socially, green non-environmental protection organisations remain marginal in the shaping of environmental policies and the cultivation of green communities. Economically, local governments keep on providing polluting businesses with investment opportunities to enable them to achieve their economic priorities. All these combine to constrain effective implementation of pollution control policies and regulations, making it difficult to close the gap between legislative intent and regulatory reality in the environmental legal regime. Looming over this dark side of the picture are that the progressive initiative of introducing the green GDP has been shelved, environmental protection bureaus have been denied active allies to support their environmental regulatory endeavors, and industrial pollution control focus has been limited to securing minimum compliance from polluting enterprises.

Despite all this cause for discouragement, there are sources of inspiration. The green development hypothesis finds increasing empirical support in the Chinese context for the notion that a polluting development path will gradually give way to a greener trajectory once the nation has achieved economic take-off. In particular, environmental governance as an evolving concept finds theoretical and practical enrichment in this third world and developing country's quest for socialist modernisation. It allows the adoption of contending approaches of environmental governance with room for a high degree of originality to examine governmental efforts dealing with challenging ecological and pollution issues and the problems that emerge out of rapid industrialisation and urbanisation. Different stages of economic development, geographical diversity, and the growing complexity of pollution structures arouse inspiration and pose exciting intellectual challenges.

Central to this new wave of intellectual effort is the rapid growth of academic research on environmental governance in China at the national and local levels, resulting in contrasting theoretical perspectives to test the validity of a host of newly formulated yet insightful hypotheses and propositions through the application of various research methods in different domains of environmental protection. The outcome is the quick replacement of paradigms of environmental governance, facilitating hundreds of contending schools of thought. This process will continue and flourish as China's effort to improve and rationalise its environmental governance at all levels progresses with periodical refinement of its development strategy through the growing incorporation of ecologically friendly components.

Given this research background, this analysis of case studies in three different Chinese cities to mitigate the disturbance of climate changes is particularly significant and timely: the two authors make a serious scholarly and inspiring effort to advance theoretical frontiers and enrich knowledge by examining this rapidly emerging topic in environmental protection in China from a governance perspective. Trained in political science and public administration, Maria Francesch-Huidobro and Qianqing Mai have displayed a commendable degree of academic sensitivity to environmental politics by undertaking an organised and systematic investigation into the growing governmental concern over climate change, which recently became a strategic focus of China's environmental governance. At the Asia-Pacific Economic Cooperation Economic Leaders Meeting in 2009, Premier Wen Jiabao advanced the strategic idea of building a "low carbon economy" to exercise active environmental stewardship in response to demands from the international community that China reduce the ecological footprint of its fast-pace economic growth.

Francesch-Huidobro and Mai's study is inspirational since it demonstrates a high degree of originality in three aspects. Firstly, they aptly captured the dynamic of climate change as a transboundary and cross-sectoral environmental governance issue in domestic politics at the city level in defining the core research questions for their study. Secondly, they made a respectful scholarly effort to formulate collaborative municipal networks (CMNs) as the theoretical framework to organise their cross-sectoral examination of climate change governance. They attempted to tackle the governance puzzles of "how intragovernmental mechanisms facilitate the coordination of climate change issues, and whether they can overcome functional fragmentation to achieve horizontal synergy". Thirdly, the research design is unconventional in selecting Hong Kong, alongside Guangzhou and Shenzhen, in their study of climate change governance in Chinese cities. This posed a number of theoretical challenges in cross-city analysis, given their respective unique constitutional, political and economic characteristics. In addition, transportation and building as the two energy end-use sectors are properly selected as the contexts for CMNs analysis, as both are major targets of climate change mitigation that have undergone a greening process involving a complex network of political and policy actors at the local, national and international levels.

The findings of in depth analyses of the use of green buildings and electric vehicles to mitigate climate change in these three cities are illuminating but less encouraging. The trend of social technical shift in these two sectors has exposed a limited government capacity, leading to the growing dynamic of climate mitigation governance with expanding space for social and business involvement. However, the future of cross-sectoral collaboration is uncertain because of its limited practice at this stage, which has undermined its legitimacy and engagement. In sum, this study has effectively achieved the objective of advancing a proper understanding of the governance dynamics and challenges of urban climate mitigation in Chinese cities, making an important contribution to the theoretical inquiry into environmental governance. This book will inspire further theoretical and empirical research efforts on climate change governance in general and in China in particular.

Carlos Wing-Hung Lo
Professor and Head
Department of Management and Marketing
The Hong Kong Polytechnic University
Hong Kong, August 2014

Foreword

This book is timely and compelling in addressing alignments of governance and climate change mitigation in the urban environment in China. There is a keen appreciation of governance as a multi-dimensional phenomenon, with the focus on climate change and strategies of mitigation providing considerable scope to explore the coalescence of people, place, function and action in government and governance. Of underlying significance are power, politics and legitimacy in various intersections of the state, market and civil society.

The analysis has conceptual and empirical relevance in three interlocking arenas, thus ensuring its appeal to a wide range of readers. The first arena is global, the second national comprising urban China, and the third the particular Chinese cities selected for comprehensive assessment. All three arenas are astutely recognised as comprising complex contextual forces and demands, coupled with the dynamics of political and institutional manoeuvring and the challenges inevitably inherent in policy design, action and evaluation concerning climate change and its many consequences. The result is a rich array of information, insights and lessons of considerable academic and practical significance.

The global arena of cities confronting and seeking to cope with climate change embodies critical issues about the bases, control and accountability of public and private power in global governance. As the analysis appreciates, opportunities abound, while challenges remain daunting, concerning informed policy learning, adaptation and review from city to city across state borders. Especially pertinent are ideas and actions about the use and efficacy of various mixes of coercive, contractual and consensual power in environmental governance. These matters highlight the ongoing need for sharpened governance capacities of contextual appreciation, institution building and legitimation in, and well beyond, urban planning and environmental policy and politics.

The national arena sees cities throughout China facing enormous interlocking challenges of massive urban growth, sprawl and environmental degradation. The analysis acknowledges that context obviously matters from city to city, without diminishing the overarching role and power of the political regime. There are numerous drivers and restrainers of responsible change and reform in environmental management. Essential responses, all of which pose significant challenges, include the need for much more effort and innovation in transforming key institutions, in harnessing private resources for public good, and in fostering appropriate forms of city–civil society collaboration.

The local arena comprises three cities in southern China: Guangzhou, Shenzhen and Hong Kong. The detailed analysis of their environmental management networks and experience in the transportation and building sectors is rightly context-specific while also cognisant of similarities. Electric vehicles and green buildings are sound targets of attention, especially given their immediate relevance to the reduction of carbon emissions. There are important lessons concerning the significance of political and economic incentives, the dispersal of planning authority, the utilisation of land and other resources, and the limits to innovative capacity – all of which are affected by, while at the same time affecting, networks of people and power.

The analysis as a whole, with arenas appreciated and in depth city assessments addressing contexts, dynamics and challenges in sectors so pertinent to climate change mitigation, neatly threads together theory and practice in making sense of urban environmental governance. In doing so, it is to be commended as adding considerably to a vital and ever-expanding body of literature and knowledge.

Ian Thynne
Australian National University and
University of Hong Kong
Canberra, August 2014

Preface

Urban climate governance is one of the least studied issues in contemporary political science and policy studies in China. Although commonly used and analysed, the concept of governance and its understanding in the urban politics research area remains not only imperfectly understood but also inadequately applied to geopolitical contexts outside the Global North and to concrete policy problems. This book is the first attempt to fill in this dearth of knowledge. It establishes a theoretical dialogue with the governance literature and carefully conceptualises, contextualises, and explains the dynamics and challenges of the governance of climate change in Chinese cities. This has enabled us to come up with a conceptual framework to explain contemporary urban climate governance practices and changes and what, how and why these changes occur and identify future research and practice prospects. What we have tried to do in this book is to look at what goes on beyond the radar of the official accounts on climate mitigation policies and the prevailing model-based, quantitative-orientated studies of climate change in China.

The two of us began this collaboration in 2010 when we were discussing common research interests, especially on the way governance was conceptualised and understood across the discipline of political science and the research areas of urban and China studies. Through the past four years, we have spent time in the field and at conferences and have collaborated in publications. In 2013, we spent a month attached to Humboldt University of Berlin. There, away from regular teaching and research commitments, we had time to discuss and outline the connections of our research and to come up with a book proposal and a couple of draft chapters, which we presented to Routledge in August. The proposal and draft chapters were then reviewed by peers in the field and a book contract was presented to us in December. Before the final draft of the manuscript was submitted in 2014, two experts in the field reviewed the penultimate manuscript. In the end, this process has produced a manuscript that reflects very critical and careful deliberation about the governance of climate mitigation innovations in China.

Besides thanking each other for intellectual and personal camaraderie in the process of researching and writing the book, we are also grateful to colleagues who have provided their advice. These colleagues include Ray Forrest, Ting Gong, Ian Scott, Philip J. Ivanhoe, Peter Hills, Shi Han, Wanxin Li, Xiaohu Wang, Simon

Yau and Sara Fuller. Andreas Thiel of Humboldt University of Berlin, Harriet Bulkeley and Simon Marvin of Durham University, as well as Kristine Kern and Timothy Moss of the Leibniz Institute for Regional Development and Structural Planning (Erkner, Germany) have been great hosts and a source of inspiration. We finally also thank three anonymous reviewers and Helena Hurd of Routledge (Abingdon, Oxford) for her support and advice in the process of producing the manuscript.

Qianqing MAI
Maria FRANCESCH-HUIDOBRO
Hong Kong, May 2014

Abbreviations

APAS	Automotive Parts and Accessory Systems Research and Development Centre
BCA	Building and Construction Authority (Singapore)
BEAM	Building Environmental Assessment Method (Hong Kong)
BEC	Business Environment Council
BEEMO	Management Office of Building Energy Efficiency (Guangzhou)
BYD	Bi Ya Di Auto Company Limited
C40	C40 Cities Climate Leadership Group
CAN	Clean Air Network
CARE	Centre of Architectural Research for Education
CCBF	Climate Change Business Forum
CCPG	Chinese Central People's Government
CEPA	Closer Economic Partnership Arrangement between Hong Kong and China
CEPAS	Comprehensive Environmental Performance Assessment Scheme for Building (Hong Kong)
CGBC	China Green Building Council
CGBC (HK)	China Green Building (Hong Kong) Council
CIC	Construction Industry Council (Hong Kong)
CITPS	Construction Inno-Tech Promotion Station (Guangzhou)
CMNs	Collaborative Municipal Networks
CO_2-eq	Carbon Dioxide Equivalent
COP	Conference of Parties
CRIAS	Carbon Reduction Implementation and Assessment Strategy
DC	District Cooling
DSM	Demand-Side Management
EB	Environment Bureau (Hong Kong)
EE	Energy Efficient
EES	Energy Efficient Systems
ENGO	Environmental Nongovernment Organisations
EPD	Environmental Protection Department (Hong Kong)
EU	European Union

EVs	Electric Vehicles
FoE	Friends of the Earth
GARC	Governance in Asia Research Centre
GBL	Green Building Labeling (China)
GD-BEEA	Guangdong Building Energy Efficiency Association
GDP	Gross Domestic Product
GFA	Gross Floor Area
GHG	Greenhouse Gas Emissions
GONGO	Government–Organised Nongovernmental Organisation
GZ-BEEA	Guangzhou Association of Energy Efficiency and Building Technology
GZ-DRC	Guangzhou Development and Reform Commission
GZ-ETC	Guangzhou Economy and Trade Commission
GZ-FD	Guangzhou Finance Department
GZ-SDA	Surveying and Design Association in Guangzhou
GZ-TC	Guangzhou Transport Commission
GZ-URCC	Guangzhou Urban–Rural Construction Commission
HKET	*Hong Kong Economic Times*
HKGBC	Hong Kong Green Building Council
HKSAR-ERM	Hong Kong Special Administrative Region-Environmental Resources Management [study]
ICLEI	International Council for Local Environmental Initiatives
IEA	International Energy Agency
IPCC	Intergovernmental Panel on Climate Change
IWGCC	Interdepartmental Working Group on Climate Change (Hong Kong)
LEED	Leadership in Energy and Environmental Design
MOHURD	Ministry of Housing and Urban–Rural Development
MOT	Ministry of Transport
MP	Ming Pao
MURE	Measures d'Utilisation Rationelle de L'Energie [Measures for the Rational Use of Energy]
NDRC	National Development and Reform Commission
NG	Natural Gas
NGOs	Nongovernmental Organisations
NP	Nuclear Power
OCTS	One Country, Two Systems
OTTV	Overall Thermal Transfer Value
PGBC	Professional Green Building Council (Hong Kong)
PLDRC	Planning and Land Development Research Centre (Shenzhen)
PRD	Pearl River Delta
RE	Renewable Energy
REA	Real Estate Association (Shenzhen)
REDA	Real Estate Developers Association (Hong Kong)
SCMP	*South China Morning Post*

SDC	Council for Sustainable Development (Hong Kong)
SGBA	Shenzhen Green Building Association
SIBR	Shenzhen Institute of Building Research
SOC	Scheme of Control
SZ-DRC	Shenzhen Development and Reform Commission
SZ-HCD	Shenzhen Housing and Construction Department
SZ-SDA	Surveying and Design Association in Shenzhen
SZ-TC	Shenzhen Transport Commission
SZ-UPLRC	Shenzhen Urban Planning, Land and Resources Commission
UNFCCC	United Nations Framework Convention on Climate Change
VAs	Voluntary Agreements
WAC	Water Cooled Air-Conditioning
WGO	World Green Organisation
WtE	Waste-to-Energy
WWF	World Wildlife Fund
WWF-Arup	World Wildlife Fund for Nature-Arup [study]

Part I

Conceptions and context of climate change governance in China

1 Governing climate mitigation in Chinese cities

Introduction

Since opening up in the 1980s, China has encountered fundamental challenges in governance. Those challenges are becoming even more pressing as the country attempts to streamline the relationship between the government and the market while maintaining a semi-authoritarian, one-party system and transforming the role of government by making it more law based and service oriented. China's need to consider how it should govern itself while addressing these challenges has been emphasised by a number of transformations in contemporary China, especially urban China.

One major transformation has been the recognition of the increasing complexity of policy problems (climate change being a case in point) and the increasing interdependency of policy areas (urban development, the built environment, urban infrastructure), policy levels (national, provincial, municipal) and policy actors (public, private, people). Further, globalisation has increased the dependency and influence of actors, to which China has not been immune (Bai 2007; Balme 2011). Nongovernmental actors (businesses, professional organisations, civil society organisations), whether independent or co-opted within the political system, have also become more directly involved in policy making and are increasingly affecting policy choices (Wu 2003). At the same time, citizens are generally wearier, less trustful and more mindful of what the Chinese government does and does not do (He and Thøgersen 2010; Ho 2007; Ho and Edmonds 2007). Reforms in the Chinese public sector have brought additional challenges concerning how best to organise the task of governing (Appendix II: CPC Central Committee 2013; Lieberthal 1997; Lieberthal and Lampton 1992; Lieberthal and Oksenberg 1988).

In responding to similar challenges elsewhere, the concept of *governance* – the way collective action steers and controls society to achieve collective goals – has become one of the most thought-about and researched in the social sciences, with policy makers and academic scholars trying to make sense of how to govern societies (Kooiman 1993; Levi-Faur 2012; Pierre 2000; Pierre and Peters 2000; Rhodes 1997; Torfing *et al.* 2012). The interest in governance reflects the ongoing attempt to understand public leadership, public institutions and social patterns of behaviour.

What, though, do we mean by governance? How is governance approached in the context of urban politics? What does governance mean in China?

In their recent, highly analytical and thought-provoking account of governance, four of the best intellectual minds in public administration, Jacob Torfing, Guy Peters, Jon Pierre and Eva Sorensen (2012), reopen once more the debate about governing contemporary societies. They encourage us to think not only about the traditional forms of governing through formal state actors and bureaucratic procedures but also about the reality of governing in the contemporary world through informal, multi-actor, multi-interest, multigoal processes. They term this style of governing 'interactive governance' and formally define it as 'the complex process through which a plurality of actors with diverging interests interact to formulate, promote and achieve common objectives by means of mobilising, exchanging and deploying a range of ideas, rules and resources' (Torfing *et al.* 2012: 14). Their thinking acknowledges contrasting explanations of governing as 'governance without government' (Rhodes 1997), which conceives the state as *primus inter pares* with no special privileges for decision making, and 'governance beyond government' (Kooiman 1993), which sees governing as a bottom-up decision-making process based on interactions among various actors, of which the state may or may not be the initiator. Without subscribing to these more extreme views of governance, Torfing et al. provide a highly analytical account of 'how we should think about, empirically observe, as well as interpret and evaluate these open-ended, fluid, complex, multiparty approaches to solving problems that are not controlled by state actors nor confined to mere conventional policy networks' (Hart 2013: 1071).

Yet while interactive governance depicts the way governing is conducted in many advanced democracies of the Global North (for example, Scandinavia, UK, US), it does not fully explain the way of governing in most democracies in transition (for example, Thailand, South Korea, Indonesia, Taiwan) and nondemocratic societies (China) of the Global South, nor does it prescribe how to govern these polities and/or all of their policy areas (see, for example, He's 2012 account of the Chinese practice of complex, deliberative governance). Thus, interactive governance may be contrasted and compared with other forms of governing that are deducted from in depth theorisation and empirical observations of comparative case studies across societies (of the Global North and South), policy sectors (environment, employment, welfare, etc.), disciplines (sociology, economics, anthropology, etc.) and research areas (urban studies, development studies, legal studies, etc.).

In this book, our aim is to reduce this dearth of knowledge by improving our understanding of the urban governance of climate change mitigation in China through conceptualising collaborative municipal networks (CMNs) and observing their dynamics in the process of greening the building and transport sectors of Chinese cities. We formally define CMNs as follows:

The complex process of interaction among a plurality of state and nonstate actors with the objective of achieving climate mitigation goals through functions such as knowledge transfer and resource exchange while maintaining municipal autonomy in vertical or horizontal coordination, and enhancing municipal capacity in horizontal collaboration. These interactions are manifested in collective decision making, which includes consensus building and joint rule making.

Urban studies and urban politics is an area of research with an institutional system (that of the city) that is struggling to maintain its capacity to govern (Torfing *et al.* 2012: 38). Three features depicting the challenges of governing at the subnational level can be identified from the urban governance literature, most of which is based on studies of cities in the Global North conducted by geographers, planners and political scientists. First, there are political and institutional constraints imposed on cities. Cities have limited autonomy in relation to higher levels of government. Urban politics tends to include a dialogue with higher institutional levels in the region, the province and the nation. This dialogue of bargaining also includes transnational institutions, those that cut across jurisdictional borders and that are established based on cities' shared objectives. Various models of multilevel governance, originally depicting governance in the European Union (EU) (Hooghe and Marks 2001), can also be applied to the role of urban areas with complex systems of governance (Francesch-Huidobro 2012). This literature is based on empirical observations of cities in the Global North. As we argue in Chapter 2, the political and institutional constraints imposed on Chinese cities are very different from those in their Western counterparts. Administratively, for instance, Chinese cities have higher levels of autonomy due to the decentralisation of the system, although politically they are more constrained by the authoritarian state structure than cities in the West.

The second feature depicting the challenges of governing at the city level is the predominance of private capital and corporate resources. This is another feature of urban governance in the Global North but also increasingly of societies in economic and political transition such as China. Political leadership forges coalitions with the corporate leadership, which boosts the city's governing capacity.

Third, the relationship between the city and civil society is another feature of urban governance that is broadening cooperation across the public–private divide. This feature is also found in the modes of urban governance, CMNs, as we propose in this volume. In considering these features, one must be mindful of the literature on collaborative planning (Booher and Innes 2010; Healey 2007) that has emerged from recognising the limitations of the rational and linear planning models.

In addition to these three features challenging urban governance, urban governance research is concerned with the role of local political institutions and their particular roles in the process of governing (Bulkeley 2005; Bulkeley *et al.* 2009). It queries, more broadly, how cities are coordinated and planned through decentred forms of governance, a result of the need of local governments to mobilise private resources and secure consent (legitimacy; Francesch-Huidobro 2012). Finally, research in urban politics points to the importance of vertical interaction through some form of multilevel governance (Corfee-Morlot *et al.* 2009). All of these features depicting the challenges of governing cities are discussed in depth in Chapters 2 and 3, in which we address the challenges of urban climate governance in China.

The *urban governance of climate change* has been mostly researched in relation to the transnational coalitions between cities – that is, the International Council for Local Environmental Initiatives (ICLEI), the Cities for Climate Protection (CCP) programme and the Large Cities Climate Leadership Group (C40) (Bulkeley 2005, 2010, 2012; Bulkeley and Newell 2010; Levi-Faur 2012). Increasingly, comparative studies are considering how climate-mitigation responses and low carbon transitions are governed in cities across the globe (Bulkeley *et al.* 2013).

More recently, we and other scholars have analysed various aspects of climate governance in Chinese cities. For example, Chu and Schroeder (2010) discuss the Hong Kong corporate sector's attempt at driving climate change mitigation; Dhakal (2009, 2011) analyses the internal dynamics of several Chinese cities in relation to energy use and CO_2 emissions; and Schroder (2011) discusses the operationalisation of the clean development mechanism by Chinese local authorities. The textbook by Harris (2012) contains a succinct analysis of the governance of climate change in Hong Kong, while Miao and Lang (2010) offer a broad discussion on China's climate change trajectory and energy profile, national responses to climate change and the evolution of state institutions (Li *et al.* 2011). Our more recent work has focused on the issues of legitimacy and the role of epistemic communities in climate governance in several Chinese cities (Francesch-Huidobro 2012; Francesch-Huidobro and Mai 2012). Finally, a group of Chinese scholars working for state-sponsored research institutions has compiled an edited volume on issues affecting climate change policies in China, with reflections on experiences elsewhere (Wang *et al.* 2013).

Against the backdrop of this scholarship, we recognise the importance of studying the context, dynamics and challenges of the governance of climate change in Chinese cities and of developing adequate conceptual frameworks to explain contemporary practices and changes, what, how and why these changes occur and the implications for future research and practice. In other words, we need to examine what goes on beyond the radar of the official accounts and the model-oriented studies of climate mitigation in China and the vacuum around studies of how climate mitigation is being governed.

As mentioned, *Chinese cities* have been important sites for rapid economic development and have contributed to the transformation of the nation into a modern state. At present, China has 665 million urban dwellers, up from 191 million 30 years ago. By 2030, 350 million new rural migrants will need to be accommodated in China's cities (Liu and Salzberg 2012). UN projections indicate that by 2050, 73 percent of China will be urbanised, with cities populated by 1.02 billion dwellers (Dhakal 2011: 73). Meeting their needs will put pressure on resources, which in turn will bring problems related to air pollution, energy efficiency, transport, solid waste and water pollution. In addition to these challenges, China has made a commitment to lower the carbon intensity of its economy by 40 to 50 percent by 2020, compared with 2005 levels. In the current 12th Five-Year Plan period (2011–15), a 17 percent reduction target for carbon intensity has been set. As Chinese cities contribute more than 70 percent of energy-related carbon emissions, addressing cities' emissions is an essential element of this plan.

In light of these developments, and within the conversations about urban governance and governing broadly in contemporary China, this book provides the first critical introduction to the challenges that Chinese cities are facing with regards to the governance of climate change mitigation. It considers these challenges from the perspective of the development of three intertwined issues driving carbon emissions and reduction in Chinese cities: municipal governance, including the effects of globalisation and decentralisation, financial and personnel incentives for municipal authorities and their legitimacy and emerging forms of governance; the policy design of low carbon development plans, including financial and land concessions and sectoral innovations; and urban energy use.

The book focuses on the dynamics of a specific arrangement of climate change governance, municipal networks, which have largely emerged in response to Chinese municipalities' international and national commitments and to the increasing scientific evidence of the seriousness of the climate change problem. While transnational municipal networks (those across cities of different national jurisdictions) have been widely studied in Western cities (e.g. ICLEI and C40; Lee and van De Meene 2012), the study of networks that bring together municipality actors from the public, private and people sectors has received no attention in the context of Chinese cities. We therefore study *collaborative municipal networks* (CMNs) specifically in the Chinese context.

Throughout the book, CMNs are discussed in relation to how the nongovernmental space is/is not expanding in the Chinese environmental state. The definition of CMNs contrasts with that of governance networks in the Western-centred, democratically grounded literature (see Sørensen and Torfing 2005a, 2005b, 2007; Torfing *et al.* 2012), which conceives state actors in the dynamics of interaction as the shadow under which the interactions of interdependent but operationally autonomous (i.e. nonstate) actors occur. If we were to ignore the horizontal articulation of interdependent, though not operationally autonomous actors (i.e. GONGOS, state and nonstate), that get involved in negotiating climate change mitigation actions within the institutional framework of the Chinese political system and the bureaucratic machinery for the purpose of facilitating policy making, we would miss not only current sectoral mitigation developments but also changes in the Chinese environmental state.

We study CMNs in *two energy end-use policy sectors – transportation and building* – of three cities, Guangzhou, Shenzhen and Hong Kong. We selected these cities not only because they are megacities at the centre of the economically thriving Pearl River Delta Region, which contributes to high GHG emission, but also because of their unique constitutional, political and economic characteristics. Hong Kong is a special administrative region, Shenzhen is a special economic zone and Guangzhou is a provincial capital of the People's Republic of China

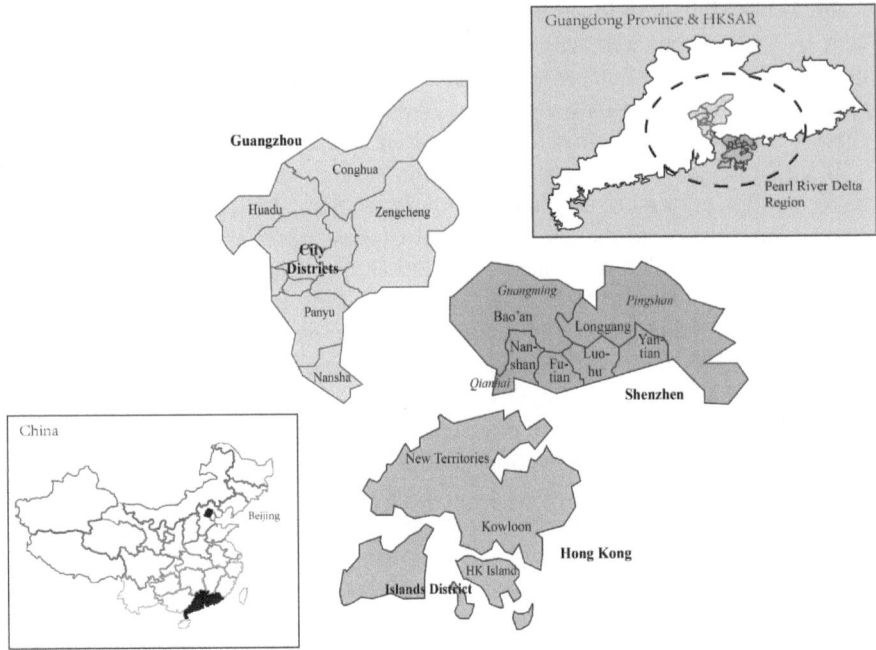

Figure 1.1 Map of study cases
Source: Compiled by authors

(Figure 1.1). Their distinctive features allow the three cities to enjoy a high degree of vertical autonomy in policy making and planning and, potentially, to develop 'new forms' of governance. These three cities also carry relative instrumentality in connecting to world city networks, as recognised in a recent inventory of globalisation and world cities (GaWC 2011). Instrumental world cities are often positioned by their national state and their own collaborative networks as contexts for developing innovative responses and urban transitions to emerging complex issues such as climate change (Hodson and Marvin 2010). Intensive interactions among network actors occur when the developed innovative responses are vertically passed down and horizontally transferred to other urban systems within the nation (Torfing *et al.* 2012).

A further reason for choosing these three cities is that they have the greatest mitigation potential and thus potentially the GHG highest reduction impact. The transport and building sectors were chosen because hard data on these sectors are available both in official documents and from players that are involved and willing to talk about these developments. There is a significant potential for reduction in China's transport and building sectors, which together account for more than 60 percent of the nation's total energy consumption. With the decentralisation of state functions (see Chapter 2), city governments have been proactively engaged in the sectoral integration of carbon reduction and low carbon innovations. In this process, the governing of mitigation responses – in particular of *electric vehicles* and *green buildings* that rely heavily on technological development and policy support – is met with incentives but also confronted with barriers such as resource mobilisation, knowledge transfer and legitimation and ongoing uncertainties over technical and market feasibility. To analyse the interactive relationship between the governance of low carbon responses in cities and the institutional transformation of environmental state functions in China, this book focuses on the dynamics of coordination and collaboration in the process of integrating mitigation innovations derived from the transport and building sectors.

The *transport* sector consumes more than 20 percent of China's total energy end-use allocation (CASS 2013). The number of privately owned vehicles has rapidly increased in tandem with the process of urbanisation: from 2,499,600 in 1995 to 35,014,000 in 2008 (Qin *et al.* 2011). The reduction potential in road transport is much larger than in rail, sea and air transport. In line with current technological developments and mitigation actions, a CO_2 reduction of up to 1.03 billion tons in road transport can be achieved by 2030, whereas the potentials for rail (0.1 billion tons), sea (0.05 billion tons) and air (0.01 billion tons) are mere fractions of that (Qin *et al.* 2011: 398). One of the optimal mitigation approaches in road transport is turning the technological innovation of *electric vehicles* into a more widely used form of private transport.

The *building* sector consumes 40 percent of China's aggregate end-use energy, including energy used in the production of building materials and during the entire construction process (CASS 2013). Given the substantial reduction potential over a building's life cycle, design guidelines and mandatory policies for *green buildings* are being adopted as an essential sectoral approach to mitigation. Green building practices carry both baseline capacity and long-term potential for emissions reduction. In terms of the estimated baseline, applying energy-efficient designs in new buildings could achieve a CO_2 reduction of 0.34 billion tons in nationwide emissions by 2030, substantially larger than the 0.02 billion ton reduction from utilising renewable energy in buildings or the 0.14 billion ton reduction from energy efficient lighting (Qin *et al.* 2011: 404). The passive design of green buildings – for example, considering southern China's subtropical climate conditions, an early stage architectural design that encourages natural ventilation and daylight with flexible solar controls – would prolong carbon reduction over a longer timeframe. However, this would require more fundamental behavioural changes in both the public and private sectors, in comparison with the active design of applying energy-efficient lighting or cooling techniques during the later stages of building operations and management.

With sectoral focuses on electric vehicles and green buildings, this book integrates the analysis of sociotechnological responses (innovation transfers) and political-economic transformations (low carbon responses and environmental state restructuring). The dynamics and complexities of sectoral integration are reflected in the interactions of cross-sector organisational actors in the process of networking to initiate and institutionalise innovative urban responses to climate change mitigation (Chapters 5 and 6).

Drawing on an innovative framework (Figure 2.2) and richly supported by detailed case studies (Figure 2.1), this book will enable scholars, practitioners and students of climate change to understand the drivers and barriers to effective climate governance in Chinese cities and to explore the consequences for Chinese cities in the future. The book will be essential reading for researchers and practitioners across disciplines, including China studies, environmental politics and policy, urban studies and planning, environmental science, management and technology and geography.

In reading this book, it is crucial to remember that the discussion of CMNs must always be considered within the wider framework of governing and governance. The link between traditional state-centric governance and the CMNs of governance discussed in this volume is articulated clearly in Chapter 2, which deals with the concepts, definitions, main arguments, theoretical framework and methodology of this book. In terms of political thought, it is crucial to recognise that the arguments presented in this volume do not derive from a neostatist or neoliberal philosophy: we consider the state neither as the sole nor the most important actor in governing climate change mitigation in contemporary urban China; hence our attention to the dynamic interactions between state and nonstate actors in CMNs (Torfing *et al.* 2012).

Urban climate governance in China

Most urban responses to climate change worldwide have focused on *climate change mitigation* – the actions taken to reduce the impact of human activity on the climate system, primarily through reducing net greenhouse gas (GHG) emissions (IPCC 2007, 2013). The objective of the United Nations Framework Convention on Climate Change is to achieve 'stabilisation of greenhouse gas concentrations in the atmosphere at a level that would prevent dangerous anthropogenic interference with the climate system' (UN 1994: 9). The Intergovernmental Panel on Climate Change (IPCC) suggests that the plausible limit of 'dangerous anthropogenic interference' set out in policymaking is beyond the work of science because normative judgments are inherently involved (IPCC 2007, 2013). The policy and scientific communities have basically agreed that a 2°C increase (from preindustrial levels) in the global annual mean surface temperature is the maximum acceptable threshold of risk to the ecosystem (CEU 2005; Hansen *et al.* 2006; IPCC 2007, 2013; Rijsberman and Swart 1990). Figure 1.2 presents China's per capita GHG emissions (only CO_2) from 1960 through 2010 in comparative perspective with other BRICS countries[1] and major economies.

While there is recognition that we are already locked into climate change due to past emissions and that there is a need to adapt to stress and change and to build resilience, efforts remain focused on policies to measure and monitor emissions, set targets and develop action plans. This is also the case in China. The 2°C threshold requires nations to follow the 'common but differentiated' principle by which the reduction of emissions is offset by a country's current growth needs.

In the past three decades, government policies in China have created an 'energy-intensive' and 'high-carbon' urban transition through their efforts to

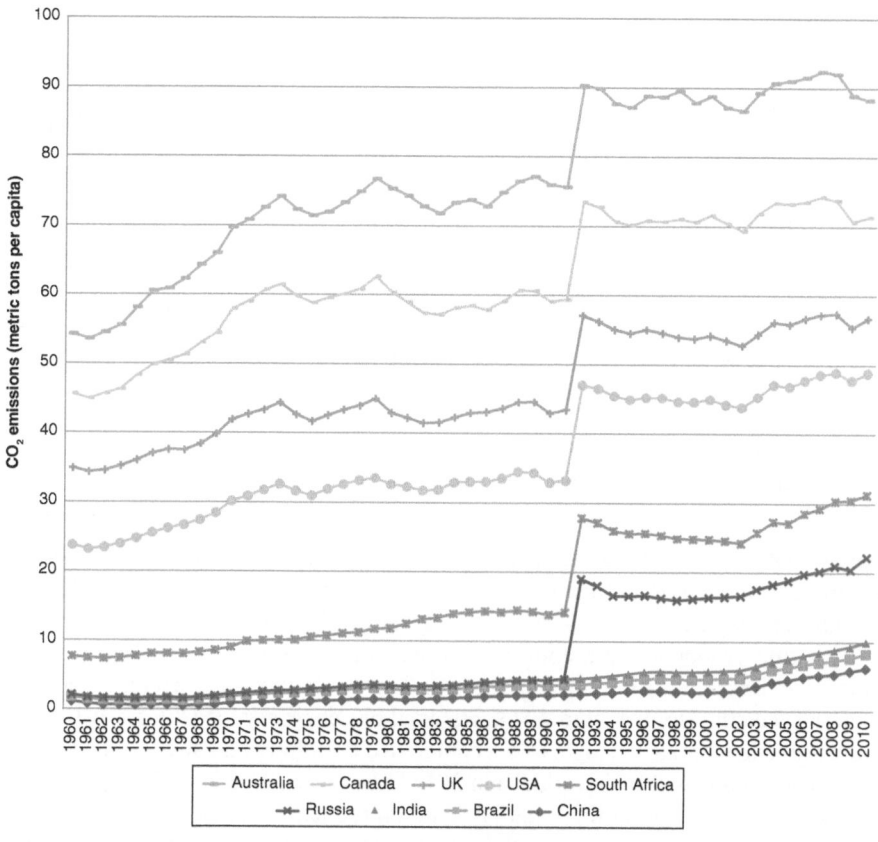

Figure 1.2 Historical per capita GHG (CO$_2$) emissions
Source: Compiled by authors from World Bank Databank 2014

increase urbanisation and economic growth (Dhakal 2011: 73). Although China's per-capita GHG emissions remain the lowest among the BRICS countries and the Global North (Figure 1.1), its total CO$_2$ emissions are on a rapid growth trend since economic reforms were launched in 1978. Figure 1.3 illustrates China's historical CO$_2$ emissions (1960–2010).

In 2009, shortly before the Copenhagen Conference of Parties (COP15), China announced a national commitment to a 40 to 45 percent reduction in carbon dioxide (CO$_2$) per unit of GDP by 2020 from 2005 levels (17 percent for the 2011–2015 period) (Appendix II: Central Government 2009). Moreover, the 12th Five-Year Plan, which charters the social and economic development of China from 2011 to 2015, states that the main priorities during this period are sustainable growth, industrial upgrading and the promotion of domestic consumption (Appendix II: Delegation of the European Union in China 2012). These priorities explain why certain sectors, including energy, automotive, IT infrastructure and biotechnology, receive a high degree of focus (KPMG Insight Series 2011). In particular, *energy security and climate change mitigation* have emerged as key policy priorities. This

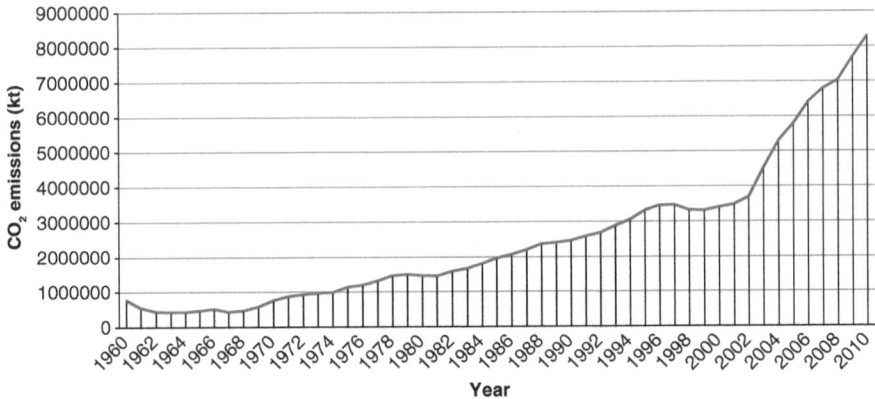

Figure 1.3 Historical GHG (CO₂) emissions in China
Source: Compiled by authors from World Bank Databank 2014

demonstrates that the Chinese central government (CCG) is committed to carbon control and low carbon transition, despite the fact that such transition challenges current development patterns and poses great implementation difficulties, such as changing production and consumption patterns and generally altering what is currently considered by many in China as quality of life – the accumulation of material wealth. On 3rd June 2014, He Jiankun, chairman of China's Advisory Committee in Climate Change, told a conference in Beijing that an absolute cap on carbon emissions will be introduced in 2016 (*The Guardian* 3 June 2014).

The implementation of these national carbon reduction policies is prescribed to take place at the municipal level through regular policy implementation avenues and through low carbon experiments (pilot projects) in metropolises such as Baoding, Shanghai, Hangzhou, Jilin, Zhuhai, Nanchang, Xiamen and 100 other cities. According to the official account, these localities have developed blueprints for building low carbon cities that are spontaneously initiated and experiment oriented (Zhuang 2012: 216). The approaches these cities have taken to the low carbon transition are determined by their own choice of pathways. For example, Shanghai used the building of the Expo Park as a means to implement a low carbon strategy through site selection, planning, design, construction and operation of the park and of the transport and accommodation infrastructures. Nevertheless, the city was unable to extend the same development pattern to other structures or to the whole city due to financial and land-use constraints. Baoding branded itself as China's Power Valley and devised its transition by investing in photovoltaic, wind, power storage and power transmission, with 105 districts powered by solar energy, although it was unable to make the transition in other sectors of its economy (Zhuang 2012: 216–17).

While it is difficult to evaluate and determine the success of urban carbon reduction actions and activities and how much they contribute to absolute or relative GHG emissions in China, it is essential to recognise three intertwined *issues affecting carbon reduction in Chinese cities: municipal governance* (including the effects of decentralisation, personnel incentives for municipal authorities and their legitimacy/accountability), *policy design of development plans* (including financial resources and land concessions, urban planning and sectoral innovation; Liu and Salzberg 2012; Qiu 2008; Wang and Tang 2005; World Bank 2009) and *urban contribution to energy consumption* (Dhakal 2009; 2011).

Municipal governance

The CCG establishes broad national policies and targets for urbanisation, rural-to-urban conversion, energy saving and CO_2 emissions reduction. The CCG also reviews and approves urban master plans, large investment infrastructural and development projects and applications for rural-to-urban land conversion. In addition, it provides technical guidelines and standards for cities in specific areas such as public transport and utility services. Direct financial support is limited to *inter*governmental transfers to provincial and municipal governments, primarily in less developed provinces.

At the local level, municipal governments have functions that range from developing the local economy and employment to the provision and management of municipal services (waste collection, leisure and culture, etc.). This devolution of government functions puts Chinese municipalities in the enviable position of being able to make decisions and to implement them swiftly. One only needs to look at the past three decades of rapid urban development to ascertain Chinese municipalities' governing capacity. However, in the planning of urban low carbon transitions that require looking beyond short-term economic growth towards longer-term planning for sustainable development, this capacity is being driven by a combination of decentralisation, incentives for municipal authorities, questions about their legitimacy (accountability) and emerging forms of urban governance.

Decentralisation

China has devolved a range of functional and fiscal responsibilities from the national to the subnational governments, particularly municipalities. Beyond the review and approval of master plans (including urban spatial plans, national socio-economic development five-year plans and state land utilisation plans), investment in large urban infrastructure projects, setting technical standards, providing policy guidelines and facilitating knowledge transfer and capacity building, the central (national) government's capacity to guide and control subnational governments is limited by the rapid economic and spatial changes at the local level and by the limited budget available for urban management (not development).

Spillovers from rapid urban development such as CO_2 emissions and energy security have been the concern of local governments. The cobenefits of transiting to a low carbon economy at the local level – such as energy saving, reducing the environmental and health risks posed by degradation and improving quality of life – are not sufficient incentives for cities to take action without central government support. Chinese mayors are held accountable not only for the provision of urban public services but also for the performance of the urban economy, investment and employment. They face difficult choices between higher GDP growth and more sustainable urban development.

Personnel, incentives and accountability

Since the opening up in the 1980s, China's local authorities – Communist Party secretaries of provinces and cities, provincial governors and city majors – have been evaluated almost exclusively by their contribution to annual GDP growth. This has created a culture of competition across time and space (with preceding and with neighbouring authorities), driven by a system that rates individuals on their technical competence and good economic performance. As such, local authorities focus on the single issue of economic development. Nevertheless, as we argue in this volume, this focus is now beginning to change due to the formation of climate collaborative municipal networks and policy directions such as 'green GDP' (2004) and 'people-centred development' (2003), which are starting to be implemented even if their effects are only just beginning to be felt.

Policy design of development plans

Financial resources and land concessions

Since 1994, China has adopted a tax-sharing system that stipulates that the 'ratio of subnational revenues to the total revenues averages 50 per cent, while the ratio of local fiscal expenditures to the total is at about 70 per cent' (Liu and Salzberg 2012: 103). However, local governments do not have tax collection powers over residential property tax and land value incremental tax, although pilot projects were launched in Shanghai and Chongqing in 2011. As such, local governments increasingly face the burden of rapidly increasing expenditures without the power to raise revenues at the required scale. This gap is filled by two sources of funds: revenue-sharing transfers and tax rebates (Shah and Shen 2006) and specific-purpose funds usually designated for poor and rural communities. These funds tend to be insufficient to meet capital expenditures for urbanisation and industrial development, given that most of these expenditures are overestimated, as they are used to attract investments (FDI) and jobs. The gap is then filled using off-budget funds: revenues from land concessions, borrowing through municipal government-owned urban development investment corporations (UDIC), municipality-imposed surcharges, public–private partnership (PPP) financial arrangements and build-operate-transfer schemes.

UDICs are established by local governments to bypass the constraints imposed by the Chinese Budget Law for municipalities to borrow from commercial banks. UDICS are given plots of public land as starting assets. With these as collateral, UDICs borrow on behalf of municipalities to fund infrastructure investments. Some UDICs are also created as majority shareholders for municipally owned urban utility companies (water supplies, central heating, town gas), thus generating cash flows. Other UDICS are simply fiscally backed 'empty shell' companies able to borrow either commercial loans from commercial banks or policy loans from the China Development Bank. Despite these extra resources, municipal governments are under tremendous pressure to continue raising funds to deliver municipal services. The only way they can do this is through the appropriation of rural land within municipal boundaries.

The conversion of rural land to urban use is undertaken through a land acquisition plan based on an urban master plan, a process that is closely monitored by the Ministry of Land and Resources (MLR). The MLR sets the price based on agricultural revenues and the cost of relocating farmers (usually very low in cost), services the land with urban infrastructure and then sells or auctions the serviced land to property developers. The revenues from land sales (concessions) are significant. According to the MLR, total revenue from land sales nationwide in 2010 amounted to RMB2.7 trillion or RMB2,000 (US$300) per capita (Liu and Salzberg 2012: 105). The prospect of huge earnings continues to drive large developments on the understanding that the rapid growth of the urban population, incomes and property prices will maintain the demand for more flats and houses. Nevertheless, in the past few years, 'ghost towns' made up of empty but perfectly finished villas and towers of flats have become increasingly common. This relentless conversion of land and construction is also creating significant amounts of local debt.

Rural land is also converted into urban land by leasing rural land in suburban villages for residential, commercial or industrial development. This entails *in situ* urbanisation without relocation. Estimates based on the World Bank's Sustainable Development on the Urban Fringe study (World Bank 2007) suggest that *in situ* urbanisation accounted for almost 40 percent of the growth in China's urban population during the 1990s, of which 33 percent was attributed to natural population increase (births exceeding deaths). These developments occur outside of the urban land acquisition plan and outside of the supervision and monitoring of the municipal government. As such, the developments tend to be fragmented and unregulated with consequential inefficient land use (Qiu 2008). In sum, local governments' excessive reliance on off-budget funds for development plans is creating a series of hidden financial liabilities that are the concern of the central government. More sustainable sources of local revenue such as property taxes, betterment charges and municipal access to loans are either underdeveloped or not permitted by law. Moreover, the rate of urban expansion that sustains such development plans cannot continue indefinitely. This financing system has a huge effect on the nature of urban development and, in turn, on the creation of planning and implementation strategies that can support carbon reduction and low carbon growth in Chinese cities.

Urban planning process

Urban planning is another aspect of the policy design of development plans that impinges on China's ability to make the transition to low carbon development by mainstreaming it into the urban planning process. While the formulation of urban planning in China is rather rigorous, its implementation is less so. Most master plans are not fully implemented within their twenty-year timeframe due to two interrelated issues. First, cities have been growing at an unprecedented rate and most are unwilling or unable to overcome the rigidity imposed by the master plans to enable them to accommodate the growing urban population and demand for urban services. While the urban planning process produces twenty-year master plans, five-year implementation plans and a number of associated sectoral master and implementation plans for a given jurisdiction, urbanisation has occurred so rapidly that the actual urban population often exceeds the planned population target for the entire twenty-year time horizon of the master plan. Often, the urbanisation area goes beyond the delineated jurisdictional boundaries. Any amendment to the original master plan requires lengthy and cumbersome processes (Wang and Tang 2005).

The second issue affecting the implementation of master plans is the lack of a system of checks and balances that hold municipal authorities accountable to provincial authorities, the People's Congress of the city and to citizens. While planners carry out their work, mayors often use the rapid pace of urban growth and demand for urban services as a reason to deviate from the master plan and to raise additional GDP growth (Qiu 2008; Wang and Tang 2005; World Bank 2009). The People's Congresses in many cities are increasingly playing a supervisory role by checking the actions of municipal authorities against urban master plans, but the lack of technical expertise and insufficient representation from all stakeholders diminish their effectiveness. There is also no established procedure for expert witness testimony over issues relating to the public interest in urban planning (Liu and Salzberg 2012; Wand and Tang 2005).

Sectoral innovation

Sectoral innovation is a crucial aspect of the policy design of development plans. Low carbon transition can be regarded as a system innovation, as it is a radical but gradual shift agglomerating various technological innovations across different socio-technical systems. A socio-technical system is a cluster of 'technology, regulations, user practices and markets, cultural meanings, infrastructure, maintenance networks and supply networks' that encompasses 'production, diffusion and use of technology' (Elzen *et al.* 2004: 3; Geels 2004: 900). A system innovation is defined as a 'large-scale transformation in the way societal functions such as transportation, communication, [and] housing . . . are fulfilled', which can also be understood as a change from one socio-technical system to another' (Elzen *et al.* 2004: 19).

Municipal governance and the design of development policies have implications for carbon emissions because the rapid growth of low-density areas at the urban periphery (urban sprawl) increases GHG emissions (mostly CO_2) from a variety of sources such as density, land-use mix, transport, commuting distances and so forth. (O'Toole 2009). Urban sprawl means longer journeys by public and private transport and more private vehicle ownership, all of which are sources of GHG emissions. It also means lower density (people per square metre), which leads to higher GHG emissions per person for home heating, cooling and power generation (Norman *et al.* 2006). Low-density development produces infrastructure that is less intensively used than that in denser core areas such as suburban access highways, thereby raising emissions per capita (Kamal-Chaoui and Robert 2009; Brown and Logan 2008).

Urban energy consumption

Examining the urban contribution to China's energy uses helps in understanding how important urban areas in China are for key ongoing national concerns such as improving energy security, mitigating climate change and substantially reducing the energy intensity of the Chinese economy. Dhakal (2009, 2011) estimated that the urban contribution to China's total energy use was 84 percent of the total energy consumed in 2006. The International Energy Agency (IEA, 2007) estimated that renewable sources contributed to 10.6 percent of total energy use in China in 2006, and this was assumed to be mainly due to the high use of biomass in rural areas. Taking these IEA figures and assuming that urban areas only use nonrenewable energy sources, China's urban input to the total energy use in 2006 would have been about 75 percent (Dhakal 2011: 75). There is a huge gap in energy use between urban and rural populations: the ratio of urban to rural energy use per capita is 6.8, and that of urban to national use is 1.9.

Dhakal (2011) analyses data from cities that are highly urbanised and explicitly designated in China's National Plan as important cities for economic development and concludes that there are thirty-five cities,[2] including provincial capitals, provincial cities and special economic zones (including Guangzhou and Shenzhen), that are important for economic growth and infrastructure development in China. These are the places through which urban transition is forged. While this transition includes the growing energy intensity of urban life, it also includes the implementation of key policy measures for urban energy efficiency and climate change mitigation. The average carbon and energy intensity (energy/carbon per unit GRP) of the province to which the city belongs is used as a proxy for the city's energy and carbon intensity.

Table 1.1 shows that while these cities represent less than one fifth of the population, they produce a large share of the nation's GDP. Collectively, they consume 40 percent of the total commercial energy of the nation and emit a similar proportion of CO_2. The wide disparity between these cities and the national GDP per capita, energy consumption per capita and CO_2 emissions per capita shows that their influence in shaping the national energy and carbon profile is disproportionate to their population.

There are also large differences within these cities. Shenzhen, a manufacturing centre and China's gateway to the Hong Kong financial centre, stands out as having an extraordinarily high GDP per capita (US$38,000 per person). There are three clear pathways in China's thirty-five key cities (Dhakal 2011): the low

Table 1.1 Key indicators and estimated energy and CO_2 for the key thirty-five cities of China, 2006

	China	*Frontrunner cities*	*Cities' contribution*
Total population, million	1, 314	237	18%
GRP (market price), billion US$	2,719	1,109	41%
Total commercial energy consumption, million TJ	65.7	26.2	40%
Commercial energy consumption per capita, MJ/person (registered permanent population)	50,000	110,771	2.2 times more
GDP/GRP per capita, US$/person (registered permanent population)	2,068	4,681	2.3 times more
CO_2 emissions (commercial energy-related), million tons	5,645	2,259	40%
CO_2 emissions per capita (commercial energy-related), tons/person	4.30	9.54	2.2 times more

Source: Dhakal 2011: 76

energy consumption and high economic output path adopted by cities such as Ningbo, Beijing, Guangzhou and Shanghai (eastern coast, with a strong presence of service industries, maritime warm climate), the high energy consumption and low economic output path adopted by cities such as Urumqi, Taiyuan and Hohhot (central and western region, with a strong presence of energy-intensive industries and continental extreme climate), and the third somewhere in between these two. While these differences certainly affect the criteria by which one can compare and evaluate energy performance and prescribe energy policies, local authorities have little control over the criteria and indicators used in the ratings.

Following Dhakal (2011: 77–79), it is possible to evaluate the implications of urban energy consumption for CO_2 emissions by looking at the available data on total energy-related CO_2 emissions (million tons) of three megacities: Beijing, Shanghai and Tianjin. The data show rapid growths in energy use and CO_2 emissions from each of these cities, particularly since the early 2000s. In Shanghai, emissions increased by 5 times between 1985 and 2006, in contrast to 2.6 times for Beijing and 2.8 times for Tianjin in the same period. In the 2000 to 2006 period alone, these cities increased their emissions by a factor of 1.5 to 1.7. In 2006, the gap in emissions among the three cities was large. In 1995, energy use in Beijing, Tianjin and Shanghai was equivalent to 2 tons of oil in each city, with 8 tons of CO_2 emissions per capita (registered population); however, by 2006, CO_2 emissions per capita for Beijing and Shanghai had reached 9 tons and 12.6 tons, respectively.

The sectoral analysis indicates that the industrial sector has dominated CO_2 emissions in these cities. In particular, Beijing and Shanghai underwent rapid transformations between 1985 and 2006, characterised by a rapid decline in the industrial sector's share of total CO_2 emissions – 65 to 43 percent for Beijing and 74 to 64 percent for Shanghai – and increasing shares for the commercial and transportation sectors. These figures indicate a rapid urban transition. The residential sector (buildings) has also grown tremendously in energy use and carbon emissions. However, due to the rapid growth and expansion of other energy-consuming sectors, the residential sector's proportion of energy use and carbon emissions has remained more or less unchanged over the past two decades in Beijing, Shanghai and Tianjin. Although the transportation sector's share in 2006 was relatively small – 7 percent for Tianjin and 16 percent for Beijing – its growth rate has been very high due to the rising car ownership rates in these cities. Despite strong control over vehicular ownership in Shanghai, CO_2 emissions from the transportation sector increased eight fold between 1985 and 2006. Beijing registered close to a seven-fold increase in the same period. In the past decade alone, CO_2 emissions in Tianjin have increased by a factor of almost 3.5.

Fuel usage has brought a rapid change in CO_2 emissions. A rapidly declining trend in the contribution of coal burning to both energy consumption and CO_2 emissions is evident. Between 1985 and 2006, coal's share of CO_2 emissions declined from 58 to 26 percent, 51 to 18 percent and 61 to 33 percent in Beijing, Shanghai and Tianjin, respectively, thus indicating a rapid transition. In contrast, the contribution of electricity and oil to CO_2 emissions and energy consumption is rising, which is compensating for the declining share of coal burning.

What these factors do not fully reflect is that urban responses to climate change in Chinese cities is experiencing a governance transition – from the traditional state-centric mode of governing to a more pluralistic and collaborative approach. Through the theoretical lenses of different conceptual streams in the governance paradigm (i.e. network governance, collaborative governance and interactive governance), this book examines how such transition dynamics – involving interactions of state and nonstate actors – are leading to changes in China's climate governance and in the overall evolution of the Chinese environmental state.

The contents of the book

Building on the basic point concerning the challenge of governing climate change in Chinese cities, this book aims to fill some of the gaps that are still to be found in the growing research literature on subnational climate networks, the Chinese environmental state in transition, governance networks, social networks and policy networks, organisational forms of governance networks, innovation, information and incentives, and the issues surrounding collaboration, coordination and the legitimisation of climate networks.

The volume is structured in three parts: Part I, Conceptions and context, provides an analysis of the essential concepts and current debates surrounding urban climate governance in China. Chapter 1 presents a succinct yet thorough examination of the current thinking about climate change policies that informs climate mitigation governance in urban China and the factors impinging on it. The chapter strongly justifies the governance approach to climate mitigation and CMNs as a form of urban governing and as a novel approach to political inquiry. Chapter 2 grounds the discussion, the arguments and the theoretical framework proposed in this volume (Figure 2.2) in a thorough review of the literature regarding the concepts of governance, coordination, collaboration and networks, with the purpose of providing clarity about the theoretical and empirical gaps left in the literature and the questions addressed in this study. This chapter also includes the rationale for the research design and methodology. Urban climate networks are embedded in given institutional frameworks and sociopolitical contexts. In the case of the three cities studied in this volume, Guangzhou, Shenzhen and Hong Kong, their contrasting constitutional contexts are also factors that impinge on their dynamic interactions. These contextual considerations are presented in Chapter 3.

Having established the concepts and context, the book turns to the thematic chapters in Part II, Dynamics of climate collaborative municipal networks. The discussion begins with what is happening within the networks themselves, with Chapter 4 considering how *intra*governmental mechanisms facilitate the coordination of climate change issues and whether they can overcome functional fragmentation to achieve horizontal synergy. The chapter theoretically discusses how the coordination dynamics are configured in the network prototype stage of CMNs. Cross-case comparative analysis and within-method triangulation are used in this chapter.

The dynamics and explanations of how green buildings can be integrated as a sectoral mitigation innovation in an expanding nongovernmental space are discussed in Chapter 5. Given the relative success of green building policies, this chapter also addresses the question of why this integration has been so successful. Cross-case comparative analysis and within-method triangulation are used in this chapter. Chapter 6, in contrast, highlights the problem of integrating green transportation technologies (electric vehicles, specifically) as a sectoral mitigation innovation. The chapter focuses on the drivers of and barriers to legitimating the uptake of electric vehicles in Hong Kong, highlighting the potential for (future) collaboration in this sector. Single-case analysis based on the unique case principle, together with cross-method triangulation, are the methods used in this chapter.

Chapter 7 addresses the entire preparation, implementation and impact generation process of sectoral climate mitigation experiments conducted in the built environment and transport sectors. Through a discussion of several experiments in these sectors and how these are institutionalised in CMNs, this chapter examines how the nexus of coordination and collaboration configures the CMNs. Diverse case analysis, rather than cross-case comparative analysis and within-method triangulation, is the design adopted in this chapter. In Part III, Challenges of climate change governance in China, Chapter 8 highlights that since sectoral *intra*governmental coordination structures display fragmented *intra*governmental capacity in low carbon transitions in Hong Kong, collaborative municipal networks will necessarily encounter obstacles at the implementation stage of this transition. As such, the chapter proposes a carbon reduction implementation strategy (CRIAS) firmly grounded in the political and technical feasibility of uptaking specific carbon reduction policy tools. Finally, Chapter 9, which highlights the governance of climate mitigation challenges, discusses CMNs in light of the empirical findings and summarises the research findings. This chapter concludes that CMNs are adjusting municipal governance practices to accommodate sociotechnical innovations and/or mitigation responses. It also addresses the theoretical and empirical implications of the study, acknowledges its limitations and offers suggestions for future research. These include but should not be limited to empirical studies that compare and contrast climate governance in other Chinese cities (first, second and third tier) in the east and west of China, which would allow concept development, hypothesis testing and theory building. Comparisons of cities across various Asian countries and other policy areas (energy for example) are also needed. The governance approach to governing is being transferred to policy problems in both the Global North and the Global South, as societies continue to be challenged on how to govern themselves. If it is here to stay, we should continue perfecting governance ideas and empirics. We hope this volume provides an initiation to such thinking in the Chinese context.

Summary

This chapter contextualises the study of climate mitigation governance in China by outlining the background on China's approach to governing complex issues and how this approach fits within the literature on governance as a trend and object of study, not just in the field of public administration but also in the field of urban politics. We have outlined our arguments concerning the relevant issues: first, regarding the climate change mitigation policies that are being formulated and implemented, highlighting their significance and direction, and second, regarding the factors that affect carbon reduction in Chinese cities, including urban governance, the policy design of development plans and urban energy consumption. Although these factors are essential in understanding the drivers and challenges of climate mitigation in Chinese cities, they do not recognise the newly emerging, informal and less institutionalised changes to urban governance, such as the formation of what we term collaborative municipal networks (CMNs), which we formally define in this chapter. Understanding how these new urban governance forms affect climate mitigation policies, urban transitions and the capacity of the Chinese environmental state is essential for interpreting and evaluating climate governance trends in China.

The chapter also provides a strong justification for taking a governance approach to the study of climate change mitigation in China. Chapter 2 grounds the discussion, the arguments and the theoretical framework in a thorough review of the literature regarding the concepts of governance, coordination, collaboration and networks with the purpose of providing clarity about the theoretical, empirical and methodological gaps left in the literature and the questions that need to be addressed.

Notes

1 BRICS is an acronym for a grouping of five major emerging national economies: Brazil, Russia, India, China, and South Africa.

2 These cities are Beijing, Tianjin, Shijiazhuang, Taiyuan, Hohhot, Shenyang, Dalian, Changchun, Harbin, Shanghai, Nanjing, Hangzhou, Ningbo, Hefei, Fuzhou, Xiamen, Nanchang, Jinan, Qingdao, Zhengzhou, Wuhan, Changsha, Guangzhou, Shenzhen, Nanning, Haikou, Chongqing, Chengdu, Guiyang, Kunming, Xi'an, Lanzhou, Xining, Yinchuan and Urumqi.

References

Bai, X. 2007. Integrating Global Environmental Concerns into Urban Management: The Scale and Readiness Arguments. *Journal of Industrial Ecology*, 11, 15–29

Balme, R. 2011. China's Climate Change Policy: Governing at the Core of Globalization. *CCLR [The Carbon & Climate Law Review]*, 5, 44–56

Brown, M. A. & Logan, E. 2008. *The Residential Energy and Carbon Footprints of the 100 Largest US Metropolitan Areas.* School of Public Policy, Georgia Institute of Technology, Working Paper 39

Bulkeley, H. 2005. Reconfiguring Environmental Governance: Towards a Politics of Scales and Networks. *Political Geography*, 24, 875–902

Bulkeley, H. 2010. Cities and the Governing of Climate Change. *Annual Review of Environment and Resources*, 35, 229–253

Bulkeley, H. 2012. Governance and the Geography of Authority: Modalities of Authorisation and the Transnational Governing of Climate Change. *Environment and Planning-Part A*, 44, 2428

Bulkeley, H., Broto, V. C. & Maassen, A. 2013. Low-Carbon Transitions and the Reconfiguration of Urban Infrastructure. *Urban Studies*, 51, 1471–1486

Bulkeley, H. & Newell, P. 2010. *Governing Climate Change*, Abingdon, Oxon, Routledge.

Bulkeley, H., Schroeder, H., Janda, K., Zhao, J., Armstrong, A., Chu, S. Y. & Ghosh, S. 2009. *Cities and Climate Change: The Role of Institutions, Governance and Urban Planning.* 5th Urban Research Symposium: Cities and Climate Change. Marseille, France.

CASS. 2013. *Reconstruction of China Low-Carbon City Evaluation and Indicator System: A Methodological Guide for Application*, Beijing, Social Sciences Academic Press.

CEU. 2005. *Presidency Conclusions – Brussels, 22 and 23 March 2005.* European Commission, Council of the European Union. Brussels, Belgium. Available: www.eu2005.lu/en/actualites/conseil/2005/03/23conseileuropen/ceconcl.pdf [Accessed 27 September 2013]

Chu, S. Y. & Schroeder, H. 2010. Private Governance of Climate Change in Hong Kong: An Analysis of Drivers and Barriers to Corporate Action. *Asian Studies Review*, 34, 287–308

Corfee-Morlot, J., Kamal-Chaoui, L., Donovan, M., Cochran, I., Robert, A. & Teasdale, P.-J. 2009. *Cities, Climate Change and Multilevel Governance.* OECD Environmental Working Papers (14). Paris, France, OECD Publishing.

Dhakal, S. 2009. Urban Energy Use and Carbon Emissions from Cities in China and Policy Implications. *Energy Policy*, 37, 4208–4219

Dhakal, S. 2011. Urban Energy Transitions in Chinese Cities, in Bulkeley, H., Broto, V. C., Hodson, M. & Marvin, S. (eds.) *Cities and Low Carbon Transitions*, New York, NY, Routledge.

Elzen, B., Geels, F. W. & Green, K. (eds.) 2004. *System Innovation and the Transition to Sustainability: Theory, Evidence and Policy*, Gloucestershire, UK, Edward Elgar Publishing.

Francesch-Huidobro, M. 2012. Institutional Deficit and Lack of Legitimacy: The Challenges of Climate Change Governance in Hong Kong. *Environmental Politics*, 21, 791–810

Francesch-Huidobro, M. & Mai, Q. 2012. Climate Advocacy Coalitions in Guangdong, China. *Administration & Society,* 44, 43S–64S

GaWC. 2011. *The World According to GaWC 2010* [Online]. Available: www.lboro.ac.uk/gawc/world2010t.html [Accessed 5 September 2013]

Geels, F. W. 2004. From Sectoral Systems of Innovation to Socio-Technical Systems: Insights about Dynamics and Change from Sociology and Institutional Theory. *Research Policy,* 33, 897–920

Hansen, J., Sato, M., Ruedy, R., Lo, K., Lea, D. W. & Medina-Elizade, M. 2006. Global Temperature Change. *Proceedings of the National Academy of Sciences,* 103, 14288–14293

Harris, P. G. 2012. *Transport, Environmental Policy and Sustainable Development in China: Hong Kong in Global Context,* Bristol, UK, The Policy Press.

Hart, P. T. 2013. Interactive Governance: Advancing the Paradigm by Jacob Torfing, B. Guy Peters, Jon Pierre and Eva Sørensen. *Public Administration,* 91, 1071–1073

He, B. 2012. Western Theories of Deliberative Democracy and the Chinese Practice of Complex Deliberative Governance, in Leib, E. J. & He, B. (eds.) *The Search for Deliberative Democracy in China,* Basingstoke, UK: Palgrave Macmillan.

He, B. & Thøgersen, S. 2010. Giving the People a Voice? Experiments with Consultative Authoritarian Institutions in China. *Journal of Contemporary China,* 19, 675–692

Healey, P. 2007. *Urban Complexity and Spatial Strategies: Towards a Relational Planning for Our Time,* London, Routledge.

Ho, P. 2007. Embedded Activism and Political Change in a Semiauthoritarian Context. *China Information,* 21, 187–209

Ho, P. & Edmonds, R. L. 2007. Perspectives of Time and Change Rethinking Embedded Environmental Activism in China. *China Information,* 21, 331–344

Hodson, M. & Marvin, S. 2010. Can Cities Shape Socio-Technical Transitions and How Would We Know If They Were? *Research Policy,* 39, 477–485

Hooghe, L. & Marks, G. 2001. *Multi-Level Governance and European Integration,* Boulder, CO, Rowman & Littlefield.

IEA. 2007. *World Energy Outlook 2007,* Paris: International Energy Agency.

Innes, J. E. & Booher, D. E. 2010. *Planning with Complexity: An Introduction to Collaborative Rationality for Public Policy,* Abingdon, Oxon, Routledge.

IPCC. 2007. Climate Change 2007: Mitigation of Climate Change, in Metz, B., Davidson, O. R., Bosch, P. R., Dave, R. & Meyer, L. A. (eds.) *IPCC Fourth Assessment Report – Working Group III,* Cambridge, UK, and New York, NY, Intergovernmental Panel on Climate Change.

IPCC. 2013. *Climate Change 2013: The Physical Science Basis – (Approved) Summary for Policymakers.* IPCC Fifth Assessment Report: Working Group I. Stockholm, Sweden, Intergovernmental Panel on Climate Change.

Kamal-Chaoui, L. & Roberts, A. 2009. *Competitive Cities and Climate Change.* OECD Regional Development Working Paper No. 2.

Kooiman, J. 1993. *Modern Governance: New Government–Society Interactions,* London; Thousand Oaks, CA, SAGE.

KPMG Insight Series. 2011. *China's 12th Five-Year Plan (2011–2015) – KPMG Insight Series.* Hong Kong [Online]. Available: www.kpmg.com/cn/en/issuesandinsights/articlespublications/publicationseries/5-years-plan/pages/default.aspx [Accessed 8 May 2013]

Lee, T. & Van De Meene, S. 2012. Who Teaches and Who Learns? Policy Learning through the C40 Cities Climate Network. *Policy Sciences,* 45, 199–220

Levi-Faur, D. 2012. *The Oxford Handbook of Governance,* Oxford, Oxford University Press.

Li, Y., Miao, B. & Lang, G. 2011. The Local Environmental State in China: A Study of County-Level Cities in Suzhou. *The China Quarterly,* 205, 115–132

Lieberthal, K. 1997. China's Governing System and Its Impact on Environmental Policy Implementation. *China Environment Series,* 1, 3–8

Lieberthal, K. & Lampton, D. M. 1992. *Bureaucracy, Politics, and Decision Making in Post-Mao China,* Berkeley, University of California Press.

Lieberthal, K. & Oksenberg, M. 1988. *Policy Making in China Leaders, Structures, and Processes,* Princeton, NJ, Princeton University Press.

Liu, Z. & Salzberg, A. 2012. Developing Low-Carbon Cities in China: Local Governance, Municipal Finance, and Land-Use Planning – the Key Underlying Drivers, in Baeumler, A., Ijjasz-Vasquez, E. & Mehndiratta, S. (eds.) *Sustainable Low-Carbon City Development in China,* Washington, DC, World Bank.

Miao, B. & Lang, G. 2010. China's Emissions: Dangers and Responses, in Lever-Tracy, C. (ed.) *Routledge Handbook of Climate Change and Society,* Abingdon, Oxon, Routledge.

Norman, J., MacLean, H. L. & Kennedy, C. A. 2006. Comparing High and Low Residential Density: Life-Cycle Analysis of Energy Use and Greenhouse Gas Emissions. *Journal of Urban Planning and Development,* 132, 10–21

O'Toole, R. 2009. The Myth of the Compact City: Why Compact Development Is Not the Way to Reduce Carbon Dioxide Emissions [Online]. *Cato Policy Analysis Series,* Washington, DC: Cato Institute. Available: http://www.cato.org/sites/cato.org/files/pubs/pdf/pa653.pdf [Accessed 15 April 2014]

Pierre, J. 2000. *Debating Governance: Authority, Steering, and Democracy,* Oxford, Oxford University Press.

Pierre, J. & Peters, G. B. 2000. *Governance, Politics and the State,* Basingstoke, Hampshire, UK, Macmillan Press.

Qin, D., Ding, Y., Lin, E., He, J., Zhou, D., Wang, H., Luo, Y., Ding, Y., Wu, S., Pan, J.-H., Ge, Q. & Yu, G. 2011. *Second National Assessment Report of Climate Change,* Beijing, China, China Meteorological Administration, Ministry of Science and Technology of the PRC.

Qiu, B. 2008. *Responding to Opportunities and Challenges: Key Issues and Policies for Urbanization Strategies in China,* Beijing, China Architecture Industry Press.

Rhodes, R. A. 1997. *Understanding Governance: Policy Networks, Governance, Reflexivity and Accountability,* Bristol, UK, Open University Press.

Rijsberman, F. R. & Swart, R. J. 1990. *Targets and Indicators of Climatic Change,* Stockholm: Stockholm Environment Institute.

Schroder, M. 2011. *Local Climate Governance in China: Hybrid Actors and Market Mechanisms,* Basingstoke, Hampshire, UK, Palgrave Macmillan.

Shah, A. & Shen, C. 2006. *The Reform of the Intergovernmental Transfer System to Achieve a Harmonious Society and a Level Playing Field for Regional Development in China.* Policy research working paper, Washington, DC, World Bank.

Sørensen, E. & Torfing, J. 2005a. The Democratic Anchorage of Governance Networks. *Scandinavian Political Studies,* 28, 195–218

Sørensen, E. & Torfing, J. 2005b. Network Governance and Post-Liberal Democracy. *Administrative Theory & Praxis,* 27, 197–237

Sørensen, E. & Torfing, J. 2007. *Theories of Democratic Network Governance,* Basingstoke, UK, Palgrave Macmillan.

Torfing, J., Peters, B. G., Pierre, J. & Sørensen, E. 2012. *Interactive Governance: Advancing the Paradigm,* Oxford, Oxford University Press.

UN. 1994. *The United Nations Framework Convention on Climate Change,* Rio, Brazil, United Nations.

Wang, J. & Tang, M. 2005. The Romance of the Three Kingdoms in the Formulation of City Master Plan. *The Outlook Newsweek,* 45.

Wang, W., Zheng, G. & Pan, J. 2013. *China's Climate Change Policies,* Abingdon, Oxon, Routledge.

World Bank. 2007. *EAP Sustainable Development on the Urban Fringe,* Washington, DC, World Bank.

World Bank. 2009. *The Spatial Growth of Metropolitan Cities in China: Issues and Options in Urban Land Use,* Washington, DC, World Bank.

Wu, F. 2003. Environmental GONGO Autonomy: Unintended Consequences of State Strategies in China. *The Good Society,* 12, 35–45

Zhuang, G. 2012. Towards a Low-Carbon Economy: International Experience and Development Trends, in Wang, W., Zheng, G., Pan, J., Luo, Y., Chen, Y. & Chen, H. (eds.) *China's Climate Change Policies.* Abingdon, Oxon, Routledge.

2 Climate governance through collaborative municipal networks (CMNs)

The arguments and theoretical framework proposed in this volume regarding the challenges of climate change governance in Chinese cities are grounded on a review of the contemporary literature on the concepts of governance, coordination, collaboration and networks. The purpose of this review is to clarify the theoretical and empirical gaps in the literature and to address the questions of this study. In this chapter, we shall first identify the notions that define CMNs and review the existing research on subnational climate networks to identify the gaps in this literature. The chapter then proceeds to a survey of the concepts that are integral to the conceptualisation and theorisation of subnational climate networks, such as the environmental state, innovation, information, incentives and legitimation, with the purpose of devising a suitable theoretical framework (Figure 2.2) to explain how the dynamic interactions within CMNs configure and transform the institutional structure of subnational governance for climate change. This chapter also includes a justification for the research design and methodology (Figure 2.1).

Research on subnational climate networks

Existing studies on urban climate governance can be summarised into four categories, each with its own distinctive research focus and objectives.

The first category comprises *review articles* that examine the development of subnational climate governance in relation to various policy initiatives and theoretical debates and also set forth the future agenda for research (Betsill and Bulkeley 2007; Biermann and Pattberg 2008; Bulkeley 2010; Bulkeley and Newell 2010; Bulkeley *et al.* 2009; Corfee-Morlot *et al.* 2009; Okereke *et al.* 2009). For example, Betsill and Bulkeley (2007) point out that after a decade of researching climate change governance, it is widely acknowledged that cities form a critical arena in which various interactions in the process of governance take place. They also highlight the significance of conducting research on whether and how local authorities are planning for the impacts of climate change (Betsill and Bulkeley 2007). Okereke, Bulkeley and Schroeder (2009) further emphasise the '*how* of governance' (or *governmentality*), calling for renewed research attention on the governance process and the actual dynamics of governance. In a recent review, Bulkeley (2010) examines various case studies, elaborates the multilevel governance context adopted in previous studies and assesses the structural and political factors that shape subnational climate governance. Several aspects of this research are highlighted to provide new insights: i) although existing studies recognise multilevel governance as the context and process that provides and structures opportunities for municipal actors, few studies pay real attention to how municipal-based governance of urban climate change problems may be reconfiguring the traditional state-based political authorities; and ii) the failing political rhetoric of urban commitments to climate change is a result of weak institutional capacity (i.e. the resources and portfolios of municipal authorities) and limited political opportunities and leadership (Bulkeley 2010). Future research directions are suggested, one of which is to engage with issues such as the processes of urban governance and the reconfiguration of political authority to understand the 'complex problem' of urban climate governance (Bulkeley 2010).

These agenda articles also pinpoint certain theoretical deadlocks in the field. Biermann and Pattberg (2008) identify the problem of how to apply existing theories to empirical studies of local climate governance, as most of the established theoretical approaches, such as governance theories, are not specifically tailored for such a newly emerging problem. They suggest that the theories need refinement before they can fit into explanatory research foundations, although they do not propose a feasible theoretical framework to resolve such deficiencies. Okereke et al. (2009) diagnose a stagnated theorisation of governance in the non-nation-state climate regime and consider the potential for thinking about the political leverage of cities, multinational corporations and carbon offset firms. Further research is needed to fill this theoretical gap and generate new insights to supplement established theories.

The second category contains *municipal-based research* and *regional-specific climate governance* studies. Most of the significant research on the subnational governance of climate change has been conducted in North America and Europe, such as Betsill's studies in American cities (Betsill 2001; Betsill 2007), Bulkeley's studies in the United Kingdom (Bulkeley and Kern 2006; Bulkeley and Schroeder 2008) and Kern's studies in Germany (Alber and Kern 2009; Bulkeley and Kern 2006). Betsill's (2001) study of the Cities for Climate Protection campaign suggests that to achieve actual policy change, globally framed climate change issues need to be localised to municipal decisions, despite the possible institutional barriers in city governments such as limited administrative capacity and budgetary constraints. A comparative analysis of local climate change policies in Germany and the UK also stresses the local dimension of climate protection policy while identifying that the contingent factors in the local capacity of European cities are the effects of financial crises, European Union policies and the political challenges of implementation (Bulkeley and Kern 2006). By analysing the emergence of regional governance institutions on the other side of the Atlantic, for example the North American Commission for Environmental Cooperation, Betsill (2007) argues that the multilevel governance of climate change has been institutionalised not only in the European Union but also in North America. All of these studies have strengthened theoretical development in the field, and their empirical findings can be generalised to cities with similar institutional settings and levels of economic development. However, they also have a 'geographical bias towards cities in more economically developed countries' (Castán Broto and Bulkeley 2012: 92). Given the radically different institutional setting, it is vital to study Chinese cases and build up a research base relevant to Chinese cities to enable further comparison.

The third category of review and research agenda articles involves *policy studies*, either specifically focusing on mitigation measures and adaptation strategies to climate change or looking at policies spanning various policy sectors. Most research efforts concentrate on energy, a strategically critical policy area (Bulkeley and Kern 2006; Bulkeley and Schroeder 2008; Dhakal 2004, 2009, 2011; Schroeder and Bulkeley 2008). Our study, in contrast, focuses on two end-use energy sectors: building and transportation. Although municipal authorities have more leverage to support change in these sectors, there has been relatively little climate policy and governance research in this area (Dodman 2009). Besides research on different policy sectors, policy studies continue to focus on either mitigation or adaptation (Alber and Kern 2009; Francesch-Huidobro 2011; Satterthwaite 2008; Pauw and Francesch-Huidobro 2010). However, one of our recent studies on climate policy advocacy coalitions in Guangdong (China) underlines the role of government-led nongovernmental organisations in the local policy subsystem of climate mitigation in China (Francesch-Huidobro and Mai 2012).

The fourth category consists of *institutional studies* that take a multilevel governance approach. These studies regard cities as non-nation-state actors within the governance networks for climate change. Cities as actors have varying degrees of horizontal autonomy through networks connecting them with other local actors and changing levels of vertical autonomy through their hierarchical relations with the national state (DeAngelo and Harvey 1998). Hooghe and Marks (2001) classify multilevel governance into two types, leading to two dimensions in the institutional studies of subnational climate governance: i) vertical coordination and ii) horizontal collaboration through subnational/transnational networks (Bulkeley *et al.* 2009). Vertical coordination conceives governance as the negotiation of power and the distribution of resources between different vertical levels of government (Hooghe and Marks 2001). Because national authorities have a set of functions distinct from those of their subnational counterparts, policy makers promoting low carbon development must incorporate a national strategy to steer the development of climate change policies while proactively engaging with subnational governance issues (Bulkeley *et al.* 2009). In the horizontal dimension of multilevel governance, a variety of often overlapping and interdependent spheres of state and nonstate authority are involved in allocating power and resources (Hooghe and Marks 2001). Institutional studies thus tend to be concerned with how to manage the subnational networks and partnerships among public and private actors. Recent attention has also been drawn to an emerging form of urban climate-change governance – transnational municipal networks (or 'transmunicipal networks'), which connect cities beyond national borders (Bulkeley 2005; Kern and Bulkeley 2009; Lee and Van De Meene 2012).

Recognising the research gaps

The review of recent studies on subnational climate networks indicates three major research gaps: i) potential theoretical incompatibility with and inapplicability to empirical research, ii) a lack of exploration into the process of governance and iii) limited theoretical and empirical application of existing research to China.

Compatibility and applicability

The first gap lies in the intertwined issues of *theoretical compatibility and generalisability*. First, the internal inconsistencies in existing theoretical frameworks may lead to theoretical inapplicability. For example, conceptual frameworks related to collaborative governance – one of the essential theoretical components used in theory building later in this chapter – 'lack generalisability' given their 'inapplicability across different settings, sectors, geographic and temporal scales, policy arenas, and process mechanisms' (Emerson *et al.* 2012: 3). Without an integrative framework of core concepts and theories, it is difficult to apply them in different and changing empirical settings.

Second, existing theoretical frameworks are also limited in their applicability to studies of urban governance in China because they originated in radically different contexts. For instance, the multilevel framework for the study of subnational climate change governance originated from Collier's (1997) research on the institutional development of climate protection in the European Union, and the two-type multilevel governance framework of Hooghe and Marks (2001) was based on reflections about federalism.

Third, there is another possible incompatibility in the theoretical assumptions. Democratic theory is the foundation for network governance and governance networks, the two core concepts commonly applied in researching the urban governance of climate change (Koliba *et al.* 2010). Network governance encompasses a greater number of actors involved in horizontal network relationships and is regarded as an extension of 'stretched liberal democratic processes' (Bogason and Musso 2006: 3). Governance studies have likewise presumed a close relationship between representative democracy and governance networks (Bogason and Musso 2006; Klijn and Skelcher 2007; Sørensen and Torfing 2005b). Scholars have sought to measure the democratic implications of these networks, particularly those of interactive network governance (Bogason and Musso 2006; Sørensen and Torfing 2005a).

However, China is a regime without representative democracy. To what extent can we empirically apply the explanatory power of existing network-governance theories to the case of China? As demonstrated in the thematic chapters of this volume (Chapters 4, 5, 6 and 7), collaborative network arrangements in the governance process of climate change in Chinese cities involve an extensive empirical reality, which has been grossly overlooked and understudied. Some China scholars have argued that China is a state undergoing a revised form of deliberative democracy (He 2012; He and Thøgersen 2010; Zhou 2012). Meanwhile, network governance is recognised as moving society away from traditional representative democracy to more 'negotiated and deliberative models', which indicates the strong potential for network governance to be theoretically compatible with the Chinese context (Bogason and Musso 2006). Diamond (2002) classifies China as a hybrid regime that is 'politically closed authoritarian', given its lack of an open and competitive electoral system at almost all government levels. However, he also emphasises that despite being politically closed in terms of its electoral system, China has 'quasi-constitutional mechanisms to limit power and consult broader opinion' and 'has taken some steps to rotate power and to check certain abuses of corrupt local and provincial officials'; therefore, 'significant steps toward a more open, competitive, pluralistic, and restrained authoritarian system can emerge in arenas other than electoral ones' (Diamond 2002: 33). This is an important insight supporting the study of China's network governance in non-electorally related problem areas, such as the mitigation of climate change, in which a more pluralistic system is encouraged by the state's developmental needs and empirically exists in the regime context.

Unexplored local climate governance

The second gap in current studies on subnational climate networks is the lack of attention to the *local governance process of climate change*. Bulkeley and Kern (2006) propose four governing modes, reflecting their particular interest in 'engaging with the processes and power dynamics through which governing is orchestrated'. These modes are *self-governance* of local government's own activities, *governing by provision* of services and resources, *governing by authority* through regulation and direction and *governing through enabling* the participation of other stakeholders. This line of thinking has been adopted by subsequent studies (Alber and Kern 2009; Bulkeley and Schroeder 2008). However, governing modes are not the same as the governance process (or *governmentality*). The latter relates not only to the modes of action initiated by governing authorities but also to the *dynamic interactions* between the government and other actors in civil society and the private sector. Both concepts are core elements that are integrated into the conceptualisation and theoretical framework of our study (Figure 2.2).

Transfer of knowledge and best practices, an essential function performed by governing through enabling, is pertinent to the horizontal subnational network governance of climate change. There are multiple channels of information transfer: vertical policy coordination by transferring national guidelines to local governments; cross-sector integration of technological innovations; horizontal exchange of governance practices and experience across nearby localities; and transnational exchange of best practices. The transnational networking of municipalities appears as a focus of research in knowledge transfer given its large scope of influence (Alber and Kern 2009; Collier 1997). However, the process of municipal governance and how it configures the influence of transnational and horizontal municipal networks remains largely unexplored. Therefore, the focus should be drawn back to the local governance sites. This is particularly the case when studying the Chinese urban governance of climate change because established transnational municipal networks concerned with climate change issues (such as ICLEI and C40) have exerted only marginal influence on their city members, let alone local climate change policies in China.

Drivers of and barriers to local governance for climate change are also discussed in current studies of climate change governance at the city level. For instance, through comparative case studies of several cities in EU member states, Collier (1997) suggests several explanatory factors for local implementation capacities: the history of local engagement, institutional establishment and staffing, local competencies, reduction potential and the availability of financial resources. Betsill (2001) proposes that bureaucratic structures, administrative capacity and budgetary constraints have created institutional barriers to municipal action. Along the same lines, Satterthwaite (2008) argues that particularly for cities in developing countries, institutional capacity is a determining factor in developing climate adaptation measures. Conversely, the interface of cities' political, business and global leaderships driven by policy entrepreneurs is a driver of local climate change governance (Bulkeley and Schroeder 2008).

While these studies are certainly informative, they do not provide an integrative theoretical framework consolidating all of the relevant factors that shape the governance process and that are applicable to cities of different political structures and economic scales. Such an integrative theoretical framework requires a comprehensive research base that includes case studies in cities not only in the Global North but also in the Global South. The 'geographical bias' of the current research field suggests that a theoretical approach to systematic comparative analysis is not widely available, considering the lack of case studies of climate governance process in developing countries (Castán Broto and Bulkeley 2012). In depth case studies of cities in developing regions, especially China, are therefore indispensable.

Limited research on China

This leads to the third gap in current studies on subnational climate networks – *the paucity of theoretical and empirical application of existing research to China.* Betsill and Bulkeley (2007, 2010) find that current research is concentrated on networks in cities of industrialised countries in the Global North and on a small number of cases. Therefore, the evidence base needs to be expanded. Considering that a growing number of cities all over the world have become engaged in climate change solutions, further comparative research in this area is needed. Moreover, as a large proportion of existing case studies focus on cities embedded in the relatively stable environmental states of European and other OECD countries, it is of theoretical significance to examine the dynamic trajectories of cities belonging to restructuring environmental states, of which China is a case in point, taking into account their empirical uniqueness.

The few studies that focus on Chinese cities are also limited both empirically and theoretically. First, much of the research is concentrated in the governance of the energy sector, such as energy-related policies and emissions reduction (Dhakal 2009, 2011; Heggelund 2007; Levine and Aden 2008; Richerzhagen and Scholz 2008). It is necessary to expand the scope of research to the governance of other policy areas that are significant to emissions reduction, including the transportation and building sectors.

Second, most of the existing governance studies analyse China as a nation. These national-level studies heavily focus on exploring the power struggles and regime legitimation of China as a national player in the international negotiations of the United Nations Framework Convention on Climate Change (UNFCCC) (Balme 2011; Delman 2011; Harris 2010; Howard 2011; Wang *et al.* 2012; Zang, 2009b). Although the governance and institutional development of climate policy and coordination has been studied with regards to China, such studies are limited to the role of the national leadership (Heggelund 2007; Tsang and Kolk 2010; Yu 2004; Zang 2009a, 2009b). Moreover, discussions on the governance of low carbon development in the Chinese context relate to measures appropriate to a national strategy – developing and deploying clean energy and supporting technologies, establishing management and regulatory institutions and enhancing public awareness of greenhouse gas mitigation (Jiang *et al.* 2010; Pan 2010; Pan *et al.* 2010). However, the specific interactive process of how to implement these measures at the subnational level remains unexplored. Although the foregoing analyses and findings are limited to the national level, the study conducted by Richerzhagen and Scholz (2008) recognises that climate capacities are distributed not only among governmental actors but also among nongovernmental actors. The step we are taking in this volume is to investigate the interactive process among these identified actors, not only at the national but also at the local level.

Third, a few case studies have explored the dynamic process of interaction among network actors for climate policy at the local level. Initial attempts examined the responses of selected Chinese cities to climate change (Bai 2007; Qi *et al.* 2008). Zhao (2011) conducted a local case study in Hong Kong; however, it merely reviews the proposed local mitigation measures and their future potential without discussing the interactive process among network actors.

Our recent study of climate advocacy coalitions in Guangzhou sought to investigate the interactions among advocacy coalitions and local government agencies (Francesch-Huidobro and Mai 2012). This study extends the previous research scope and examines the dynamic interactions within a broader range of governance networks.

Nevertheless, important insights have emerged from studies on China, which this volume builds on and takes forward. First, it is now agreed that nongovernmental and external actors should be incorporated in studies of multilevel governance structures as a basis for the analysis of network governance for climate change (Eichhorst and Bongardt 2009; Richerzhagen and Scholz, 2008). Most significantly, Eichhorst and Bongardt's (2009) analysis posits that the cooperative, voluntary approach developed in Chinese cities has evolved into a tool of local implementation responses to the stringent national mandates on energy efficiency, that is a domestic driving force structured in the *vertical hierarchy*. Second, cooperative arrangements have become a way to improve relations between local enterprises and local authorities. Such arrangements have the potential to activate the formation of *horizontal networks* at the local level.

A knowledge-based process of policy specialisation and norm making imposed by the international climate change regime is identified as a significant driver of national policy change in China, while bottom-up policy innovations (in addition to the traditional top-down commands) are also found in the process (Balme 2011; Schreurs 2008; Yu 2004). Our study thus further explores how global knowledge is localised and how the local experience is globalised in the process of urban climate governance.

Furthermore, obstacles to successful implementation are observed in existing studies on policy implementation in China. The capacity for implementation – in terms of enforcement power and the available resources of local governments' mitigation actions – is questioned (Francesch-Huidobro *et al.* 2012; Li 2011; Lin 2012). In terms of social norms, there is a lack of incentives or a conflicting set of incentives at the local level and likewise an absence of social infrastructure – composed of trust, consensus, effective social organising and collective action – that distances citizens from pressing local governments to take action for climate change (Li 2011). This is partly due to the stringent state control of grassroots environmental nongovernmental organisations (ENGOs) in China imposed by the social-organisation licensing requirement (Yang 2008).

Institutionally, local responses are also constrained by i) the fragmented sectoral approach conventionally adopted by policy makers; ii) the lack of democratic governance, with unaccountable processes and undeveloped avenues for citizen participation; iii) the weakness of the rule of law and supporting legal institutions; and iv) the inherent 'institutional design bias' arising from putting the NDRC at the lead of Chinese climate change policies, allowing the dominant economic growth agenda to override climate-relevant objectives (Francesch-Huidobro 2012; Li 2011; Yang 2008).

Integrating concepts

Based on the review of the literature on subnational climate networks and identification of the gaps in the existing research on China, we shall review in this section the concepts that are integral to the conceptualisation and theorisation of subnational climate networks with the intent of devising a suitable theoretical framework (Figure 2.2).

Chinese environmental state in transition

'Environmental state' refers to the organic composition of instruments and arrangements, laws and regulations and institutional authorities in the state structure, intended to achieve the ends of the ecological metafunction – one of the three major functions of the state in addition to its economic and social metafunctions (Jänicke 2006; Mol and Carter 2006). Unlike developed countries in the Global

North, where the environmental state has been largely 'stabilised', China is experiencing a major restructuring of the institutional configuration of its environmental state. To make the transition to a low carbon economy, the Chinese national and local climate-governing regime has continuously shifted its leadership from the environmental protection agencies to the meteorological administration and the state development planners. The transformation of the environmental state is the *outcome variable* of this study. By analysing how the state and nonstate actors interact to achieve this transformation, our study aims to identify the key governance mechanisms driving such transformations.

As mentioned, the environmental state serves the ecological metafunction of the state, including 'environmental research and education, as well as the workings of the policy subsystem comprising environmental bureaucrats, political advocates, including environmental non-governmental organisations (ENGOs), the media, the scientific community and green parties' (Francesch-Huidobro and Mai 2012: 47S; Jänicke 2006). The Chinese environmental state is rapidly evolving, witnessing 'increased state involvement in environmental tasks through regulatory or enabling policies' (Francesch-Huidobro and Mai 2012: 47S; Li *et al.* 2011; Mol and Buttel 2002).

The process of transitioning the environmental state in climate governance terms involves a common trajectory of low carbon restructuring in the state's environmental regulation (Mol and Carter 2006; While *et al.* 2010). The restructuring has led to reforms of institutional arrangements and policy instruments concerned with climate change issues at various levels of government, including both national and subnational (provincial, prefectural and county/district). The core focus of our study is on the subnational level.

Governance networks, social networks and policy networks

Governance *networks* can be defined as a dynamic process of interaction among multiple social actors that takes place within a relatively institutionalised framework regarding issues of negotiation and self-regulated policy making in the shadow of a hierarchy, together with implementation and service delivery, with or without the direct involvement of government (O'Toole 1993; Pierre and Peters 2000; Sørensen and Torfing 2007). Studies of governance networks focus on the interactive style of governance, 'the complex process through which a plurality of social and political actors (organisations and/or individuals) with diverging interests interact to formulate, promote, and achieve common objectives by means of mobilising, exchanging, and deploying a range of ideas, rules, and resources' (Torfing *et al.* 2012: 2–4), which ultimately has the potential to transform the role and functioning of the state.

The 'relatively institutionalised framework' of governance networks is further specified by Knoke (2001), who sets out three basic conditions for mutual influences between networks and actors in a given policy domain: i) there are stable patterns of repeated interactions among actors in the social structure of any complex system; ii) the social relations among actors are the primary explanatory units of analysis; and iii) actors are embedded in multiple structural networks and thus their perceptions, attitudes and actions are shaped by these networks, while their behaviour can, in turn, change the network structure. The social relations and links among actors in network governance can be extended to the scope of multiple *inter*organisational ties.

There are various types of network relations – including resource exchange, information transmission and knowledge sharing, power relations and boundary penetration – that are embedded in *coordination enhancement* to maximise joint efforts and *facilitate collaboration* through developing joint definitions of and solutions to emerging problems (Knoke 2001; Torfing *et al.* 2012). Governance networks are also found in different forms: networks mandated from above and self-grown from below; formal and relatively close networks; informal and relatively open networks; *intra*organisational networks formed within public organisations; and joined-up government networks established between public organisations (Torfing *et al.* 2012). Taking into account the various forms and types of governance networks for urban responses to climate change, our study focuses on analysing the dynamic interactive processes in the *inter*organisational and *intra*governmental networks at the municipal level of governance, as depicted in the theoretical framework of CMNs (Figure 2.2). The concept of governance networks is incorporated in the study of CMNs as a notion and methodological tool that is distinct from social and policy networks. Governance networks are more specific to public administration (including policy studies) than are social networks.

The concept of *social networks* has been broadly applied in sociology, economics and management studies and in interdisciplinary social science studies such as human ecology. The fundamentals of social network analysis are still applied in governance network analysis, such as the logic of social structures taking precedence over individual nodes, and the ties of coordination and resource exchange that connect those nodes (Koliba *et al.* 2010). However, what distinguish governance networks from social networks are 'the characteristics of network actors and the kinds of functions and collective actions the actors take on' (Koliba *et al.* 2010: 43–44). Sørensen and Torfing (2005a: 197) further define this particular feature of governance networks as 'a relatively stable horizontal articulation of interdependent but operationally autonomous actors who interact through negotiations that involve bargaining, deliberation, and intense power struggles', where the networking 'take[s] place within a relatively institutionalised framework of contingently articulated rules, norms, knowledge, and social imaginaries, that is self-regulating within limits set by external agencies', and it ultimately 'contribute[s] to the production of public purpose in the broad sense of visions, ideas, plans, and regulations'.

Governance network analysis is also more appropriate than the notion of a *policy network*, which places a restricted emphasis on the governance process. Moreover, governance network theory has a broader definitional scope, combining policy network frameworks with system analysis and state theories (Koliba *et al.* 2010: 55). Policy networks tend to emphasise resource interdependency and the process of resource exchange in policy making while neglecting the structural effects of governance (Compston 2009; Hudson 2004). Governance network analysis positions itself more closely with the interaction dynamics of collaboration and coordination. It thus looks beyond the mere process of resource exchange to the dimensions in which the interactions affect network members and the network structure in the governance process.

Governance networks and policy networks are, nevertheless, two closely related and overlapping concepts, and therefore choosing the former does not mean abandoning the virtues of the latter. As governance networks and policy networks are basically two forms of social network (Carlsson and Sandström 2008, Rhodes 2006), in this study we attempt to integrate the essence of the governance process into the perspectives of both the policy network and social network research traditions and thus build a broader view of network structures and interactions regarding the governance of climate change.

Contextualising governance networks

Understanding the interactive process of governance networks is based on *institutional* and *dimensional contextualisation* (Torfing *et al.* 2012). Institutionally, the CMNs in China are embedded in a semi-authoritarian, hybrid regime (Diamond 2002). Because the interactions of the governance networks are also contextualised vertically and horizontally, the CMNs are structured within the multilevel governance framework and the bureaucratic politics of fragmented authoritarianism.

Semi-authoritarian, hybrid regime context

China is a *hybrid regime* comprising both democratic and authoritarian elements in its political system, with power and resources centralised upwards to the state, while flexibility in local governance has become decentralised downwards as part of the marketisation process. As a hybrid regime, China requires a much higher level of 'opposition mobilisation unity, skill and heroism', compared to that in liberal democratic regimes (Diamond 2002: 24). In the environmental sector, mobilisation has become an essential tool of environmental governance in China, while environmental nongovernmental organisations are socially embedded as a stabilising force in this *semi-authoritarian context* (Ho 2007). Given the embeddedness of environmental nongovernmental organisations, Ho and Edmonds (2007: 334) highlight two critical features relevant to the Chinese experience: the *contextualisation* of human action in the semi-authoritarian system and the interaction and negotiation within the *networks.* As Ho (2007: 189) points out, through 'a web of informal ties', the transforming social structures are capable of 'effectively mobilising resources, appealing to citizens' newly perceived or desired identities and building up a modest level of counter-expertise against state-dominated information on social cleavages and problems', to moderate environmental conflicts such as energy safety and incinerator construction. Such a web of informal social ties therefore forms the broader structural basis of the collaborative municipal networks that we conceptualise at the end of this chapter.

The three cities we study in this volume are located in diverse institutional contexts. Hong Kong, as a Special Administrative Region of the People's Republic of China, enjoys constitutionally independent political institutions agreed upon in the Sino-British Joined Declaration enshrined in the Basic Law (Basic Law of the HKSAR 4 April 1990; Article 31 of the Constitution of the People's Republic of China). Shenzhen, as a Special Economic Zone, is directly under the jurisdiction of the central government and enjoys stronger vertical autonomy in policy making and planning than do other Chinese cities (Number 41 Document, National People's Congress, May 1980). Guangzhou, as a provincial capital city, interacts closely with the provincial authority – a mediator in the hierarchical influence of the central government. Therefore, despite their geographical proximity, the three cities have unique institutional characteristics (Figure 3.5).

Multilevel governance and fragmented authoritarianism

As the intellectual tradition and the institutional context of the *multilevel governance* (MLG) framework are quite different from the institutional situation in China, its explanatory power is not fully applicable to research on Chinese cities. It is thus necessary to incorporate the logic of *fragmented authoritarianism* into the MLG framework to devise an explanation that fits China's context.

In the analytical framework of multilevel governance, municipalities are regarded as important sites of interaction and as having an important role to play in the governance of climate change. Type I multilevel governance emphasises *inter*governmental relations in a hierarchical nation-state system, with multitask jurisdictions of mutually exclusive territorial boundaries. Type II multilevel governance recognises the existence of a vast number of jurisdictions operating in a flexible manner over diverse territorial scales on functionally specific tasks (Hooghe and Marks 2001). The multilevel governance framework is applicable to the analysis of the Pearl River Delta region, given the diffusion of decision making in this subnational, cross-boundary region (Yang 2005).

Type I multilevel governance can be depicted as vertical or horizontal coordination (Bulkeley *et al.* 2009). Vertical coordination denotes the relations among municipal, regional and national authorities, in which municipal authorities need to maintain vertical autonomy under the enabling or constraining effects of national planning. Horizontal coordination refers to the relations between environmental agencies and other sectoral agencies at the municipal level, where administrative fragmentation may emerge. Conversely, type II multilevel governance is a manner of governing whereby cities or municipal actors need to maintain horizontal autonomy among a number of powerful stakeholders, such as multinational corporations and resourceful pressure groups (Bulkeley 2005).

Pertinent issues of fragmentation in governance networks are further discussed by Biermann and Pattberg (2008), who identify increasing segmentation of different layers and clusters of rule making and rule implementing. They clarify the concept of vertical fragmentation and horizontal fragmentation. Vertical fragmentation exists among supranational, international, national and subnational layers of authority and thus is a by-product of multilevel governance. Conversely, horizontal fragmentation emerges from the vacuum between different parallel rule-making systems maintained by different groups of actors. Thus, it results from *multipolar governance.*

In the Chinese context, the fragmented authoritarianism of bureaucratic politics results in discrete environmental policies and disjointed authority in both the vertical and the horizontal structures of the state system (Lieberthal 1997; Lieberthal and Lampton 1992). Accordingly, the Chinese environmental state in transition is conditioned by the design of its administrative system, particularly the multilayered territorial structure and the administrative ranks of public agencies and local governments. This fragmented authoritarianism results in a disjointed, protracted and incremental policymaking process (Lieberthal 1997; Yu 2004). Even though national climate change policy making takes place in a highly coordinated manner at the ministerial level (Yu 2004), the decentralised action plans and localised implementation may experience conflict between the vertical and horizontal lines of authority and between the competing functions in the administrative system. Decision making in this context is also based on the consensus built between government agencies that share the same administrative rank, which enables the effective operation of governance networks for climate change (Eichhorst and Bongardt 2009; Lieberthal 1997; Lieberthal and Oksenberg 1988; Ma and Ortolano 2000).

In China, the government structure is stratified vertically across the central state, provinces, cities, counties, townships and villages. Each administrative agency (such as the Ministry at the national level, the Commission at the provincial level and the Department at the prefectural level) has an administrative rank (or 'bureaucratic rank'). This ranking system enables all agencies (units) to assess their status with respect to all other units (Lieberthal and Oksenberg 1988). The division of cities is in accordance with their size and strategic importance to the country, and city status is aligned with administrative rank across government agencies. Territories can find their equivalent administrative agencies in the state structure. The lines of authority in China's administrative system are mapped out in Chapter 3 (Figure 3.5), linking up the organisational actors in the governance network.

Organisational practices and space in governance networks

The *inter*organisational networks within CMNs comprise organisational actors such as government agencies, intermediary organisations such as political parties and nongovernmental organisations (NGOs), independent research institutions, other community groups and business associations, trade unions and corporations in the private sector. Incorporating the notion of publicness (for an extended discussion of this concept, see what follows) into these organisational forms, an organisational typology is summarised and presented in Table 2.1.

Table 2.1 Organisational typology

	Organisation type	Examples
Government involvement/dimensional publicness	Government agencies at multiple levels of governance	National Development and Reform Commission (NDRC), China Meteorological Administration (CMA), etc.
	*Inter*governmental organisations	Intergovernmental Panel on Climate Change (IPCC)
	Transnational network organisations	C40
	Government organised nongovernmental organisations (GONGOs)	Guangdong Low Carbon Association, etc.
	Non-profit third-sector organisations	Business associations: Business Environment Council, etc., Sectoral associations: Professional Green Building Council, etc.
	Policy think tanks	Civic Exchange, etc.
	Research/academic network organisations	South China Climate Change Network, etc.
	Social organisations	Organisations subject to licensing requirements for social management
	Grassroots community organisations	
	Independent environmental organisations in civil society	Environmental nongovernmental organisations (ENGOs): Worldwide Fund for Nature Hong Kong (WWF), Greenpeace, etc.
	Private entities	Environmental consultancy firms, real estate developers, property management companies, etc.

GONGO – a unique organisational form

The governance networks of mitigation responses to climate change span the sector boundaries among the government, the market and civil society located in the three municipalities in our study. Special forms of civil society organisations – government-organised nongovernmental organisations (GONGOs, or 'officially organised NGOs') – have emerged in China, which adopt a nonconfrontational position on environmentalism with regards to the existing state system (Francesch-Huidobro and Mai 2012).

GONGOs play a significant role in *inter*organisational networks for environmental governance in general and for governance of climate change in particular, given their potential to straddle and even bridge governmental agencies and independent grassroots environmental NGOs (Wu 2002). The idea of setting up a GONGO often originates from a government entity, which makes GONGOs intrinsically different from independent NGOs. However, GONGOs are not extended organs of government, because they carry neither the legal mandate nor the political authority to directly implement projects through formal administrative systems (Wu 2002). They have their own expertise and closer relationships with society than with the government itself and are 'pervasive within both national and local level environmental policy making' in China (Wu 2003: 39).

GONGOs involved in environmental governance are not only co-opted by the state system but are also gaining autonomy and building up capacity in terms of access to resources and connections with other organised groups in society (Wu 2003). Contrary to the common assumption, studies indicate that GONGOs enjoy significant *de facto* autonomy, comparable to that of grassroots NGOs in China (Wu 2002, 2003). As state control of established organisations is not always effectively enforced, GONGOs are capable of pursuing their own independent agenda by using their personal relations with businesses and local governments and seeking financial independence and powerful patrons (Lu 2007). Such *de facto* autonomy ensures that GONGOs have the capability to generate and cooperate with local responses to climate change.

Organisational publicness

Among different organisations participating in the governance networks of climate change, there are diverse modalities of legal status, policy influence and access to the resources of the local governing authority and of international entities, including funding, partnership and information. An essential concept that captures a wide range of these organisational features is the *publicness* of organisations. Publicness is 'a characteristic of an organisation which reflects the extent the organisation is influenced by political authority' (Bozeman and Bretschneider 1994: 197). Traditionally, the core of organisational publicness is ownership, funding sources and political control.

Dimensional publicness and environmental publicness are also discussed in the literature. Dimensional publicness considers that organisations are affected by the relative extent of externally imposed authority; hence, publicness is determined by the constructs of political and economic authority that enable or constrain organisational behaviour (Bozeman and Bretschneider 1994; Wamsley and Zald 1973). Economic authority is built on the economic paradigm of property rights, including the right to the economic returns of public resources, the partitioning of demarcated resource domains, the decision-making process to determine the use of resources and the problems of transaction costs (Alchian 1977; Alchian and Demsetz 1973). Political authority refers to the coerced or constrained engagement of the organisation to serve the regime objectives (Bozeman and Bretschneider 1994; Miller and Moulton 2013; Wamsley and Zald 1973).

Environmental publicness refers to the political influence in a policy environment or in the multilevel context of governance networks, which indirectly shapes organisational publicness by altering organisational behaviour and interaction outcomes (Lynn *et al.* 2001; Miller and Moulton 2013). It is measured by i) the aggregate publicness of organisations in the multilevel governance environment, for instance, the respective proportions of public agencies, nonprofit organisations and privately owned organisations in the policy area of climate change; and

ii) the public priority of the particular policy area in the governance context, as reflected by state budget allocations to or statutory support of climate change policy (Miller and Moulton 2013). In this regard, we shall discuss in Chapter 3 the budgetary support and regulatory context of climate change mitigation in the three cities, whereas in Chapters 5 and 6 we shall analyse how the aggregate publicness of organisations in the sectoral governance networks is driving the integration of green buildings and electric vehicle innovations.

In addition, stronger multilevel publicness in the policy environment leads to higher institutional pressures for organisational policy learning, given the uncertainties of technologies and best practices (Miller and Moulton 2013). Best-practice transfer and policy learning are essential functions in the CMNs for the mitigation of climate change in Chinese cities. Understanding environmental publicness facilitates our analysis and lays the groundwork for the calculation of publicness scores in Chapter 5, in particular to identify the drivers for the implementation of green building technologies.

Nongovernmental space

To contextualise nongovernmental network organisations as distinct from public agencies, we propose the notion of nongovernmental space. To distinguish it from civil society and the third sector, nongovernmental space partly captures the described notion of environmental publicness while incorporating and expanding the concepts of epistemic community and political space.

Compared with the concepts of civil society and the third sector, nongovernmental space is only a structural concept in relation to the extent of governmental authority. It does not imply civil society's emphasis on the value of citizen participation or the third sector's focus on nonprofit service delivery (Axtmann 1996; Brandsen and Pestoff 2006). Nevertheless, nongovernmental space does share the value of diversity with civil society and the third sector (Blunkett 2008; Gilchrist 2009). It aims to create 'strong, active and connected communities', and to promote 'enterprising solutions to social and environmental challenges' (Blunkett 2008: 5). Nongovernmental space is an essential place for networking and interactions, because 'for diversity to flourish, communities need neutral communal spaces, which are neither private nor public, where the integrative process of community and civil society can be continually renewed' (Gilchrist 2009: 133). Analysing the expansion of nongovernmental space and its driving forces captures the aggregate publicness of organisations (i.e. environmental publicness) in a dynamic state.

The nongovernmental space also accommodates the epistemic community of climate change at multiple levels of influence. An epistemic community is often formed by groups of self-recruited, like-minded professionals who gather around a particular problem such as climate change to generate and channel consensual knowledge into public bureaucracies and legislative entities, with the ultimate objective of producing public policies and driving change (Haas and Haas 1995).

Nongovernmental space is, in essence, an integral part of the urban political space. The municipality is 'an especially significant political space', offered to ordinary people to participate in the business of the state (Magnusson 1996: 8, 302). The practical search for political space in a municipal context requires an understanding of 'the concrete activities' that people 'can enter' and 'in which people are engaged' (Magnusson 1996: 8). The political space is a space for action and is fluidly fragile in sheltering political freedom (Magnusson 1996: 3–5; Magnusson and Shaw 2003). Distinguishing the nongovernmental sector from the governmental arena in discussing the political space of municipalities reemphasises and specifies the nature of organisations and their interactions in the collaborative networking setting.

Innovation, information and incentives

A major function of climate governance networks is the diffusion of innovation, information and incentives across a large group of societal actors. 'Innovation' here means the creation of science and technology and the diffusion of best practices and knowledge. 'Information' refers to communication among network actors and public disclosure, while 'incentive' refers to the motivation offered by resources such as funding, partnership and reputation to encourage mitigation practices for climate change.

Innovation and governance networks

A sociotechnical system is a cluster of 'technology, regulations, user practices and markets, cultural meanings, infrastructure, maintenance networks and supply networks', which encompasses the 'production, diffusion and use of technology' (Elzen *et al.* 2004: 3; Geels 2004: 900). A system innovation is defined as a 'large-scale transformation in the way societal functions such as transportation, communication, [and] housing . . . are fulfilled', which can also be understood as 'a change from one socio-technical system to another' (Elzen *et al.* 2004: 19).

Hodson and Marvin (2010) apply a multilevel perspective to analyse the role of cities in the governance of sociotechnical transitions. To initiate systemic transitions at the urban scale, it is necessary to achieve effective coordination of capacity and capability, particularly in governing visions, sustaining research capability and necessary political processes, encouraging participation and mobilising resource capacities in world cities (Hodson and Marvin 2010). Instrumental world cities are 'positioned by national states and their own collaborative networks as contexts for developing innovative responses (i.e. urban transitions)' to emerging complex issues such as climate change, 'with an aligned view that these responses can collectively be "passed down" or transferred to other cities in the urban system' (Hodson and Marvin 2010: 478; Hodson and Marvin 2009). The exemplary role of the three instrumental world cities (Hong Kong, Shenzhen, Guangzhou) in the political and sociotechnical system of low carbon responses to climate change is analysed and discussed in depth in Chapter 7.

In addition to innovation production, the technology use and technology diffusion aspects of sociotechnical transitions are integral to the analysis of the sectoral integration of green buildings and electric vehicles. This makes user organisations, societal groups, public authorities and research institutes essential players in the transition process as they turn *inter*organisational networks into the necessary infrastructure for innovation distribution (Geels 2004).

Ways to systematically analyse sociotechnical system transformations and innovations have been sought through various models of innovation diffusion (Wejnert 2002). Network governance is currently identified as an essential approach to make innovation diffusion feasible and as an analytical tool for conducting research on the diffusion process (Caniëls and Romijn 2008; Elzen *et al.* 2004; Geels 2004; Schot and Geels 2008; Wejnert 2002). One of the various diffusion models is defined as 'strategic niche management', which is concerned with 'how governments can help initiate experiments within protected small-scale (technological) niches and then encourage these innovations to spread through the introduction of new supportive policies and regulations' (Lovell 2007: 35).

Technological niches refer to 'protected spaces that allow nurturing and experimentation with the co-evolution of technology, user practices, and regulatory structure', which precede the development of a 'market niche' in the transition of sociotechnical systems (Schot and Geels 2008: 538, 539). Potential innovations have emerged to achieve sociotechnical transitions in response to climate change challenges. Technical niches are being created by government and quasigovernment research institutions. The market niches that closely follow are needed to achieve broader engagement and the sectoral integration of low carbon technologies and user demands (Schot and Geels 2008). At this stage, networking is a critical part of the interactive process of learning, engagement and expectation convergence (Caniëls and Romijn 2008). However, the dilemmas of networking engagement may emerge in the selection of actors and the assessment of their resources, competencies, innovative ideas and vested interests (Schot and Geels 2008).

A specific focus in our analysis of CMNs is on how governance networks facilitate and how organisational actors interact to realise the integration and mainstreaming of technological innovations for broader system innovation. In this regard, Chapters 5 and 6 consider the technological and market niches of green building development and electric vehicles, respectively, while the interactive process of low carbon experimentation in the three cities is analysed in Chapter 7.

Information and communication

The complexity of managing governance networks determines that 'no single actor holds all the necessary information or the necessary resources for efficient policy' (Toikka 2010: 135). 'In the informal sense, networks are assumed to be simple membership structures, but real life governance systems are complex communication structures, where the interplay of institutions produces policy. Policy depends on collecting information dispersed in various organisations. As organisations seek to mobilise the information required for policy-making, they form communication links. As individual organisations establish communication links, a complex structure of interwoven links is born' (Toikka 2010: 136).

Information regarding policy making and implementation is therefore an essential feature of such a network's structure. Complex communication structures are the interactive outcomes of the institutions and the participating organisational actors, which produce policies by collecting the dispersed information. Our analysis of CMNs thus focuses on the positions of actors and on 'who talks to whom' in the interaction process to deliver network outcomes.

The public disclosure of emissions data is a major issue pertinent to information and communication in the network governance of climate change. Since the implementation of the Measures on Open Environmental Information in 2008, with some disclosure initiatives carried out by ENGOs in China, local governments and industrial suppliers of multinational corporations have responded to a more transparent and open network context with altered environmental behaviour (Tan 2012). To conduct research on CMNs, it is pertinent to investigate the hidden patterns of communication and information transmission among the organisational actors.

Incentives and resource exchange

Institutional incentive structures in network governance 'shape societal interdependency structures in desired ways' by 'giving certain actors important resources and capacities that grant them a key role in the network' (Sørensen and Torfing 2005b: 229). Conversely, network actors have incentives to exchange resources on a continuous basis and thus have the incentive 'to establish mutually recognised procedures including legal rules and informal norms to facilitate their interactions' (Compston 2009: 731). For members participating in the governance networks of climate change, different resources – including investment and funding, access and information and political support – are controlled by different actors but at the same time are transferrable across sectors of government, market and civil society (Compston 2009; Li 2006).

Incentives are thus necessary in governance networks to mobilise resource exchange. In studying the interaction dynamics of CMNs, it is pertinent to find out how the incentives are incorporated in the network structure and how they operate to influence specific actors and the governance structure itself.

Coordination

Coordination and collaboration are two instrumental components of the interactive process of governance. Early management studies defined coordination as an 'orderly arrangement of group effort to provide unity of action in the pursuit of common purpose' (Mooney, 1947, quoted in Camerer and Knez 1997: 158). Coordination partly constitutes the initial stage of collaboration, while collaboration is the outcome of coordination (Gray 1989; Thomson and Perry 2006). Coordination therefore precedes collaboration. For local governments dealing with complex problems such as climate change, coordination requires three types of interactions: vertical *inter*governmental hierarchy, horizontal *inter*governmental relations and *intra*governmental management.

Inter*governmental coordination: vertical hierarchy and*
horizontal relations

Coordination in *inter*governmental relations is a 'process of solving [*inter*govern-mental] problems under conditions of high uncertainty and complexity through the creation and use of governmental and nongovernmental networks' (Wright and Krane 1998). To initiate joint actions in governance networks, coordination in *inter*governmental relations includes the skills necessary for mobilising resources in the nongovernmental space and the capacity for engaging external actors.

Several distinctive elements characterise *inter*governmental coordination – leaders and actors, participating jurisdictions, dominant values, conflict resolution mechanisms and authority relationships (Wright 1990). *Inter*governmental coor-dination functions within the structures of all governmental entities at multiple levels of the state hierarchy, and these entities interact with an array of public and private actors involved in the implementation of designated policies (Mandell 1990; Wright and Krane 1998). The primary participants in the interactions within *inter*governmental coordination are policy professionals and administrative gen-eralists, who are in charge of authority, resources and information. Their decision-making capacities, administrative roles and institutional responsibilities are thus the focus of the analysis undertaken in Chapter 7.

In governance processes involving various actors, conflicts between goal ori-entations and values are inevitable among specialists, generalists, clients and the general public. In this sense, conflict resolution and consensus building through active networking, together with 'coordination and orchestration of the discordant organisation' in the network, are the major tasks for *inter*governmental coordina-tors (Mandell 1990; Wright and Krane 1998: 1166). Conflict resolution mecha-nisms are conventionally found in bargaining and negotiation, informal personal connections, joint dialogues and trust-building platforms. Chapter 7 provides an analysis of different modes of negotiation and consensus building in the *inter*gov-ernmental context.

*Inter*governmental relations are structured along two dimensions. The vertical *inter*governmental hierarchy denotes the notion of hierarchical control. A hierar-chy is an instrument of coordination. Although hierarchical structures are increas-ingly being replaced by collaborative arrangements, the concept of a hierarchy is persistent in public administration. In the governance context, a hierarchy is regarded as 'a means of coordinating social activity through the mechanism of the rank ordering of authority within a society' rather than the traditional institution of bureaucratic controls (Lynn 2011: 228). A hierarchy can serve the function of an instrument used to organise and carry out the work of a complex social system and to achieve coordination by minimising transaction costs (Lynn 2011; Simon 1962; Williamson 1996). The instrumental view of hierarchy considers it closer to coordination, as a hierarchy is 'a mechanism for coordination and control of productive activity, preferable to other mechanisms, such as markets or networks, under certain circumstances' (Lynn 2011: 229).

While a vertical *inter*governmental hierarchy is characterised by hierarchi-cal influence, horizontal *inter*governmental relations emphasise the network relationships between governments on a horizontally equal footing. The arenas of horizontal *inter*governmental relations are situated across different jurisdictional localities and thus connect municipal governments with similar administrative status and power. With various actors on a level playing field, horizontal *inter*governmental relations are built upon a network of authority relationships with loosely coupled patterns of influence directed by each actor's interactions (Mandell 1990; Wright and Krane 1998). Unlike vertical boundary spanning, horizontal boundary-spanning activities in sustainable development require further illumination (Zeemering and Romero 2012), a key justification for the selection of three different city-level cases in this book (Chapter 3).

Coordinating horizontal *inter*governmental relations is necessary for local governmental officials and stakeholders to navigate through *inter*governmental complexity in response to climate change challenges (Zeemering and Romero 2012: 286). One of the challenges of horizontal *inter*governmental relations is to 'identify and coordinate the complex array of actors that have a stake in the governance of sustainable development', including climate change policies (Zeemering and Romero 2012: 289). Having identified the stakeholders, the process of coordinating horizontal *inter*governmental relations involves reaching shared values horizontally with other agencies, identifying the correct scale on which to address the natural resource problem and incorporating the collaborative process into governance (Boone 2012). Practitioners find that the networking activities of informal horizontal *inter*governmental relations are often more effective and responsive than formal activities, particularly when they are initiating innovative or collaborative projects on sustainable development (Boone 2012). In studying the experimental and exemplary initiatives of climate mitigation in the three cities, Chapter 7 further investigates the networking process of *inter*governmental coordination in both hierarchical and horizontal relations.

Intra*governmental management*

*Intra*governmental management refers to the behaviour and interactions of institutions within a single branch or on a single tier of government (Koliba *et al.* 2010). Existing studies of governance in the context of climate change seldom focus on *intra*governmental mechanisms (except López-Santana 2009; Ward and Rodriguez 1999; Koliba *et al.* 2010). Although research has been conducted on similar concepts of 'interagency networks' and 'joined-up government', a gap remains in the study of *intra*governmental management and its coordination and collaboration dynamics among network actors and institutions (6 2004; Koliba *et al.* 2010; Lyall 2007). In this regard, in Chapter 4 we shall provide an in depth analysis of *intra*governmental coordination mechanisms in the three cities.

Despite being considered a variation of governance network configuration, *intra*governmental management is often theorised as a modified version of *inter*governmental management. Following Wright's (1990) characterisation, the defining characteristics of *intra*governmental management networks are i) leading and primary actors who hold resources and authority, ii) participating agencies or institutions, iii) dominant values and consensus building, iv) coordination mechanisms to mediate conflicts and carry out networking activities and v) authority relationships with flows of influence.

Horizontal coordination in an *intra*governmental setting can be defined and analysed from either a process or an outcome perspective. From the process perspective, coordination in *intra*governmental networks is an 'organised attempt to optimise the coherence and consistency of political decisions and policy implementation' across actors, levels of government and sectoral policies to achieve a holistic response to complex problems such as climate change (Wollmann 2003: 594). *Intra*governmental coordination can be defined as 'the process whereby two or more [governmental] organisations create and/or use existing decision rules that have been established to deal collectively with their shared task environment' (Mulford and Rogers 1982: 12). The coordination process is constantly involved with the balance between dealing with arising conflicts and achieving cooperation and between imposing coercion and obtaining consent (Dunsire 1978; Pressman and Wildavsky 1973). Networking interactions, such as 'the development of ideas about joint and holistic working, joint information systems, dialogue between agencies, process of planning, and making decision', are therefore necessary in this process (6 2004: 106). From the outcome perspective, coordination is analysed as 'an end state in which government policies are characterised by minimal redundancy, incoherence and lacunae', which facilitates the evaluation of successful or failing coordination (Peters 1998: 296). The outcome perspective of coordination specifies four problems: i) redundancy, in which more than one actor or institution performs the same task; ii) lacunae, in which no actor performs the core task; iii) overlapping clientele groups, whereby different policies or public services are delivered to the same group of people; and iv) incoherence in the goals and requirements of institutions (Peters 1998).

Problems in horizontal coordination in *intra*governmental management are often caused by the fragmentation of responsibilities and governance functions, colloquially named the 'silo effect'. In practice, this includes phenomena such as 'the separation of functions and authority among different agencies', a lack of communication among departments of the same organisation and 'the inclination of agencies to protect their functions and authority from possible intrusion by other agencies' (Mitchell 2005: 1341). An alternative perspective views problems of interagency coordination as a principal–agent problem to be solved by strengthening institutional capacity. In this context, setting up interagency mechanisms of coordination can enhance management and enforcement capacities (Li and Chan 2009). The presence of horizontal fragmentation in the resolution of complex policy problems such as climate change thus provides strong motivation to 'search for a way to achieve integration through coordination and collaboration' (Mitchell 2005: 1341).

Despite the perceived dichotomy between outcomes and processes in analysing *intra*governmental management, the end-state outcomes and the dynamic processes must be considered together. When evaluating the success of a particular outcome of coordination, a retrospective view of the process leading to such an outcome is necessary. Similarly, when analysing how network actors engage in the process of coordination, predicting the likely outcomes is necessary to enable benchmarking against an evaluation framework.

Coordination not only spans agencies within *intra*governmental networks but also configures the internal structure of a given agency. The capacity of managers to operate collaboratively within their own organisations is dependent on the levels of internal coordination and internal capabilities needed to form and sustain external links in a boundary-spanning role (6 *et al.* 2006; Mcguire and Silvia 2010). This internal capability of a given agency is determined both by its institutional and power configuration (leadership) and by the current trends of institutional and network change in its external organisation, that is, the nongovernmental space defined above (6 *et al.* 2006).

Collaboration and collaborative governance

Changing definitions of collaboration

To study collaboration in the governance networks of climate change, it is necessary to look inside the 'black box' of subnational networks, coalitions and partnerships (Thomson and Perry 2006). The concept of collaboration is distinguishable from cooperation and coordination. It occurs 'when a group of autonomous stakeholders of a problem domain engage in an *interactive process*, using shared rules, norms, and structures, to act or decide on issues related to that problem domain' (Wood and Gray 1991: 146). Collaboration is perceived as 'a process through which parties who see different aspects of a problem can constructively explore their differences and search for solutions that go beyond their own limited visions of what is possible' (Gray 1989: 1, 5); thus collaboration emphasises the notion of 'constructive management of differences'.

Thomson (2001) clarifies the understanding of 'interactive processes' as collaboration that can be either formal or informal while mutually benefitting participants. He integrates the ideas of joint decision making and self-imposed governing rules into a revised definition of Gray's conceptualisation (1989). Thomson (2001: 83) defines collaboration as 'a process in which autonomous actors interact through formal and informal negotiation, jointly creating rules and structures governing their relationships and ways to act or decide on the issues that brought them together, involving shared norms and mutually beneficial interactions'.

Given the complexity of the climate change problem, cross-sector collaboration has emerged as the major approach to spur action at various levels of governance. Drawing from the various definitions of collaboration, Bryson et al. (2006: 44) identify the 'partnerships involving government, business, non-profits and philanthropies, communities, and/or the public as a whole' as cross-sector collaboration and define it as 'the linking or sharing of information, resources, activities, and capacities by organisations in two or more sectors to achieve jointly an outcome that could not be achieved by organisations in one sector separately'.

Comparing frameworks of collaborative governance

Collaborative governance has arisen as a new paradigm in governance studies. In their meta-analysis, Ansell and Gash (2008: 544) propose the following definition of collaborative governance: 'a governing arrangement where one or more public agencies directly engage non-state stakeholders in a collective decision-making process that is formal, consensus-oriented, and deliberative and that aims to make or implement public policy or manage public programs or assets'. Emerson et al. (2012: 3) challenge this definition in their synthesis of an integrative framework for collaborative governance as 'limited to only formal state-initiated arrangements and to engagement between government and nongovernmental stakeholders'. They provide an alternative definition of the collaborative governance as 'the process and structures of public policy decision making and management that engage people constructively across the boundaries of public agencies, levels of government, and/or the public, private and civic spheres to carry out a public purpose that could not otherwise be accomplished' (Emerson *et al.* 2012: 2).

Thus, collaborative governance in this alternative definition is not limited to unidirectional engagement initiated only by public actors but also includes mutual engagement between public and private actors. 'Multipartner governance' is emphasised in this alternative definition, which covers partnerships, joined-up government and hybrid arrangements, established among and within the state, the private sector, civil society and the community (Emerson *et al.* 2012: 3). This extends the scope of collaborative governance to both *inter*governmental collaborative structures and interagency collaboration: the former refers to collaboration among actors on the vertically arranged governance levels of the nation-state, whereas the latter denotes actors at the same governance level collaborating on specific policy issues such as climate change (Emerson *et al.* 2012).

Further to the definition of collaborative governance, this conceptual development identifies other differences between the frameworks (Chapter appendix: Figure 2.1A and Figure 2.2A). First, Emerson et al. (2012) distinguish the 'drivers' of collaborative governance from its 'system context', while Ansell and Gash (2008) consider both as the 'starting conditions', through which the 'drivers' differentiate the importance of various components in the context leading to the development of collaborative governance arrangements. By adding the 'drivers' to their explanation of collaborative governance, Emerson et al. (2012) identify direct causal variables that provide greater insights into process analysis. Second, Emerson et al. (2012) extend the scope of initial engagement during the collaborative process beyond mere 'face-to-face dialogue', as identified by Ansell and Gash (2008). Emerson et al. (2012: 10) argue that 'although face-to-face dialogue is advantageous at the outset, it is not always essential, particularly when conflict may be low and shared values and objectives quickly surface'.

Third, in addition to trust, mutual understanding and shared commitment, Emerson et al. (2012) add 'internal legitimacy', which was not previously identified by

Ansell and Gash (2008). They define internal legitimacy as 'a sense of interpersonal validation' generated by mutual understanding and, more specifically, 'the confirmation that participants in a collective endeavour are trustworthy and credible, with compatible and interdependent interests, which legitimises and motivates ongoing collaboration' (Emerson *et al.* 2012: 14). Internal legitimacy should be differentiated from 'discursive legitimacy', one of the sources of power in the collaborative governance process (Purdy 2012). Discursive legitimacy is defined as 'the ability of an organisation to represent a discourse or speak on behalf of an issue in the public sphere' (Hardy and Phillips 1998; quoted in Purdy 2012: 411). The discursive legitimacy of participants in a collaborative governance arrangement may determine the leverage that they have during collaborative interactions, both formal and informal. Due to its effect on the power structure in the collaborative governance process, we opine that discursive legitimacy should be incorporated into the framework of collaborative dynamics, although it was omitted by both Emerson et al. (2012) and Ansell and Gash (2008).

Fourth, while the collaborative process depicted by Ansell and Gash (2008) remains a single-layered cycle of interactions, Emerson et al. (2012) group the variables involved in collaborative dynamics into two layers of interactive relationships. Tier One comprises three steps in the collaborative process – principled engagement,[1] shared motivation and capacity for joint action. Tier Two encompasses different variables that form an interactive cycle in each step of Tier One. Shared motivation is initiated by principled engagement, while the capacity for joint action is an outcome of the interaction between principled engagement and shared motivation (Emerson *et al.* 2012: 13, 14). Once the shared motivation and capacity for joint action are initiated in Tier One, they can respectively develop into self-sustaining cycles with restructuring of their own component elements in Tier Two and in turn can reinforce or accelerate the previous steps in Tier One of the collaborative process (Emerson *et al.* 2012: 13–15).

Fifth, Emerson et al. (2012) separate 'collaborative actions' as an intermediate output from the ultimate collaborative outcomes. Regarding the collaborative outcomes, Emerson et al. (2012) continue the discussion about the effects and future adaptation. Rogers and Weber (2010) propose three dimensions of research on the outcomes of collaborative governance: improving the problem-solving capacity of government agencies the development and transfer of technology and governance beyond compliance. These dimensions are the specific areas on which the interaction outcomes of CMNs can be evaluated based on their respective collaborative governance arrangements for climate mitigation.

Despite their differences, there are also some similarities in the frameworks proposed by Emerson et al. (2012) and Ansell and Gash (2008). Both frameworks perceive collaborative dynamics as cyclical, iterative and interactive processes. Nevertheless, neither framework fully integrates its analysis with hierarchical influences. Gibson (2011: 6) seeks to combine the models of collaborative governance with the concept of multilevel governance discussed earlier, thereby creating 'collaborative multilevel governance', defined as 'a process and institutions, distinct from government, built on the foundations of collaboration and multi-level partnerships among community-based organisations, statutory agencies, and the private sector engaged'. Although this theorisation is not entirely satisfactory, it sheds light on how different concepts from different research traditions can be combined to generate new insights.

Legitimation: convergence of coordination and collaboration

In addition to the distinction between internal and discursive legitimacy that emerges from the discussion of collaboration dynamics, the legitimation process is also classified in terms of input and output legitimacy. Input legitimacy in network governance refers to procedural demands, which involve consultation with and the participation of nonstate actors in policy making, balanced representation of different stakeholder groups, forums for multisectoral collaboration and deliberation, transparent and accessible information sharing and accountability/ reporting mechanisms (Bäckstrand 2006; Skogstad 2003).

Output legitimacy relates to the distributed benefits and regulatory advantages among the actors in the network (Héritier 1999; Skogstad 2003) and to the effectiveness of partnership agreements in terms of their problem-solving capacity (Bäckstrand 2006). The effectiveness of a network arrangement encompasses two dimensions: first, whether it leads to the expected environmental and developmental outcomes and second, its institutional effectiveness, that is, whether the institutional design is adequate and effective for achieving the expected environmental outcomes (Bäckstrand 2006). In the context of climate mitigation, the first dimension is closely related to the criteria of environmental effectiveness, cost effectiveness and distributional effects, while the second dimension relates to the criteria of institutional feasibility, as specified by the Intergovernmental Panel on Climate Change in the Fourth Assessment Report on mitigation (IPCC 2007). In a case study of the Chinese Clean Development Mechanism market, Schroeder (2009: 388) argues that the input legitimacy of Chinese carbon governance is low, despite the existing opportunities for nonstate actors to participate in the rule-setting process, while the output legitimacy is satisfactory in terms of the environmental and institutional effectiveness (Schroeder 2009).

In our study, we examine the network dynamics involved in the legitimation gap between input and output legitimacy given the semi-authoritarian context of climate governance in China. The legitimation process also sees the convergence of the outcomes of coordination mechanisms and collaborative arrangements, whereby the institutional feasibility and effectiveness of sociotechnical and policy innovations are confirmed prior to the transformation of the environmental state.

Methodology

In order to operationalise the theoretical framework of collaborative municipal networks (CMNs) and build the empirical case studies, data were collected and collated data from multiple sources. These include transcripts of semistructured interviews with informants from the networks (Appendix 1), relevant policy documents, statistical records and newspaper archival records. The analysis of the data was undertaken through a mixed-methods research design framework which includes i) case study analysis combining a within-case process tracing of the causal mechanism with a cross-case comparison; ii) qualitative and quantitative content analysis of textual data; and iii) statistical analysis of quantitative numerical data. This is diagrammatically presented in Figure 2.1. The diagram was developed following graphic examples in drawing visual representations of triangulation designs (Creswell *et al.* 2003; Ivankova *et al.* 2006: 15; Tashakkori and Teddlie 2003).

Conclusion: collaborative municipal networks (CMNs)

In sum, our core research inquiry *investigates the climate mitigation process in cities through the lens of collaborative municipal networks (CMNs) to extrapolate how such processes ultimately transform the Chinese environmental state.* The concept of CMNs is developed from governance network theories and the process characteristics offered by collaboration theories. When applied in the context of subnational governance for climate change in the transportation and building sectors, a collaborative municipal network (CMN) is conceptualised as *the complex process of interaction among a plurality of state and nonstate actors, with the objective of achieving common climate mitigation goals through functions such as knowledge transfer and resource exchange while maintaining municipal autonomy in vertical or horizontal coordination and enhancing municipal capacity in horizontal collaboration. Their interaction is manifested through collective decision making, which includes consensus building and joint rule making.*

The nonstate actors include both participants from network organisations and NGOs in civil society and corporate actors in the private sector. This concept comprises four dimensions – participants, objectives, network functions and collaborative traits. Based on the proposed definition, we shall now discuss the theoretical framework for the analysis of CMNs. The competing frameworks of collaborative governance discussed in the previous section provide, in part, a basis for this theorisation process. However, to be fully compatible with the analysis of CMNs, the existing frameworks must be modified to focus on the governance-design process and structural configuration.

Figure 2.2 depicts the interactive process of CMNs represented by two mutually reinforcing yet independently operating analytical tiers: the stages of network development and the cyclical process of network dynamics. The stages of network development represent the varied forms of *institutional design* during the entire life cycle of CMNs. Network development is thus an *external process* of CMNs, which is analytically divided into four stages: i) *network prototype*, including the initial drivers derived from the system context, and principled engagement once the network participants have been identified; ii) *network formation*, including the activities of shared motivation and capacity building prior to joint action; iii) *network in action*, spanning the process by which actual actions are taken or initiatives

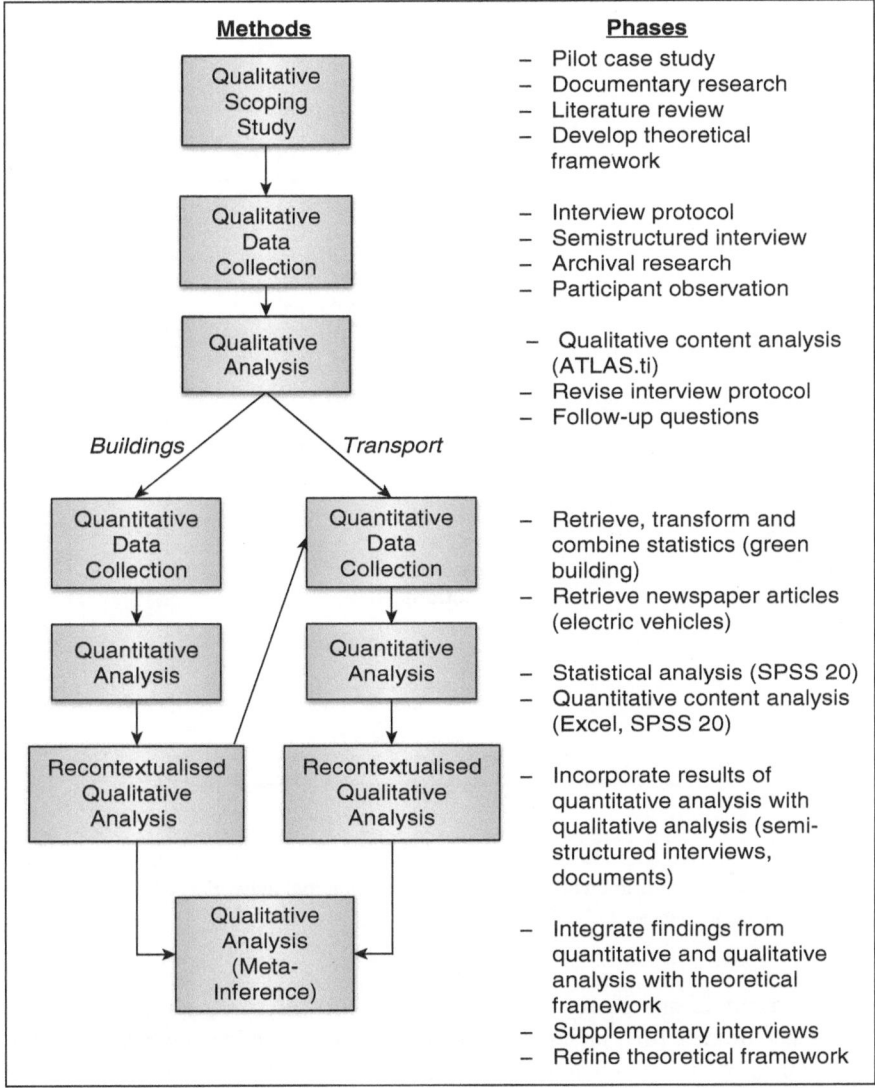

Methods

Phases

| Qualitative Scoping Study |
| Qualitative Data Collection |
| Qualitative Analysis |

- Pilot case study
- Documentary research
- Literature review
- Develop theoretical framework

- Interview protocol
- Semistructured interview
- Archival research
- Participant observation

- Qualitative content analysis (ATLAS.ti)
- Revise interview protocol
- Follow-up questions

Buildings   *Transport*

Quantitative Data Collection	Quantitative Data Collection
Quantitative Analysis	Quantitative Analysis
Recontextualised Qualitative Analysis	Recontextualised Qualitative Analysis
Qualitative Analysis (Meta-Inference)

- Retrieve, transform and combine statistics (green building)
- Retrieve newspaper articles (electric vehicles)

- Statistical analysis (SPSS 20)
- Quantitative content analysis (Excel, SPSS 20)

- Incorporate results of quantitative analysis with qualitative analysis (semi-structured interviews, documents)

- Integrate findings from quantitative and qualitative analysis with theoretical framework
- Supplementary interviews
- Refine theoretical framework

Figure 2.1 Research design and methodology
Source: Compiled by authors

are carried out by the network; and iv) *network outcomes*, the ultimate effect on and adaptation of the network and its future operation.

The second tier of analysis is the cyclical process of network dynamics embedded in CMNs. It is thus an *internal process* driving the external changes in the network design in the first tier of analysis. Iterative components are identified in the network dynamics. First, a *coordination mechanism* is formed, composed of a conflict resolution platform and the pattern of authority relationships that

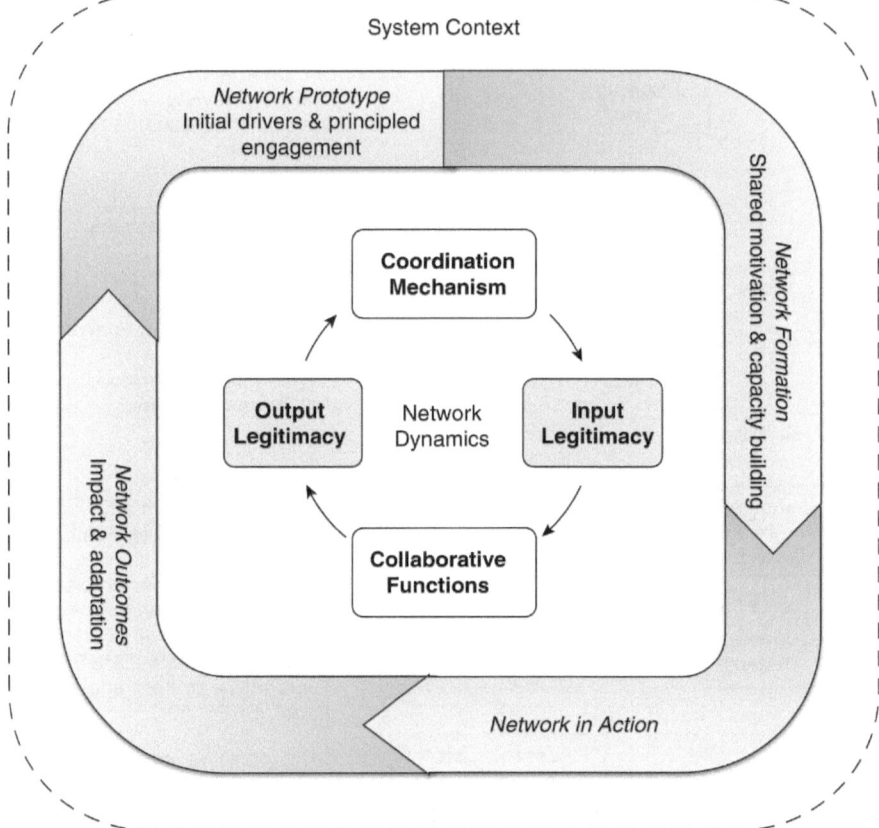

Figure 2.2 Theoretical framework of collaborative municipal networks (CMNs)
Source: Compiled by authors

transfer resources and power. Second, *collaborative functions* are integrated as the essence of the CMN, including innovation transfers, information sharing and communication and effective incentive structures. Once the outcomes of coordination and collaboration have been achieved, the internal configuration proceeds to the legitimation process. The legitimation process starts by generating *internal legitimacy* in the network then sustains *discursive legitimacy* in the public sphere by establishing facilitative leadership and shared values. Analytically, it can be divided into *input* and *output legitimacy*, combining the influence of coordination with that of collaboration.

Based on a critical review of the relevant literature, in this chapter we have identified the research gap of municipal governance for climate change and justified the theoretical necessity of focusing the research on China. The chapter conceptualises and theorises the framework of CMNs, and this theorisation underpins the empirical analysis of this study (Chapters 4–7).

Chapter appendix

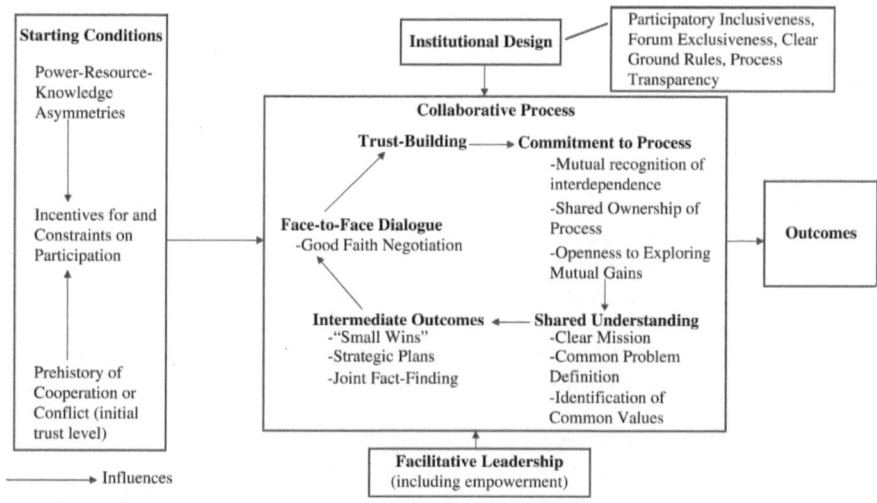

Figure 2.1A A model of collaborative governance
Source: Ansell and Gash 2008: 550

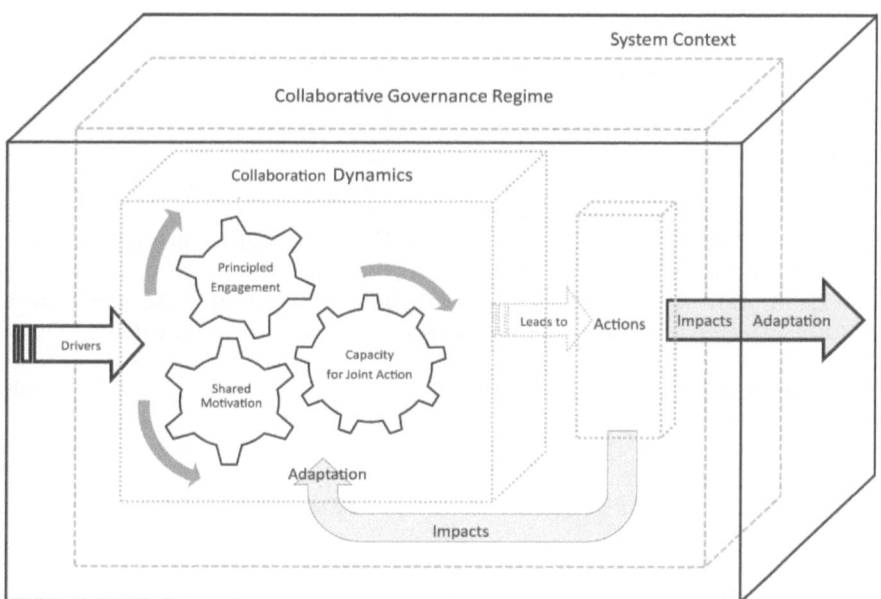

Figure 2.2A The integrative framework for collaborative governance
Source: Emerson et al. 2012: 14

Note

1 'Principled engagement' is defined as a process during which 'people with differing content, relational, and identity goals work across their respective institutional, sectoral, or jurisdictional boundaries to solve problems, resolve conflicts, or create value' (Emerson et al. 2012: 10).

References

6, P. 2004. Joined-up Government in the Western World in Comparative Perspective: A Preliminary Literature Review and Exploration. *Journal of Public Administration Research and Theory,* 14, 103–138

6, P., Goodwin, N., Peck, E. & Freeman, T. 2006. *Managing Networks of Twenty-First Century Organisations,* Basingstoke, Palgrave Macmillan.

Alber, G. & Kern, K. 2009. Governing Climate Change in Cities: Modes of Urban Governance in Multi-Level Systems, in OECD (ed.) *OECD Conference Proceedings,* Paris: OECD.

Alchian, A. A. 1977. Some Economics of Property Rights, in Alchian, A. A. & Coase, R. H. (eds.) *Economic Forces at Work,* Indianapolis, IN, Liberty Press.

Alchian, A. A. & Demsetz, H. 1973. The Property Right Paradigm. *The Journal of Economic History,* 33, 16–27

Ansell, C. & Gash, A. 2008. Collaborative Governance in Theory and Practice. *Journal of Public Administration Research and Theory,* 18, 543–571

Axtmann, R. 1996. Citizenship and Civil Society, in Axtmann, R. (ed.) *Liberal Democracy into the Twenty-First Century: Globalization, Integration and the Nation-State,* Manchester, Manchester University Press.

Bäckstrand, K. 2006. Multi-Stakeholder Partnerships for Sustainable Development: Rethinking Legitimacy, Accountability and Effectiveness. *European Environment,* 16, 290–306

Bai, X. 2007. Integrating Global Environmental Concerns into Urban Management: The Scale and Readiness Arguments. *Journal of Industrial Ecology,* 11, 15–29

Balme, R. 2011. China's Climate Change Policy: Governing at the Core of Globalization. *CCLR [The Carbon & Climate Law Review],* 5, 44–56

Betsill, M. 2001. Mitigating Climate Change in US Cities: Opportunities and Obstacles. *Local Environment,* 6, 393–406

Betsill, M. 2007. Regional Governance of Global Climate Change: The North American Commission for Environmental Cooperation. *Global Environmental Politics,* 7, 11–27

Betsill, M. & Bulkeley, H. 2007. Looking Back and Thinking Ahead: A Decade of Cities and Climate Change Research. *Local Environment,* 12, 447–456

Biermann, F. & Pattberg, P. 2008. Global Environmental Governance: Taking Stock, Moving Forward. *Annual Review of Environment and Resources,* 33, 277–294

Blunkett, D. 2008. *Mutual Action, Common Purpose: Empowering the Third Sector,* London, Fabian Society.

Bogason, P. & Musso, J. A. 2006. The Democratic Prospects of Network Governance. *The American Review of Public Administration,* 36, 3–18

Boone, J. 2012. A Practitioner Responds – Sustainability: A View from the Trenches, in Meek, J. W. & Thurmaier, K. (eds.) *Networked Governance: The Future of Intergovernmental Management,* Thousand Oaks, CA, CQ Press.

Bozeman, B. & Bretschneider, S. 1994. The "Publicness Puzzle" in Organization Theory: A Test of Alternative Explanations of Differences between Public and Private Organizations. *Journal of Public Administration Research and Theory*, 4, 197–224

Brandsen, T. & Pestoff, V. 2006. Co-Production, the Third Sector and the Delivery of Public Services: An Introduction. *Public Management Review*, 8, 493–501

Bryson, J. M., Crosby, B. & Stone, M. M. 2006. The Design and Implementation of Cross-Sector Collaborations: Propositions from the Literature. *Public Administration Review*, 66, 44–55

Bulkeley, H. 2005. Reconfiguring Environmental Governance: Towards a Politics of Scales and Networks. *Political Geography*, 24, 875–902

Bulkeley, H. 2010. Cities and the Governing of Climate Change. *Annual Review of Environment and Resources*, 35, 229–253

Bulkeley, H. & Kern, K. 2006. Local Government and the Governing of Climate Change in Germany and the UK. *Urban Studies*, 43, 2237–2259

Bulkeley, H. & Newell, P. 2010. *Governing Climate Change*, Abingdon, Oxon, Routledge.

Bulkeley, H. & Schroeder, H. 2008. *Governing Climate Change Post-2012: The Role of Global Cities – London.* Working paper (123). Norwich, UK, Tyndall Centre for Climate Change Research.

Bulkeley, H., Schroeder, H., Janda, K., Zhao, J., Armstrong, A., Chu, S. Y. & Ghosh, S. 2009. *Cities and Climate Change: The Role of Institutions, Governance and Urban Planning.* 5th Urban Research Symposium: Cities and Climate Change, Marseille, France.

Camerer, C. & Knez, M. 1997. Coordination in Organizations: A Game-Theoretic Perspective, in Shapira, Z. (ed.) *Organizational Decision Making,* Cambridge: Cambridge University Press.

Caniëls, M. C. J. & Romijn, H. A. 2008. Actor Networks in Strategic Niche Management: Insights from Social Network Theory. *Futures*, 40, 613–629

Carlsson, L. & Sandström, A. 2008. Network Governance of the Commons. *International Journal of the Commons*, 2, 33–54

Castán Broto, V. & Bulkeley, H. 2012. A Survey of Urban Climate Change Experiments in 100 Cities. *Global Environmental Change*, 23, 92–102

Collier, U. 1997. Local Authorities and Climate Protection in the European Union: Putting Subsidiarity into Practice? *Local Environment*, 2, 39–57

Compston, H. 2009. Networks, Resources, Political Strategy and Climate Policy. *Environmental Politics*, 18, 727–746

Corfee-Morlot, J., Kamal-Chaoui, L., Donovan, M., Cochran, I., Robert, A. & Teasdale, P.-J. 2009. *Cities, Climate Change and Multilevel Governance.* OECD Environmental Working Papers (14). Paris, France, OECD publishing.

Creswell, J. W., Clark, V. L. P., Gutmann, M. L. & Hanson, W. E. 2003. Advanced Mixed Methods Research Design, in Tashakkori, A. & Teddlie, C. (eds.) *Handbook of Mixed Methods in Social & Behavioral Research,* 2nd ed., Thousand Oaks, CA: SAGE Publications.

DeAngelo, B. J. & Harvey, L. D. 1998. The Jurisdictional Framework for Municipal Action to Reduce Greenhouse Gas Emissions: Case Studies from Canada, the USA and Germany. *Local Environment*, 3, 111–136

Delman, J. 2011. China's "Radicalism at the Center": Regime Legitimation through Climate Politics and Climate Governance. *Journal of Chinese Political Science*, 16, 183–205

Dhakal, S. 2004. *Urban Energy Use and Greenhouse Gas Emissions in Asian Mega Cities: Policies for a Sustainable Future,* Kitakyushu, Japan, Institute of Global Environmental Strategies.

Dhakal, S. 2009. Urban Energy Use and Carbon Emissions from Cities in China and Policy Implications. *Energy Policy,* 37, 4208–4219

Dhakal, S. 2011. Urban Energy Transitions in Chinese Cities, in Bulkeley, H., Broto, V. C., Hodson, M. & Marvin, S. (eds.) *Cities and Low Carbon Transitions,* New York, NY, Routledge.

Diamond, L. J. 2002. Thinking about Hybrid Regimes. *Journal of Democracy,* 13, 21–35

Dodman, D. 2009. Blaming Cities for Climate Change? An Analysis of Urban Greenhouse Gas Emissions Inventories. *Environment and Urbanization,* 21, 185–201

Dunsire, A. 1978. *The Execution Process, Vol. 2: Control in a Bureaucracy,* London, Martin Robertson.

Eichhorst, U. & Bongardt, D. 2009. Towards Cooperative Policy Approaches in China – Drivers for Voluntary Agreements on Industrial Energy Efficiency in Nanjing. *Energy Policy,* 37, 1855–1865

Elzen, B., Geels, F. W. & Green, K. (eds.) 2004. *System Innovation and the Transition to Sustainability: Theory, Evidence and Policy,* Gloucestershire, UK, Edward Elgar Publishing.

Emerson, K., Nabatchi, T. & Balogh, S. 2012. An Integrative Framework for Collaborative Governance. *Journal of Public Administration Research and Theory,* 22, 1–29

Francesch-Huidobro, M. 2011. Governance of Climate Change in Coastal Cities: The Example of Hong Kong, in Aerts, J., Botzen, W., Bowman, M., Ward, P. & Dircke, P. (eds.) *Climate Adaptation and Food Risk in Coastal Cities,* Oxford, Earthscan Climate from Routledge.

Francesch-Huidobro, M. & Mai, Q. 2012. Climate Advocacy Coalitions in Guangdong, China. *Administration & Society,* 44, 43S–64S

Francesch-Huidobro, M. 2012. Institutional Deficit and Lack of Legitimacy: The Challenges of Climate Change Governance in Hong Kong. *Environmental Politics,* 21, 791–810

Francesch-Huidobro, M., Lo, C. W.-H. & Tang, S.-Y. 2012. The Local Environmental Regulatory Regime in China: Changes in Pro-Environment Orientation, Institutional Capacity, and External Political Support in Guangzhou. *Environment and Planning-Part A,* 44, 2493–2511

Geels, F. W. 2004. From Sectoral Systems of Innovation to Socio-Technical Systems: Insights About Dynamics and Change from Sociology and Institutional Theory. *Research Policy,* 33, 897–920

Gibson, R. 2011. *A Primer on Collaborative Multi-Level Governance* [Online]. Canadian Regional Development: A Critical Review Of Theory, Practice And Potentials. Available: http://research.library.mun.ca/310/1/primer_collaborative.pdf

Gilchrist, A. 2009. *The Well-Connected Community: A Networking Approach to Community Development,* Bristol, Policy Press.

Gray, B. 1989. *Collaborating: Finding Common Ground for Multiparty Problems,* San Francisco, CA, Jossey-Bass.

Haas, P. M. & Haas, E. B. 1995. Learning to Learn: Improving International Governance. *Global Governance,* 1, 255

Hardy, C. & Phillips, N. 1998. Strategies of Engagement: Lessons from the Critical Examination of Collaboration and Conflict in an Interorganizational Domain. *Organization Science,* 9, 217–230

Harris, P. G. 2010. China and Climate Change: From Copenhagen to Cancun. *Environmental Law Reporter,* 40, 10858–10863

He, B. 2012. Western Theories of Deliberative Democracy and the Chinese Practice of Complex Deliberative Governance. *Search for Deliberative Democracy in China,* 133–148

He, B. & Thøgersen, S. 2010. Giving the People a Voice? Experiments with Consultative Authoritarian Institutions in China. *Journal of Contemporary China,* 19, 675–692

Heggelund, G. 2007. China's Climate Change Policy: Domestic and International Developments. *Asian Perspectives,* 31, 155–191

Héritier, A. 1999. Elements of Democratic Legitimation in Europe: An Alternative Perspective. *Journal of European Public Policy,* 6, 269–282

Ho, P. 2007. Embedded Activism and Political Change in a Semiauthoritarian Context. *China Information,* 21, 187–209

Ho, P. & Edmonds, R. L. 2007. Perspectives of Time and Change: Rethinking Embedded Environmental Activism in China. *China Information,* 21, 331–344

Hodson, M. & Marvin, S. 2009. 'Urban Ecological Security': A New Urban Paradigm? *International Journal of Urban and Regional Research,* 33, 193–215

Hodson, M. & Marvin, S. 2010. Can Cities Shape Socio-Technical Transitions and How Would We Know If They Were? *Research Policy,* 39, 477–485

Hooghe, L. & Marks, G. 2001. *Multi-Level Governance and European Integration,* Boulder, CO, Rowman & Littlefield.

Howard, P. 2011. 'Harmony' in China's Climate Change Policy, in Hossain, M. & Selvanathan, E. (eds.) *Climate Change and Growth in Asia,* Cheltenham, Glos, UK, Edward Elgar.

Hudson, B. 2004. Analysing Network Partnerships: Benson Re-Visited. *Public Management Review,* 6, 75–94

IPCC. 2007. Climate Change 2007: Mitigation of Climate Change, in Metz, B., Davidson, O. R., Bosch, P. R., Dave, R. & Meyer, L. A. (eds.) *IPCC Fourth Assessment Report – Working Group III,* Cambridge, UK and New York, NY, Intergovernmental Panel on Climate Change.

Ivankova, N. V., Creswell, J. W. & Stick, S. L. 2006. Using Mixed-Methods Sequential Explanatory Design: From Theory to Practice. *Field Methods,* 18, 3–20

Jänicke, M. 2006. The Environmental State and Environmental Flows: The Need to Reinvent the Nation-State, in Spaargaren, G., Mol, A. P. J. & Buttel, F. H. (eds.) *Governing Environmental Flows: Global Challenges to Social Theory,* Cambridge, MIT Press.

Jiang, B., Sun, Z. & Liu, M. 2010. China's Energy Development Strategy under the Low-Carbon Economy. *Energy,* 35, 4257–4264

Kern, K. & Bulkeley, H. 2009. Cities, Europeanization and Multi-Level Governance: Governing Climate Change through Transnational Municipal Networks. *JCMS: Journal of Common Market Studies,* 47, 309–332

Klijn, E. H. & Skelcher, C. 2007. Democracy and Governance Networks: Compatible or Not? *Public Administration,* 85, 587–608

Knoke, D. 2001. *Changing Organizations: Business Networks in the New Political Economy,* Boulder, CO, Westview Press.

Koliba, C., Meek, J. W. & Zia, A. 2010. *Governance Networks in Public Administration and Public Policy,* Boca Raton, FL: CRC Press, Inc.

Lee, T. & Van De Meene, S. 2012. Who Teaches and Who Learns? Policy Learning through the C40 Cities Climate Network. *Policy Sciences,* 45, 199–220

Levine, M. D. & Aden, N. T. 2008. Global Carbon Emissions in the Coming Decades: The Case of China. *Annual Review of Environment and Resources,* 33, 19–38

Li, W. 2006. Environmental Governance: Issues and Challenges. *Environmental Law Reporter News and Analysis,* 36, 10505–10525

Li, W. 2011. Engaging with the Climate Change Regime: China's Challenges and Activities. *The China Monitor,* October, 4–9

Li, W. & Chan, H. S. 2009. Clean Air in Urban China: The Case of Inter-Agency Coordination in Chongqing's Blue Sky Program. *Public Administration and Development,* 29, 55–67

Li, Y., Miao, B. & Lang, G. 2011. The Local Environmental State in China: A Study of County-Level Cities in Suzhou. *The China Quarterly,* 205, 115–132

Lieberthal, K. 1997. China's Governing System and Its Impact on Environmental Policy Implementation. *China Environment Series,* 1, 3–8

Lieberthal, K. & Lampton, D. M. 1992. *Bureaucracy, Politics, and Decision Making in Post-Mao China,* Berkeley, University of California Press.

Lieberthal, K. & Oksenberg, M. 1988. *Policy Making in China Leaders, Structures, and Processes,* Princeton, NJ, Princeton University Press.

Lin, J. 2012. Climate Governance in China: Using the 'Iron Hand', in Richardson, B. J. (ed.) *Local Climate Change Law: Environmental Regulation in Cities and Other Localities,* Cheltenham, UK, Edward Elgar Publishing.

López-Santana, M. 2009. Having a Say and Acting: Assessing the Effectiveness of the European Employment Strategy as an Intra-Governmental Coordinative Instrument. *European Integration online Papers (EIoP),* 13

Lovell, H. 2007. The Governance of Innovation in Socio-Technical Systems: The Difficulties of Strategic Niche Management in Practice. *Science and Public Policy,* 34, 35–44

Lu, Y. 2007. The Autonomy of Chinese NGOs: A New Perspective. *China: An International Journal,* 5, 173–203

Lyall, C. 2007. Changing Boundaries: The Role of Policy Networks in the Multi-Level Governance of Science and Innovation in Scotland. *Science and Public Policy,* 34, 3–14

Lynn, L. E. J. 2011. The Persistence of Hierarchy, in Bevir, M. (ed.) *The Sage Handbook of Governance,* London, SAGE.

Lynn, L. E. J., Heinrich, C. J. & Hill, C. J. 2001. *Improving Governance: A New Logic for Empirical Research,* Washington, DC, Georgetown University Press.

Ma, X. & Ortolano, L. 2000. *Environmental Regulation in China: Institutions, Enforcement, and Compliance,* Lanham, MD, Oxford, Rowman & Littlefield.

Mandell, M. P. 1990. Network Management: Strategic Behavior in the Public Sector, in Gage, R. W. & Mandell, M. P. (eds.) *Strategies for Managing Intergovernmental Policies and Networks,* New York, NY, Praeger.

Magnusson, W. 1996. *The Search for Political Space,* Cambridge, Cambridge University Press.

Magnusson, W. & Shaw, K. (eds.) 2003. *A Political Space: Reading the Global through Clayoquot Sound,* Minneapolis, University of Minnesota Press.

McGuire, M. & Silvia, C. 2010. The Effect of Problem Severity, Managerial and Organizational Capacity, and Agency Structure on Intergovernmental Collaboration: Evidence from Local Emergency Management. *Public Administration Review,* 70, 279–288

Miller, S. M. & Moulton, S. 2013. Publicness in Policy Environments: A Multilevel Analysis of Substance Abuse Treatment Services. *Journal of Public Administration Research and Theory,* advanced access, 1–37

Mitchell, B. 2005. Integrated Water Resource Management, Institutional Arrangements, and Land-Use Planning. *Environment and Planning A,* 37, 1335–1352

Mol, A. P. J. & Buttel, F. H. 2002. *The Environmental State under Pressure,* Amsterdam: JAI Press.

Mol, A. P. J. & Carter, N. T. 2006. China's Environmental Governance in Transition. *Environmental Politics,* 15, 149–170

Mooney, J. D. 1947. *The Principles of Organization,* New York, NY, London, Harper & Brothers.

Mulford, C. L. & Rogers, D. L. 1982. Definitions and Models, in Rogers, D. L., Whetten, D. A. & Associates, A. (eds.) *Interorganizational Coordination: Theory, Research, and Implementation,* Ames, Iowa State University Press.

O'Toole, L. 1993. Multiorganizational Policy Implementation: Some Limitations and Possibilities for Rational-Choice Contributions, in Scharpf, F. (ed.) *Games in Hierarchies and Networks,* Boulder, CO, Westview Press.

Okereke, C., Bulkeley, H. & Schroeder, H. 2009. Conceptualizing Climate Governance Beyond the International Regime. *Global Environmental Politics,* 9, 58–78

Pan, J.-h. 2010. Background and Approach of Low-Carbon Transition – from the Copenhagen Conference. *Yuejiang Academic Journal* [in Chinese], 4

Pan, J.-h., Zhuang, G.-y., Zheng, Y., Zhu, S.-x. & Xie, Q.-y. 2010. Identifying the Concept of Low Carbon Economy and Analyzing Its Core Elements. *International Economics Review* [in Chinese], 4, 88–101. (*ditan jingji de gainian bianshi ji hexin yaosu fenxi*)

Pauw, P. & Francesch-Huidobro, M. 2010. Hong Kong, in Dircke, P., Aerts, J. & Moleneer, A. (eds.) *Connecting Delta Cities: Sharing Knowledge and Working on Adaptation to Climate Change,* Rotterdam, the Netherlands, City of Rotterdam.

Peters, B. G. 1998. Managing Horizontal Government: The Politics of Co-ordination. *Public Administration,* 76, 295–311

Pierre, J. & Peters, G. B. 2000. *Governance, Politics and the State,* Basingstoke, Hampshire, UK, Macmillan Press.

Pressman, J. L. & Wildavsky, A. B. 1973. *Implementation: How Great Expectations in Washington Are Dashed in Oakland,* Berkeley and Los Angeles, University of California Press.

Purdy, J. M. 2012. A Framework for Assessing Power in Collaborative Governance Processes. *Public Administration Review,* 72, 409–417

Qi, Y., Ma, L., Zhang, H. & Li, H. 2008. Translating a Global Issue into Local Priority: China's Local Government Response to Climate Change. *The Journal of Environment & Development,* 17, 379–400

Rhodes, R. A. 2006. Policy Network Analysis, in Moran, M., Rein, M., and Goodin R. (eds.) *The Oxford Handbook of Public Policy,* New York, United States, Oxford University Press.

Richerzhagen, C. & Scholz, I. 2008. China's Capacities for Mitigating Climate Change. *World Development,* 36, 308–324

Rogers, E. & Weber, E. P. 2010. Thinking Harder About Outcomes for Collaborative Governance Arrangements. *The American Review of Public Administration,* 40, 546–567

Satterthwaite, D. 2008. Climate Change and Urbanization: Effects and Implications for Urban Governance, in *United Nations Expert Group Meeting on Population Distribution, Urbanization, Internal Migration and Development,* UN/POP/EGM-URB/2008/16, New York, NY, 21–23

Schot, J. & Geels, F. W. 2008. Strategic Niche Management and Sustainable Innovation Journeys: Theory, Findings, Research Agenda, and Policy. *Technology Analysis & Strategic Management,* 20, 537–554

Schreurs, M. A. 2008. From the Bottom Up: Local and Subnational Climate Change Politics. *The Journal of Environment & Development,* 17, 343–355

Schroeder, H. & Bulkeley, H. 2008. *Governing Climate Change Post-2012: The Role of Global Cities – Los Angeles.* Working paper (122). Norwich, UK, Tyndall Centre for Climate Change Research.

Schroeder, M. 2009. Varieties of Carbon Governance: Utilizing the Clean Development Mechanism for Chinese Priorities. *The Journal of Environment & Development,* 18, 371–394

Simon, H. A. 1962. The Architecture of Complexity. *Proceedings of the American Philosophical Society,* 106, 467–482

Skogstad, G. 2003. Legitimacy and/or Policy Effectiveness? Network Governance and GMO Regulation in the European Union. *Journal of European Public Policy,* 10, 321–338

Sørensen, E. & Torfing, J. 2005a. The Democratic Anchorage of Governance Networks. *Scandinavian Political Studies,* 28, 195–218

Sørensen, E. & Torfing, J. 2005b. Network Governance and Post-Liberal Democracy. *Administrative Theory & Praxis,* 27, 197–237

Sørensen, E. & Torfing, J. 2007. *Theories of Democratic Network Governance,* Basingstoke, Palgrave Macmillan.

Tan, Y. 2012. Transparency without Democracy: The Unexpected Effects of China's Environmental Disclosure Policy. *Governance,* 27, 37–62

Tashakkori, A. & Teddlie, C. 2003. The Past and Future of Mixed Methods Research: From Data Triangulation to Mixed Model Designs, in Tashakkori, A. & Teddlie, C. (eds.) *Handbook of Mixed Methods in Social & Behavioral Research,* 2nd ed., Thousand Oaks, CA: SAGE Publications.

Thomson, A. M. 2001. *Collaboration: Meaning and Measurement.* Ph.D. 3038517, Indiana University.

Thomson, A. M. & Perry, J. L. 2006. Collaboration Processes: Inside the Black Box. *Public Administration Review,* 66, 20–32

Toikka, A. 2010. Exploring the Composition of Communication Networks of Governance – a Case Study on Local Environmental Policy in Helsinki, Finland. *Environmental Policy and Governance,* 20, 135–145

Torfing, J., Peters, B. G., Pierre, J. & Sørensen, E. 2012. *Interactive Governance: Advancing the Paradigm,* Oxford, Oxford University Press.

Tsang, S. & Kolk, A. 2010. The Evolution of Chinese Policies and Governance Structures on Environment, Energy and Climate. *Environmental Policy and Governance,* 20, 180–196

Wamsley, G. L. & Zald, M. N. 1973. *The Political Economy of Public Organizations: A Critique and Approach to the Study of Public Administration,* Lexington, MA, D.C. Heath.

Wang, W., Zheng, G. & Pan, J. (eds.) 2012. *China's Climate Change Policies,* New York, NY, Routledge.

Ward, P. M. & Rodríguez, V. E. 1999. New Federalism, Intra-Governmental Relations and Co-Governance in Mexico. *Journal of Latin American Studies,* 673–710

Wejnert, B. 2002. Integrating Models of Diffusion of Innovations: A Conceptual Framework. *Annual Review of Sociology,* 28, 297–326

While, A., Jonas, A. E. & Gibbs, D. 2010. From Sustainable Development to Carbon Control: Eco-State Restructuring and the Politics of Urban and Regional Development. *Transactions of the Institute of British Geographers,* 35, 76–93

Williamson, O. E. 1996. *The Mechanisms of Governance,* New York, NY, and Oxford, Oxford University Press.

Wollmann, H. 2003. Coordination in the Intergovernmental Setting, in Peters, B. G. & Pierre, J. (eds.) *Handbook of Public Administration,* London, Sage.

Wood, D. J. & Gray, B. 1991. Toward a Comprehensive Theory of Collaboration. *The Journal of Applied Behavioral Science,* 27, 139–162

Wright, D. S. 1990. Federalism, Intergovernmental Relations, and Intergovernmental Management: Historical Reflections and Conceptual Comparisons. *Public Administration Review,* 50, 168–168

Wright, D. S. & Krane, D. 1998. Intergovernmental Management (IGM), in Shafritz, J. M. (ed.) *International Encyclopedia of Public Policy and Administration,* Boulder, CO, Westview Press.

Wu, F. 2002. New Partners or Old Brothers? GONGOs in Transnational Environmental Advocacy in China. *China Environment Series,* 5, 45–58

Wu, F. 2003. Environmental GONGO Autonomy: Unintended Consequences of State Strategies in China. *The Good Society,* 12, 35–45

Yang, C. 2005. Multilevel Governance in the Cross-Boundary Region of Hong Kong-Pearl River Delta, China. *Environment and Planning A,* 37, 2147

Yang, T. 2008. The Implementation Challenge of Mitigating Greenhouse Gas Emissions in the Developing World: The Case of China. *Georgetown International Environmental Law Review,* 20, 1–26

Yu, H. 2004. Knowledge and Climate Change Policy Coordination in China. *East Asia,* 21, 58–77

Zang, D. 2009a. From Environment to Energy: China's Reconceptualization of Climate Change. *Wisconsin International Law Journal,* 27, 543

Zang, D. 2009b. Green from Above: Climate Change, New Developmental Strategy, and Regulatory Choice in China. *Texas International Law Journal,* 45, 201–232

Zeemering, E. & Romero, J. 2012. The Evolution of Sustainable Cities as a Metropolitan Policy Challenge, in Meek, J. W. & Thurmaier, K. (eds.) *Networked Governance: The Future of Intergovernmental Management,* Thousand Oaks, CA, CQ Press.

Zhao, Y. 2011. Responding to the Global Challenge of Climate Change – Hong Kong and 'One Country Two Systems'. *CCLR [The Carbon & Climate Law Review],* 5, 70–81

Zhou, W. 2012. In Search of Deliberative Democracy in China. *Journal of Public Deliberation,* 8, 8

3 Climate networks in Guangzhou, Shenzhen and Hong Kong

This chapter provides the empirical context for the analysis of climate governance in the three southern Chinese cities. By analysing the urban attributes, resource capacity, political structure and climate-policy development in the three cities under study, we contextualise CMNs in a loosely bounded sociopolitical system that connects the national regime and the global city networks. In the first section of the chapter, we shall present the cities' socio-economic profiles, their relationships with the national climate regime and the central–regional–municipal administrative structures. The second section reviews the recent developments in climate mitigation policy.

Cities' capacity for climate change mitigation

In an inventory of globalisation and world cities, Guangzhou, Shenzhen and Hong Kong were graded 'Beta', 'Beta-' and 'Alpha+', respectively, denoting their connections to world city networks (GaWC 2011). Leading Chinese cities have emerged as 'globalising cities', with Guangzhou and Shenzhen as typical Beta-grade examples, in the sense that although these cities 'remain functionally based on the domestic economy', they 'have demonstrated a trend of growing interaction [with orientation] toward the global economy' (Lin 2004: 144).

The responsibility for meeting China's national carbon reduction targets through sectoral mitigation actions is devolved to these globalised subnational localities, which are economically developed and have larger resource capacities than other regions (Table 3.1). These cities are also important sites for the transferral of and experimentation with sectoral mitigation innovations through the process of globalisation, thus leading to broader integration with and contributing to the transformation of the Chinese environmental state.

Cities' profile

The basic municipal demographics summarised in Table 3.1 demonstrate the comparability of the three cities as typical and diverse cases (see Chapter 2, methodology) in geographical, socioeconomic and environmental terms, as the three cities have similar geographical characteristics, sizes and populations and have

Table 3.1 Demographics for Guangzhou, Shenzhen and Hong Kong

	Guangzhou	Shenzhen	Hong Kong
Location	Mid-southern Guangdong; northern end of the Pearl River Delta	Southern Guangdong; Pearl River estuary to the west; borders Hong Kong to the south	Pearl River estuary to the west; south-eastern tip of mainland China; bordered on three sides by the South China Sea
Area (km^2)	3843.43@	1991.64	1104.4
Population, million	8,994,900@#	8,912,300#	7,067,800
Economy (2009)^	GDP 913,821 million (¥) with 10.3% annual growth; per capita GDP 89,082 (¥) with 8.7% annual growth	GDP 820,132 million (¥) with 5.3% annual growth; per capita GDP 92,772 (¥) with 3.6% annual growth	GDP 1,622,322 million (HKD) with −3.3% year-on-year change; per capita GDP 231,638 (HKD) with −3.6% year-on-year change
Energy efficiency of economy (2009)*	5.7 *(Conversion from Guangzhou Statistical Yearbook 2010)*	7.0 *(Conversion from Shenzhen Statistical Yearbook 2010)*	18.9 *(World Bank data)*
Carbon emissions (per capita, 2009)	5.8 tonnes CO_2 *(City data not available; World Bank data on China)*		5.3 tonnes CO_2
Urbanisation (m^2)	• Per capita area of roads: 10.63 • Per capita living floor space of buildings: 16.92 (2004), 21.01 (2009)	• Per capita area of roads: 9.9 • Per capita living floor space of buildings: 21.80 (2004)	• Per capita area of roads: 5.7 • Per capita living floor space: public housing 12.6 (2010); private housing 15.6 (1999)$^\Delta$

Note:

@This area only covers the urban districts, excluding two county-level cities (Zengcheng and Conghua) that are also under this municipality administration.

#Residential population (with or without residence cards/household registration) as of the end of 2009.

^Exchange rate of RMB (¥) to HKD ($) was 1.13 in 2009.

*The energy efficiency of an economy is measured by GDP per unit of energy use (constant 2005 purchasing power parity dollars per kilogram of oil equivalent).

$^\Delta$Relevant statistics for private housing aare not currently kept by the Hong Kong government.

Source: Guangzhou Statistical Yearbook 2010, Shenzhen Statistical Yearbook 2010, Chinese Urban Development Report 2003–2004, Hong Kong Energy Statistics Annual Report 2010, Population and Household Statistics analysed by District Council District 2010, Hong Kong Statistics on Key Economic and Social Indicators 2010, Housing in Figures 2010 by the Hong Kong Housing Authority, Greenhouse Gas Emissions and Carbon Intensity in Hong Kong, Shenzhen Medium and Long Term Plan of Low Carbon Development (2011–2020), Housing Standards of Private Dwellings (1999).

comparable levels of economic development and urbanisation. Despite these similarities, energy efficiency varies across the three cities. Energy efficiency is measured by the value of gross domestic product (constant 2005 purchasing power, parity dollars) per unit of energy use (kilogram of oil equivalent): the higher the value, the higher the energy efficiency level of the local economy. As indicated in Table 3.1, Guangzhou, Shenzhen and Hong Kong, respectively, generated $5.7, $7.0 and $18.9 per unit of energy consumption/efficiency in 2009.

The cities' demographics can be further compared in terms of national statistics, as shown in Figure 3.1 (World Bank 2009b). It is apparent that Guangzhou and Shenzhen are paying a higher price to energise economic growth. The high energy prices and outbound relocated industries of Hong Kong partly contribute to its lead in energy efficiency. This means that Guangzhou and Shenzhen have greater potential for mitigating climate change through energy retrofitting practices while Hong Kong, with its higher energy efficiency levels, needs to seek alternative means of mitigation.

In terms of carbon emissions, although city-level data are not fully available in Guangzhou and Shenzhen, the city figures of Hong Kong and the national figures of China are available and summarised in Table 3.1. Carbon emissions per capita at the national level and in Hong Kong are similar (World Bank 2009a).

Figure 3.2 shows the carbon emissions measured in economic terms. Although the carbon intensity of Hong Kong is drastically lower than that of China as a whole, it is still higher than the figures of other high-income developed economies such as Japan, Germany and the United Kingdom (EIA 2010). China, for its part, has the highest carbon intensity among all of the selected countries in Figure 3.2. Despite the lack of reliable statistics, the carbon intensity figures in Guangzhou and Shenzhen are expected to be lower than the national figure, given their relatively higher values of gross domestic product (GDP) compared to that of the average Chinese city.

Local budgetary support

Table 3.2 summarises the environmental expenditure of the three cities and the proportion of this expenditure to total local municipal expenditure. This indicates budgetary support for environmental activities that cover climate mitigation plans and actions. In Hong Kong, environmental expenditure increased steadily between 2006 and 2011. Before 2006, neither Shenzhen nor Guangzhou institutionalised the category of environmental protection and energy conservation as a regular budget item. After the introduction of an environmental budget in 2007, the proportion of environmental expenditure in Shenzhen rapidly increased to 5.47 percent in 2011, higher than that in Hong Kong in the same year. In contrast, the environmental expenditure in Guangzhou since 2007 has remained lower than in Hong Kong and Shenzhen.

Given the relatively stable environmental expenditure in Hong Kong, it is pertinent to examine how it is distributed among policy sectors. Figure 3.3 presents the annual changes in environmental expenditure by policy function (HK-EPD 2012). 'Coordination' refers to the proportion of environmental spending on

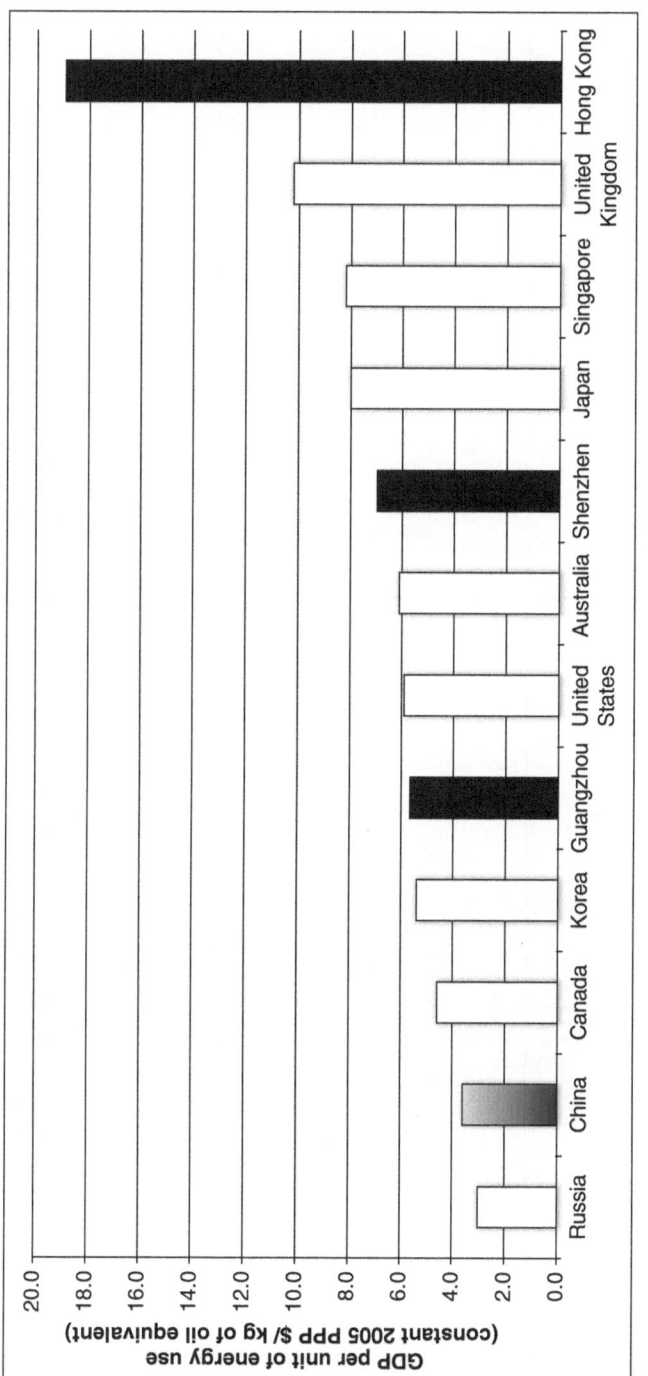

Figure 3.1 Energy efficiency of world economies

Source: Compiled by authors with data from World Bank (2011)

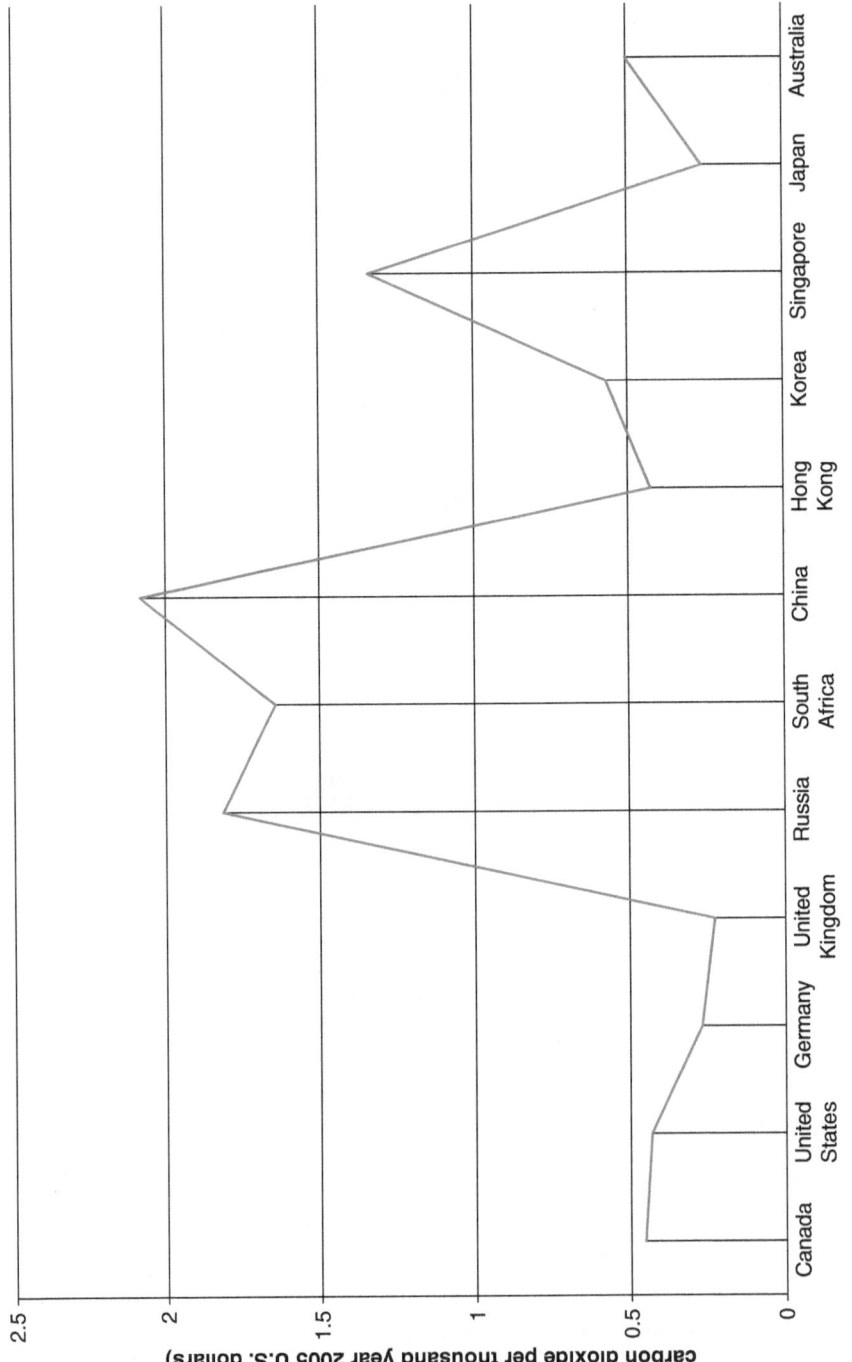

Figure 3.2 Carbon intensity of world economies

Source: Compiled by authors with data from the International Energy Statistics (EIA 2010)

Table 3.2 Environmental expenditure as percentage of total local expenditure (denoted in local currencies[#] [unit: million])

Financial year	Hong Kong			Shenzhen			Guangzhou		
	Environmental expenditure	Total expenditure	%	Environmental expenditure	Total expenditure	%	Environmental expenditure	Total expenditure	%
2006	$7,843	$248,400	**3.16%**	¥ –	¥58,445	**0.00%**	¥ –	¥52,456	**0.00%**
2007	$7,683	$226,900	**3.39%**	¥343	¥92,682	**0.37%**	¥457	¥58,305	**0.78%**
2008	$10,715	$304,600	**3.52%**	¥1,281	¥105,404	**1.21%**	¥703	¥91,278	**0.77%**
2009	$10,715	$291,200	**3.68%**	¥3,480	¥115,408	**3.02%**	¥823	¥96,794	**0.85%**
2010	$12,207	$303,500	**4.02%**	¥7,082	¥149,884	**4.73%**	¥2,376	¥136,132	**1.75%**
2011	$14,432	$366,400	**3.94%**	¥9,752	¥178,258	**5.47%**	¥2,511	¥164,586	**1.53%**

Sources: Estimates of Government Resources Devoted to Environmental Protection and Conservation Work (HK-EPD 2007-2012), Hong Kong Budgetary Reports (Financial Secretary 2007-2012), Shenzhen Statistical Yearbook (Shenzhen Statistics Bureau 2007, 2008, 2012), Guangzhou Statistical Yearbook (Guangzhou Statistics Bureau 2007-2012), Guangzhou Environmental Protection Department (GZ-EPD 2011)

Note: [#]Exchange rate for RMB (¥) with HKD ($) was between 0.9604 (1 Jan 2006) and 1.2291 (30 Dec 2011).

regular administrative activities conducted by the Environmental Bureau and the Environmental Protection Department. Environmental spending in the building sector is distributed to agencies such as the Architectural Services, Buildings and Electrical and Mechanical Services Departments. The Government Logistics Department, the Highways Department and the Transport Department are major governmental actors incurring environmental expenditures in the transport sector. Infrastructural activities are implemented by the Planning Department. 'Others' in Figure 3.3 refers to environmental expenditures in all other departments. Since 2007, there has been a substantial increase in the environmental budget allocated to the transport sector, as the government in Hong Kong has invested more revenue into decarbonising local transport. However, environmental expenditure in the transport sector is still lower than that in the building sector.

Earmarked grants

Various earmarked grants have been set up in the three cities to endorse climate mitigation projects. In 2012, the Guangzhou municipal administration injected

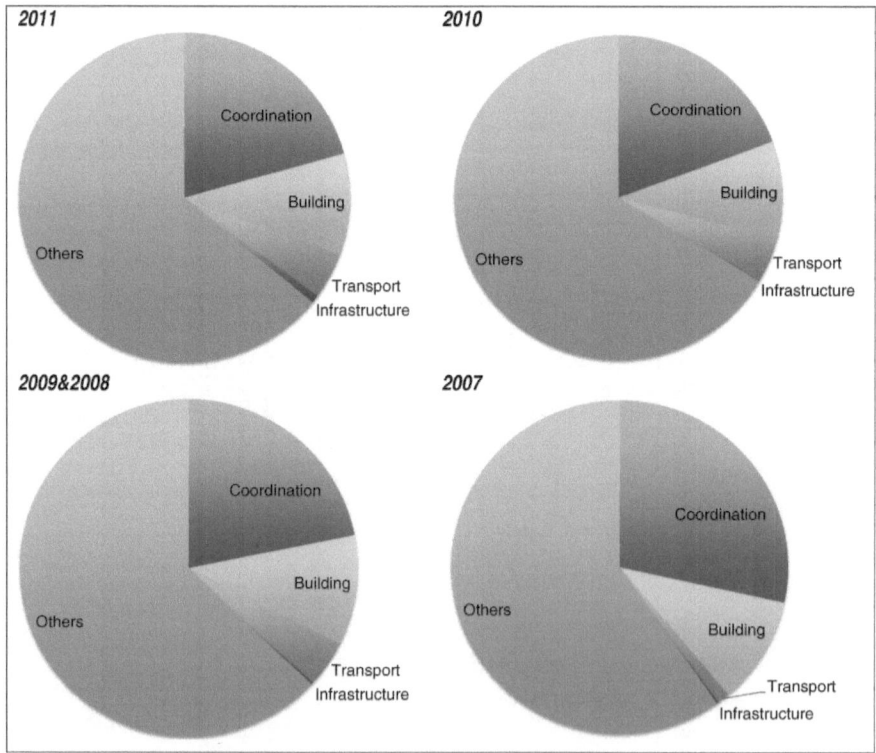

Figure 3.3 Functional distribution of environmental expenditure in Hong Kong (2007–2011)

Source: Compiled by authors

RMB60 million into its existing Energy Efficiency Special Fund, established in 2009 (Appendix I: Interviews 2 & 9). In 2011, seventeen research and industrial projects were approved in Shenzhen, with RMB37.5 million invested from its Environmental Protection Special Fund (Appendix II: Shenzhen Human Habit and Environment Commission 2011). A sectoral special fund to retrofit buildings for energy efficiency has also been created in Shenzhen. In 2011, this earmarked fund approved RMB4.7 million to support various climate-mitigation projects, including i) green building guidelines, yearbooks, technological research and training, ii) research and development of solar-powered electricity generation, iii) development and application of energy-efficient buildings and iv) a low carbon indicator system (Appendix II: Shenzhen Finance Bureau 2011).

The Environment and Conservation Fund (ECF), a Hong Kong government grant earmarked for environmental activities, has received HKD1.735 billion since 1994 (ECF 2012). The chief executive proposed a further HKD5 billion capital injection in his 2013 Policy Address, which was subsequently approved by the Finance Committee of the Legislative Council (Appendix II: Legislative Council 2013). Figure 3.4 presents the expenditure distribution of the total fiscal allocation between 2008 and 2011. The largest amount of funding went to environmental education and awareness-raising activities, in line with the ECF's funding objectives. Energy-efficient building projects, which the ECF has emphasised since 2008, also received a significant proportion of the funding. Part of the funding was allocated to local universities, research institutes and environmental nongovernmental organisations to carry out environmental research and other work related to energy efficiency.

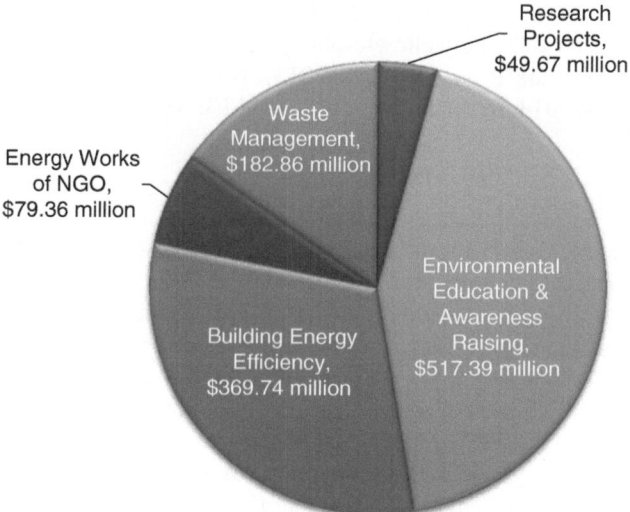

Figure 3.4 Expenditure structure of environmental conservation fund (2008–2011)
Source: Compiled by authors

National climate vision

National policies

China's *National Climate Change Programme* (Appendix II: NDRC 2007), issued by the National Development and Reform Commission (NDRC) and circulated by the State Council in 2007, formally recognised climate change as an important policy issue. The programme identified key policies related to energy conservation and the development of a circular economy. It proposed the development of climate-change mitigation and adaptation strategies in various policy sectors, including transport, building, agriculture and industry; it also emphasised the role of government in promoting public awareness of climate change, strengthening leadership and institutional development and establishing an implementation system for local climate change policies (Appendix II: NDRC 2007). However, neither specific targets for carbon reduction nor any other clearly defined climate actions were included in the 2007 national programme.

A specific carbon reduction target was set out only after a State Council Executive Meeting hosted by then Premier Wen Jiabao on 25 November 2009 (Central Government 2009). One of the important decisions made in the meeting was to establish a target of a 40 to 45 percent reduction in CO_2 emissions per unit of GDP by 2020 from the 2005 level. This target was adopted as the bottom line for accounting, monitoring and evaluating domestic energy at different levels of government.

In 2010, the NDRC organised the drafting of the *National Climate Change Plan (2011–2020)*, a state planning project prioritised in the 12th Five-Year Plan (NDRC 2011). The drafting process involved an eight-person team of climate-research scholars coordinated by the NDRC, with input from all twenty-nine national ministries (Appendix I: 34/BJ/06–08–2013). The plan was approved by the State Council in 2011 and is the only climate-policy plan mandated for nationwide implementation (Appendix I: 34/BJ/06–08–2013). However, the plan was not publicised until August 2013, in the midst of the 12th Five-Year Plan period.

Key actors and central coordination

National climate policies are mainly formulated by two state agencies, the NDRC and the China Meteorological Administration (CMA). The NDRC is a powerful ministry in China in terms of political resources and economic-planning authority. It started to deal with climate change issues after China's adoption of the Kyoto Protocol to the United Nations Framework Convention on Climate Change in 1998. In 2007, the National Climate Change Leadership Group (NCCLG) was convened by then Premier Wen Jiabao. The NCCLG's secretariat office was set up as part of the NDRC by strengthening the power of the existing National Coordination Committee on Climate Change, which at that time was China's contact point with the Intergovernmental Panel on Climate Change (IPCC).

An exclusive authority was eventually institutionalised when the Department of Climate Change was established by the NDRC in the State Council's 2008 structural reform (Appendix II: NDRC 2009). The department performs the NCCLG's secretariat tasks, organises international climate negotiations and coordinates the analysis, planning and drafting of national climate policies. During the United Nations Climate Change Conference in Durban towards the end of 2011, China established a new climate change organisation – the National Centre for Climate Change Strategy and International Cooperation (NCSC). This is a public institution (*'shiye danwei'*) directly subordinate to the NDRC, and therefore its administrative rank equals that of the Department of Climate Change. The major mission of the NCSC is to facilitate international negotiations and implement collaborative projects for research, training and publicity connected to climate change issues (NCSC 2012). More than twenty people currently work for the Department of Climate Change, while the NCSC has thirty-two employees (Appendix I: 35/BJ/08–08–2013; NCSC 2012).

The CMA is the location of the IPCC's Focal Point in China, based on a remit from the State Council (IPCC 2013; Appendix I: 33/BE/28–05–2013). It is also commissioned to organise the writing of national climate-assessment reports. The Climate Change Centre was formed in 2008 in affiliation with the National Climate Centre (NCC) and set in the CMA (NCC 2012). The NCC is the research facility where one of the two IPCC-China Technical Support Units is located. The NCC and CMA organise and recommend leading Chinese authors to the IPCC scientific community; they also organise government reviews and ministerial reviews of the draft national climate change assessment reports (Appendix I: 33/BE/28–05–2013). Around forty multidisciplinary researchers in the NCC work on climate impact in various sectors, including water, agriculture and energy generation, providing professional climate services to other government agencies.

In addition to the regular staff of the NCC, experts on climate research are appointed to the National Panel on Climate Change, with a secretariat office set in the CMA's Department of Scientific Development. In 2006, when the panel was originally formed, there were twelve expert members from different ministerial-level agencies, and in 2010 it was expanded to more than thirty members. The panel is regarded as a high-level think tank for decision making on climate change issues in the central government (Appendix II: 35/BJ/08–08–2013).

Centralised climate knowledge transfer

The National Communications on Climate Change Statistics were compiled as the initial national greenhouse-gas (GHG) inventory on which to base future carbon-reduction decisions (Appendix II: NDRC, 2005). Since 2008, the NDRC has published an *Annual Report on China's Climate Policies and Actions* with the stated aims of facilitating climate policy communication and advancing climate governance (Appendix II: NDRC, 2008–2012). Nonetheless, it is unclear to what extent these national climate communications and information disclosure documents have been effective in raising nationwide awareness and encouraging climate action by subnational governments.

Professional reviews and policy recommendations are also collected at the national level. In collaboration with the Chinese Academy of Social Sciences (CASS), the CMA founded the Joint Laboratory of Climate Change Economics Simulation, which has produced the annual *Climate Change Green Book* since the 2009 United Nations Climate Change Conference in Copenhagen. The green books collect professional views for a general audience as a way to raise public awareness of climate change issues (Appendix I: 34/BJ/06–08–2013, 35/BJ/08–08–2013).

For a professional audience, the CMA published a handbook on *China Climate and Environmental Change* in 2012, which has been distributed among different ministerial-level agencies (Appendix I: 34/BJ/06–08–2013). The *National Assessment Report on Climate Change*[1] is intended for decision makers (Appendix I: 34/BJ/06–08–2013). Steered by the Ministry of Science and Technology, its writing process involved climate specialists from the CAS, CASS and NCC. As stated by our informants, it serves as the scientific foundation for the production of the *National Climate Change Plan* and the related content in the national five-year plans (Appendix I: 34/BJ/06–08–2013). These centralised professional reviews and policy recommendations on climate policies are circulated in both the vertical *inter*governmental hierarchy and horizontal *intra*governmental networks.

As for the vertical hierarchy, authorised staff members at all levels of the meteorological administrative system in China are appointed to officially budgeted posts directly funded by the CMA (Appendix I: 34/BJ/06–08–2013). The entire meteorological administrative system in China covers more than 2,600 subnational offices at the provincial, municipal and county levels (CMA 2011). As a senior researcher in the NCC reflected:

> All the meteorological bureaus beneath are under the aegis of the meteorological administration in Beijing, meaning the CMA pays their salaries. The CMA is the only ministerial-level agency structured in this way. If you are not the one who pays their wages, they are going to *ignore* you.
>
> (Appendix I: 34/BJ/06–08–2013)

Unlike the CMA, neither the Ministry of Environmental Protection nor the NDRC directly pays for the staff at their local offices from their central budget in Beijing. The existing human resources structure of the CMA thereby ensures more intensive climate research knowledge generation and transfer through the vertical *inter*governmental hierarchy in the meteorological administrative system. Moreover, joint session systems ('*lianxi huiyi zhidu*') – regular meetings of designated departments – are institutionalised between the CMA and most provincial-level governments to facilitate *inter*governmental communication on climate service.

At the interface between policy making and scientific climate-change research, the dynamics of knowledge transfer are also found in the horizontal *intra*governmental networks among the agencies of the central government. The NCC,

CASS and CAS are the key agencies that produce scientific knowledge on climate change for the reference of central decision makers. There is close collaboration on climate research among experts from different disciplines in these agencies (Appendix I: 35/BJ/08–08–2013). However, the NDRC – the coordinating ministry for national climate policy – may 'only adopt things that are politically feasible' from the scientific reports on climate research conducted by the national-level scientific community (Appendix I: 33/BE/28–05–2013).

Furthermore, as the CMA – the parent agency of the NCC – has a less powerful position among the ministerial-level agencies in terms of resources and political significance, it is difficult for it to initiate cross-sectoral climate action. Although the CMA actively convenes cross-sectoral meetings with other ministries, it lacks the authority and resources to lead the coordination. To strengthen its influence on policy making for climate change, the CMA has attempted to initiate and promote cross-sectoral collaboration with powerful ministries such as the Ministry of Water Resources (Appendix I: 34/BJ/06–08–2013). In 2012, a Memorandum of Cooperation between the two was formulated, with mutually agreed-upon protocols on joint practices (Ministry of Water Resources 2012).

Declining political interest by the central leadership?

Nevertheless, new institutional barriers can be found in the process of centralised knowledge transfer, which diminish the influence of science on policy at the national level. In terms of governing norms, political attention towards climate change issues appears to have been diverted towards the economic agenda since the new central leadership took office in 2012. In China, an internal instruction (*'pishi'*) is an essential indicator of how important a public issue is to the political agenda. It is a type of *inter*governmental correspondence in which the upper-level government issues review notes in response to lower-level agencies' submissions of official documents discussing certain public policy issues of concern and their proposed solutions. Internally, such correspondence is regarded as indicating the will and preference of the central leadership, which determines the actions taken by officials at different levels of government (Hu 2010). By the same token, the changing political significance of climate change issues in the central leadership is implied and assessed by lower-ranking government agencies, based on the number of replies they receive during the process of *pishi* with the central leadership. As revealed by NCC staff, one year after the new national leaders assumed office, none of their documentary submissions about climate risks in China had been reviewed through internal instruction *(pishi)*. This is interpreted as a reflection of the current national leadership's declining political interest in climate change:

> Currently, unfortunately, the new government, for example, Li Keqiang, is not interested in climate change, because he insists that we have no time for climate change discussions – he is concentrating on the economy.
>
> (Appendix I: 34/BJ/06–08–2013)

Moreover, there is hardly any consensus on climate change issues even within the Central Politburo of the Party, the governing elite of the nation:

> The Politburo members are still fighting among each other; they don't know who to trust during their disputes about climate change.
>
> (Translated Appendix I: 34/BJ/06–08–2013)

Despite this declining political interest, China's climate change experts are conscious of their role in the policy process (Wübbeke 2013). As a director at a national think tank on climate mitigation expressed:

> We publish our papers, we influence our media and we submit our policy briefs to the leadership; and whether they listen to us or not, it's not our concern. We don't care at all. We just *do* what we want, we just do what we believe, so whether they take it or not, it is not our business. Our business is to research, to make the analysis, to provide the policy input – whether they take it or not, it is not our business.
>
> (Appendix I: 35/BJ/08–08–2013)

Another noteworthy institutional barrier that may exacerbate the declining political interest in climate change is an unintended outcome of the changing rules in the bureaucracy. One month after Xi Jinping's appointment as party leader at the end of 2012, he promulgated the 'eight-point bureaucracy-busting and formalism-fighting guidelines'[2] ('*baxiang guiding*') as prescriptions to maintain party discipline, reduce bureaucratic 'red tape' and erase rampant political corruption at all levels of government. One of these new prescriptions is to streamline the processing of internal written documents by prohibiting excessive document submissions from individual ministries. Submissions must be made collectively after collective decision making. The Xinhua News Agency is responsible for coordinating document submissions by gathering all ministerial documents on various topics and entering them into an internal referencing system for national political leaders. In this way, agencies such as the CMA, CAS and CASS cannot separately submit documents regarding climate change issues for the national leadership to review. In addition, after the information is filtered by the internal referencing system, a great deal of important scientific information is lost and the climate change issue may be sidelined. Thus, the proactive act of knowledge transfer among individual agencies is turned into passive reaction.

Nationally endorsed local experiments

The strong, domestically generated climate knowledge has marginalised the influence that transnational municipal networks such as C40 and the Local Governments for Sustainability (ICLEI) have on climate change issues in China. However, under the umbrella of national climate policy, domestic municipal networks remain strong.

The vital role of subnational governments in driving climate policy was emphasised in the National Communications on Climate Change and the 2009 Annual Progress Report of China's Policies and Actions on Climate Change (Appendix II: NDRC 2005, 2009). After the State Council's decision on carbon reduction targets in November 2009, actions on low carbon development were devolved to cities and provinces. In 2010, the NDRC formally designated five provinces (including Guangdong) and eight cities (including Shenzhen) as 'low carbon pilots' for conducting local experiments on climate-resilient economic growth (Appendix II: NDRC 2010).

The CASS has initiated collaborations with many subnational governments, including Shenzhen and Guangdong, to conduct low carbon projects as experimental design models for climate mitigation. The involvement of CASS in the local projects in Guangdong and Shenzhen is in relation to a policy study for low carbon development design, the China Prosperity Strategic Programme Fund (SPF) project, financed by the British Foreign and Commonwealth Office. The SPF has an ultimate goal of promoting sustainable global growth (Foreign & Commonwealth Office 2013). Low carbon pilot projects in cities or provinces that are oriented towards a partnership between the United Kingdom and the NDRC on climate change policies are given higher funding priority. Government endorsement is regarded as an integral part of project implementation, and a signed letter of support from the local DRC is required in the funding-application process.

Although the CMA has a less powerful position in terms of its overall ministerial rank, it has attempted to strengthen its policy impact by holding joint sessions with subnational governments. As an example, since 2011, the CMA has cooperated with Guangdong in carrying out laboratory and infrastructural projects on climate proofing (CMA 2012a). In 2012, a formal agreement on joint sessions was established between the CMA and the Guangdong Provincial Government, in the form of a Memorandum of Cooperation (Hu 2012). This process also serves to engage the respective authorities in Hong Kong and Macau, which as Special Administrative Regions are not within the authority of the Chinese meteorological administrative system (CMA 2012b; Dong 2013).

Vertical inter*governmental relations*

The three cities in our study have a different status from others in their vertical *inter*governmental relationships with the national government. Their unique status ensures sufficient national attention to drive policy and experiments on climate mitigation in these three cities.

Guangzhou and Shenzhen are the only two 'subprovincial' cities in southern China. Subprovincial cities enjoy more administrative autonomy in social and economic planning than do other prefectural-level municipalities, which are governed by their respective provincial governments. We should consider, then, what administrative concerns led the State Council to put Guangzhou and Shenzhen at the subprovincial level.

Guangzhou is the capital city of Guangdong province and is geographically at the centre of the fast-growing Pearl River Delta region. Guangzhou is also one of the five National Central Cities,[3] a concept proposed by the Ministry of Housing and Urban–Rural Development (MOHURD) in 2010, changed from the original terminology of 'global functional cities' in the MOHURD's *National Urban System Planning (2005–2020)*. These National Central Cities are located at the apex of the Chinese urban system and have both strong ties to the world economy and an important influence on domestic development. As the political, geographic and economic centre of its region, Guangzhou has often been chosen as a pilot for national and regional policy experiments, and the city's independence in the planning of economic and social development is leading the broader growth of the region. Meanwhile, maintaining the provincial controls on the city sets a political limit on the potential expansion of its power boundaries.

Shenzhen was constituted as a Special Economic Zone at the start of China's open-door policy in the late 1970s because of the city's geographical proximity to the then-British colonial territory, Hong Kong. After rapid changes in the city over three decades of economic development, Shenzhen is currently one of the five Chinese cities 'specifically designated in the state plan'[4] (*'jihua danlie shi'*) and has therefore been automatically turned into a subprovincial city in terms of its administrative status. The 'specifically designated' cities have a budgetary status independent from their respective provincial governments. Although administratively they are still subordinate to the province, the 'specifically designated' cities receive policy commands directly from the national government. They also enjoy autonomy in making laws about economic management and urban planning. For instance, in 2009, the Shenzhen Municipal Government implemented the Shenzhen Special Economic Zone Circular Economy Promotion Regulations, the first subnational law on the circular economy. The Shenzhen Municipal Government also incorporated the Shenzhen Measures of Green Building Promotion into its legal framework in 2013 after a year of consultation with practitioners and stakeholders in the city. Although the MOHURD–proposed Chinese urban system only classifies Shenzhen as a Regional Central City, it continues to be an experimental site for the nation.

Hong Kong is a Special Administration Region governed by its own Basic Law under the constitutional principle of 'one country, two systems' (OCTS). Hong Kong was a British colony until 1997. The Basic Law (Article 12) and the OCTS prescribe 'a high degree of autonomy' for Hong Kong 'directly under the Central People's Government' (Appendix II: Basic Law Promotion Steering Committee 1990). The pre-1997 executive-led system has been retained while the Hong Kong government has independent decision-making power except on issues related to foreign affairs and defence. Therefore, environmental policies are by constitution decided and implemented independently in Hong Kong. Nevertheless, in the 15 years since Hong Kong's return to Chinese sovereignty, the city has gradually become more integrated into State planning. In the MOHURD's *National Urban System Planning (2005–2020)*, Hong Kong was recognised as one of the 'global functional cities' of China. In the recent 12th Five-Year Plan, one chapter was set aside for the two Special Administrative Regions, Hong Kong and Macau (HKSAR Government 2011). Hong Kong's regional cooperation with the Guangdong Provincial Government has been institutionalised in a *Framework Agreement on Hong Kong/Guangdong Co-operation* since 2010, with the endorsement of the State Council and the NDRC. Thus, the decision-making process in Hong Kong is steered or influenced by the preferences of the national leadership, despite the fact that direct intervention is prohibited by the constitutional framework (Cheung 2011).

Organisational forms and multilevel climate governance

Wang and Wu (2005) suggest that emerging organisational forms, such as environmental research institutes, think tanks and nongovernmental organisations – both government-organised (GONGOs) and independent – are integrated into the changing institutional landscape of the Chinese environmental state. They recognise that the Chinese administrative reform prescribing greater decentralisation has affected local environmental management, the localisation of governing authorities and jurisdictions, the streamlining of environmental management, the incorporation of market forces into policy tools and, most importantly, internalisation of the concept of environmental governance in public participation and information disclosure (Wang and Wu 2005). However, their discussion focuses on the administrative supervisory role of the central government and thus remains at the national level rather than extending the focus to the interaction dynamics of the multilevel governance networks. Hence, they do not address both national and subnational aspects, which is one of the objects of the analysis we present in this volume. Moreover, their work is concerned only with air pollution and thus does not consider the broader and more complex issue of climate change and its mitigation.

Nevertheless, Wang and Wu's (2005) approach to the institutional analysis of the Chinese environmental state is useful. To explain the empirical context, the following paragraphs present an institutional analysis of the policy context and the agencies operating at different levels of administration in relation to the emerging organisational forms and network agglomerations.

Figure 3.5 presents the full mapping of the governmental agencies related to climate governance in China, representing the vertical and horizontal *inter*governmental relations. The municipal administrations of the three cities are included in this organisational context.

Urban climate mitigation: current policy design

In the unique context of vertical *inter*governmental relations, different sets of subnational climate mitigation policies are being implemented in the three cities. Because Guangzhou is the provincial capital city, its climate mitigation policies involving low carbon economic development plans are often intertwined with those in Guangdong province. In contrast, as a 'specifically designated city in the state plan', the design and implementation of Shenzhen's climate mitigation policy are relatively independent from provincial control. As a Special Administrative Region, Hong Kong has the widest autonomy and political space of the three localities in its decision making and implementation of climate mitigation

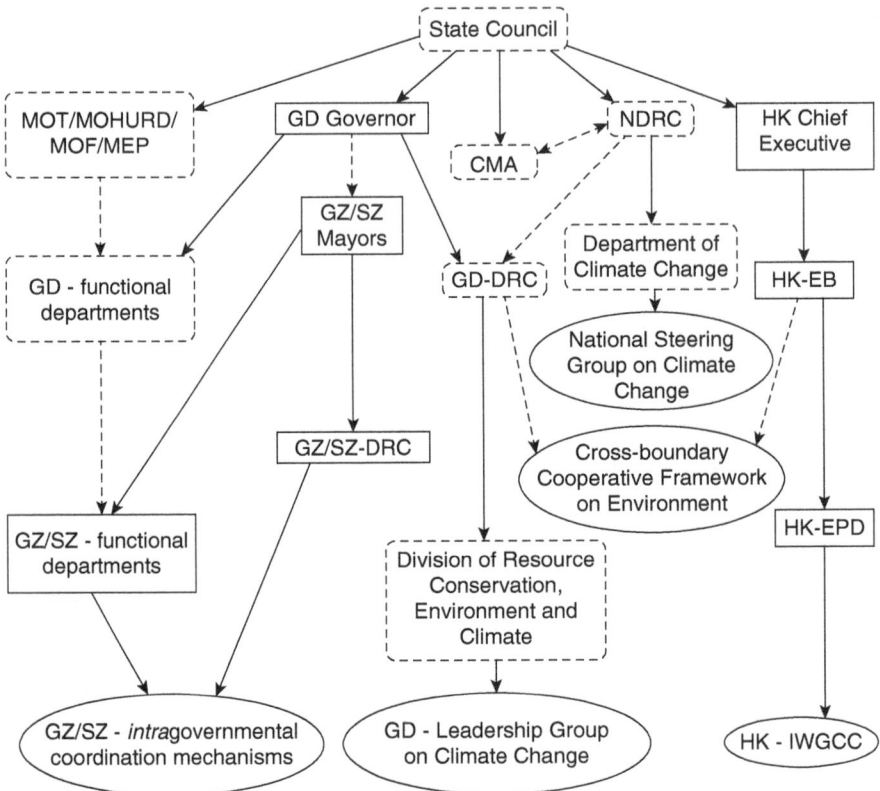

Figure 3.5 Organisational forms and networks in the central-local climate change
　　　　regime

Source: Compiled by authors

policies. Policy prescriptions can be found in the annual policy addresses made
by the chief executive, who holds the highest political authority within the Hong
Kong government. A public consultation on the city's climate action plan was carried out in 2010, although no policy plans on climate mitigation have been made
at the time of writing.

Climate mitigation through low carbon development plans

Guangdong and Guangzhou: province–city twinning

At the provincial level, two climate policy developments are noticeable. First,
the policy agenda has moved from a predominant concern with energy to a more
focused response to climate change. For example, a climate change agenda was
articulated more explicitly in Guangdong's 2010 *Climate Change Programme*
than in its 2007 *Comprehensive Strategy of Energy Conservation and Emissions*

Reduction (Appendix II: Guangdong Government 2007, 2011b). Second, provincial plans are moving from mere reliance on self-governing functions and promotional support, as in the 2008 *People's Action Plan for Guangdong Low Carbon Economy*, to a set of concrete plans on low carbon development, as in *Guangdong's 12th Five-Year Plan*, the *Implementation Strategy of Low Carbon Pilot Province* and the *Key Working Areas of Low Carbon Pilot Province*, all issued after 2010 (Appendix II: Guangdong Government 2011a, 2012; GD-DRC 2008, 2011b, 2012).

Energy issues dominated the agenda of China's national and subnational approaches to sustainable development and environmental protection before 2010. A Leadership Group on Energy Conservation and Emissions Reduction was established by the Guangdong government in 2007. In the same year, the *Guangdong Comprehensive Strategy of Energy Conservation and Emissions Reduction* was issued to monitor the energy consumption and pollutant emissions of public-sector agencies and local industries (Appendix II: Guangdong Government 2007). The top-down approach to energy efficiency monitoring was further institutionalised into provincial legislation (the Guangdong Energy Conservation Ordinance) in 2010. The concepts of low carbon development and climate-resilient growth are nowhere to be found in these provincial regulations.

Nevertheless, this focus on energy issues did provide a practical implementation basis for the *Guangdong Climate Change Programme*, issued in 2011 (Appendix II: Guangdong Government, 2011b). One important step was the 2010 establishment of a second Leadership Group on Climate Change, a reworking of the first group that examined energy conservation and emission reduction. The implementation prospects have expanded from a focus on energy towards other aspects of climate change mitigation, such as expanding carbon-sink capacity in forests and oceans and launching low carbon pilot provincial projects.

Meanwhile, governing strategies for dealing with climate change issues have changed from scattered self-governing actions to a more integrated emphasis on low carbon transition. Self-government was previously the major governing mode of local Chinese governments to address energy efficiency. The *People's Action Plan for Guangdong Low Carbon Economy* encourages self-governing activities in public-sector agencies, such as providing general education in the civil service, raising awareness of low carbon behaviour, saving energy consumption, retrofitting equipment and appliances to improve energy efficiency and strengthening energy-consumption monitoring in government offices (Appendix II: GD-DRC 2008). It is clear that the role of civil society was not made explicit in this 2008 plan. Only GONGOs were recognised, mainly to take up the responsibility of organising science-oriented activities for community outreach on energy conservation and low carbon development.

After the NDRC designated Guangdong as one of the national low carbon pilot provinces, the provincial leadership incorporated the work on climate change and carbon-emissions reduction into *Guangdong's 12th Five-Year Plan* (Appendix II: Guangdong Government 2011a). A target of 17 percent carbon dioxide (CO_2) emissions reduction was put forward in this plan, in addition to an energy-intensity[5] reduction target of 16 percent. However, the plan did not specify the base level for reduction. Interpretation of the published policy document indicates that the proposed emission cap was probably based on the final stage of the 11th Five-Year Plan, a 17 percent CO_2 reduction from 2010. To improve the control mechanism of climate change mitigation, the plan also suggests establishing a greenhouse gas emissions inventory to enhance the still-weak evidence base for any future provincial climate change policies (Appendix II: Guangdong Government 2011a).

The NDRC document on low carbon pilot provinces and cities prescribes the purpose of these localised experiments as, first, to fully mobilise green development for climate change through the hierarchy and across sectors, and, second, to accumulate good practices and operational experiences at the local level to generate bottom-up influence (Appendix II: NDRC 2010). To prepare for the low carbon pilot in Guangdong, the province's Development and Reform Commission (GD-DRC) published annual reports on the *Key Working Areas of Low Carbon Pilot Province*, which have provided timely updates on the progress of climate-policy efforts since 2011 (Appendix II: GD-DRC 2011, 2012). Eventually, the provincial government finalised and endorsed the provincewide *Implementation Strategy of Low Carbon Pilot Province* in 2012, which stipulates essential low carbon actions for various sectors including building, transportation, industry and energy (Appendix II: Guangdong Government 2012). These provincial low carbon development plans also stress the roles and specific responsibilities of the prefectural-level cities in Guangdong, especially Guangzhou, in sharing the tasks incurred in the process of transition. In the 2012 Key Working Areas, the concept of 'province–city cooperation' between Guangzhou and Guangdong was first proposed as a joint way to establish the exchange of carbon emissions rights for the Guangdong area.

Closely following the provincial policies, the lexicon of low carbon development became a recurrent theme in Guangzhou's city agenda after 2010, emerging as an integrative component in major policy documents published by the Guangzhou Municipal Government and by the city's Development and Reform Commission (GZ-DRC). Currently, Guangzhou's climate-mitigation policies are primarily found in the *Guidelines on Developing Low Carbon Economy*. A loose carbon reduction target of approximately 3 percent per annum has been set. To reach the target, the municipal authority needs to identify and review the sectoral focus in the process of low carbon transition, including the promotion of green buildings and low carbon transport (Appendix II: Guangzhou Municipal Government, 2010).

Economic incentives are institutionalised to stimulate sectoral actions. As suggested in the guidelines, to drive the integration of sectoral innovations in low carbon development, the municipal government may consider providing financial incentives for enterprises, tertiary education and research institutes. However, rather than establishing a specialised funding source, these incentives come from existing pools of technology funding, such as the Guangzhou Energy Efficiency Special Fund. The municipal authority also encourages local businesses and research institutes to seek and make good use of foreign capital, by developing technological and capital collaborations with foreign enterprises specialising in low carbon. Potential economic incentives are also embedded in the guidelines related to land-use requirements. The idea is that low carbon components in infrastructure, buildings and transport should be incorporated into the city's new construction and urban renewal projects. Because Guangzhou as a municipality has expanded its administrative jurisdiction, land resources for new development are abundant on the outskirts of the city. The old districts in the city centre are also facing redevelopment. To secure land-use and development rights, real estate developers and urban planners are under pressure to adopt low carbon technologies in their development proposals.

In addition, there is recognition of the need to strengthen organisational coordination across fragmented governmental divisions; to that end, the guidelines institutionalise a municipal Leadership Group on Low Carbon Economic Development, an essential step in preparing for cross-sectoral collaborative actions. Moreover, local GONGOs have been recruited into the transition process as an organised tool for financing carbon sinks and research and development and for demonstrating industrial projects.

The Guangzhou 12th Five-Year Plan builds on the similar focus of the guidelines and incorporates the agenda for a low carbon city (Appendix II: Guangzhou Municipal Government, 2011). Nevertheless, the low carbon strategies recognised by the municipal authorities continue to focus partly on energy rather than a low carbon approach. Only four aspects are covered: i) developing low carbon industry, ii) advocating a low carbon lifestyle, iii) promoting energy conservation and iv) building a clean and safe energy security system. Sectoral approaches to climate mitigation such as green buildings and low-emissions transport are not mentioned in this five-year plan.

Equally, municipal and provincial authorities are together incorporating the concept of 'low carbon' into the Pearl River Delta (PRD) regional development plan. Previously proposed by the NDRC in 2008, the regional plan envisions provincial ideas for future growth in the region. The initial *Outline Plan for the Reform and Development of the Pearl River Delta (2008–2020)* did not directly cover climate change issues (Appendix II: NDRC 2008). Nevertheless, in response to regional discussions, Guangzhou's Implementation Details and the Work Programme of the PRD Outline Plan prioritise the construction of a low carbon city in Guangzhou's green development (Appendix II: Guangzhou Municipal Party Committee 2010 a, b). The municipal policy design closely embraces the provincial advocacy of 'four-year grand development' for the period 2009 to 2012. In a GZ-DRC progress report on the implementation of this work programme, the administrative approach of 'province–city joint actions' ('*shengshi liandong*') was reemphasised by the municipal authority in response to the 'province–city cooperation' advocated by the provincial authority (Appendix II: GZ-DRC 2011a).

Shenzhen: municipal-based institutional framework and sectoral approaches

Given the economic and administrative privileges of being a Special Economic Zone and a 'specifically designated city', Shenzhen, more so than Guangzhou, is autonomous from provincial control in the planning and implementation of low carbon development. In addition, Shenzhen has already taken a low-emission development trajectory due to the now predominant service- and trade-oriented industries of the city. The Shenzhen municipal authority is also branding itself as an arena for experimentation and innovation. Therefore, the national leadership has always regarded Shenzhen as a suitable test ground and eventually endorsed Shenzhen as one of the eight low carbon pilot cities in the country (Appendix II: NDRC 2010).

To institutionalise the experiment, the city government recently developed a series of climate mitigation policies, grounded in the existing municipal-based legal and regulatory framework. With the power to legislate autonomously, the Shenzhen government devised and put into effect the Shenzhen Special Economic Zone Circular Economy Promotion Regulation and the Shenzhen Special Economic Zone Energy Efficiency Regulation. These regulations have prepared the city to promptly start the low carbon transition process.

The *Shenzhen 12th Five-Year Plan* (Appendix II: Shenzhen Municipal Government 2011b) dedicates a chapter specifically to low carbon and green development. A subsequent policy document, the *Shenzhen 12th Five-Year Plan on Energy Efficiency*, reiterates the policy focus on energy issues (Appendix II: Shenzhen Municipal Government 2012). Although the approach to low carbon development in Shenzhen is still heavily oriented towards energy, its two most recent five-year plans have started to take into account climate change mitigation practices in the transportation and building sectors as integral solutions to two nationally endorsed projects – 'low carbon pilot city' and 'low carbon ecological demonstration city'.

Land-use management is clearly the basis for Shenzhen and the MOHURD's *Policy Programme of Co-building Low Carbon Ecological City* with (Appendix II: Shenzhen Municipal Government, 2011a, SZ-UPLRC, 2011). Land use steers the spatial planning of the city towards low carbon infrastructural development by integrating green building design and green transport systems. To try out these sectoral low carbon innovations, the programme identifies the untapped urban outskirts of Guangming and Pingshan as key experimental zones while calling for demonstration projects in the Qianhai border cooperative zone and several other urban-renewal areas (see map in Figure 1.1). The increased availability of land is the driver for both governmental and nongovernmental actors to take action. In addition, this policy programme sets out several quantifiable sectoral mitigation actions, with specific implementation targets to be achieved by 2015, as summarised in Table 3.3.

More government-led climate change mitigation actions are found in the *Medium and Long Term Plan of Low Carbon Development (2011–2020)*. Drafted and published by the Shenzhen Development and Reform Commission (SZ-DRC) in 2012, it is an integrated low carbon development plan approved by the Shenzhen Municipal Government (Appendix II: SZ-DRC 2012). It is more forward thinking and detailed than the basic low carbon guidelines circulated in Guangzhou. The policy text elaborates the government's position on climate change issues and feasible sectoral mitigation actions while specifying concrete conditions to create financial incentives for sectoral innovations. Most importantly, the

Shenzhen plan sets a relatively specific carbon reduction target: a 39 percent reduction in carbon dioxide emissions per unit of GDP by 2015 and a minimum of 45 percent by 2020, based on 2005 emission levels. Although these sectoral mitigation actions remain the means to achieve citywide energy efficiency, each action has quantified targets that enable short-term controls on implementation outcomes. The major policy actions and quantified targets are summarised in Table 3.3.

Table 3.3 Sectoral actions of Shenzhen's low carbon policies

Mitigation actions	*Quantified targets (by 2015)*
Low carbon transport	
Strengthen energy efficiency and emission reduction controls on vehicles	(no specific targets)
Prioritise the development of public transport	– 229 km rail transport network[@#] – 400 km (two-way) public bus lanes[@#] – Public transport to take up 56% of motorised travel[@#] – 10% reduction in public vehicle fuel consumption per 100 km[@#] – 85%+ of bus stops within 500-meter radius of service coverage[#]
Promote 'new-energy vehicles'	– 50,000 'new-energy vehicles' in use accumulatively[@] – 500,000 ton carbon reduction[@]
Speed up the construction of pedestrian and bicycle transport systems	– 200 km bicycle path network[@#]
Green building	
Apply green-building principles in new buildings	– 40% of new buildings to be certified green buildings[@] – 100% green building implementation rate in government-invested new buildings[#] – 50% green building implementation rate in privately invested new buildings[#]
Carry out green-building demonstration projects	– 2 to 4 green building demonstration projects to be constructed in each district per year [#]
Adopt whole life-cycle management	(no specific target)
Enforce energy-efficiency standards in new buildings	– 100% of new buildings to comply with energy efficiency standards[@#]
Carry out energy-efficiency retrofits in existing buildings	(no specific target)
Encourage innovation in building technology	(no specific target)
Apply renewable energy in buildings	(no specific target)
Strengthen urban renewal	– 35 km^2 of urban redeveloped area[#]

Note: [#] Requirements in the Policy Programme on Co-building Low Carbon Ecological City
[@] Requirements in the Medium and Long Term Plan for Low Carbon Development (2011–2020)

'Incentive' is one of the key words in these two policies, an approach to stream-lining sectoral-mitigation actions. One of the underlining policy principles is to advance low carbon technology, particularly locally developed innovations, with policy support and financial incentives. An incentive mechanism is proposed to encourage patent applications for low carbon technologies, especially those from the research and development of local enterprises. To support the demonstration projects identified in the low carbon ecological city, the municipal authority is planning to utilise *land resources* as policy incentives by altering land-transfer conditions, increasing gross floor area ratios and giving priority to new land supplies and urban renewal projects.

Increased monetary incentives are also planned. The SZ-DRC plan indicates the municipal authority's intention to introduce policy incentives for pricing, taxation and financing mechanisms that are favourable to the city's low carbon development. The budget for investment in low carbon innovations has been increased to support technological development, industrialisation, pilot demonstrations and capacity building. The SZ-UPLRC project promises a certain amount of annual funding for studies of low carbon ecological technology. Meanwhile, the SZ-DRC has established a low carbon development fund by integrating existing policy resources and funding channels, with an annual capital injection of at least RMB200 million. The SZ-DRC anticipates that by 2015, investment in low carbon technology will take up 10 percent of the total funding for research and development in Shenzhen. The municipal government is also planning to set up a financing platform to direct domestic nongovernmental funding into low carbon city construction and to actively search for and utilise the financial support of foreign governments and international organisations.

The proposed government tasks in these two policy plans are coordinated by a ministry–city joint session system, together with the Leadership Group on Climate Change, Energy Efficiency and Emission Reduction in the SZ-DRC. The structure of the municipal leadership encompasses incorporates both vertical and horizontal *inter*governmental relations. To obtain more policy and funding support from upper levels of government, the Shenzhen leadership is actively joining its own established municipal low carbon plan to the relevant national and provincial plans (the city before the province) instead of advocating for province–city joint actions in Guangzhou (i.e. the city following the province). In Shenzhen's two city-level low carbon plans, the government leadership is engaging with non-governmental driving forces and initiating collaboration across disciplines and sectors.

In addition to the integrated low carbon plans, various functional departments of the city government have developed individual *sectoral action policies* with sectoral-specific incentives and emissions reduction targets. In August 2011, the Shenzhen Transportation Commission (SZ-TC) designed an *Implementation Programme for Constructing a Pilot Low Carbon Transportation System* (Appendix II: SZ-TC, 2011) and has secured a total investment of RMB4.3 billion for its implementation. Between 2010 and 2015, carbon dioxide emissions from road transport will be reduced by 2.9 kg CO_2 per 1,000 persons per kilometre of passenger transport and 0.2 kg CO_2 per 100 tons of goods per kilometre of freight transport.

In the building sector, a new piece of legislation – Shenzhen Green Building Promotion Measures – has been introduced into the municipal legal system with effect from August 2013. It adds to the existing environmental regulatory framework of the city, formally composed of the Shenzhen Circular Economy Promotion Regulation and the Shenzhen Building Energy Efficiency Regulations. The new law institutionalises the fiscal subsidies that provide financial incentives for green-building certification and technological innovation and lays down the legal foundation for preferential policies on priority land supplies and gross floor area rewards. It also legalises the liabilities for noncompliance with green-building verification, with enforceable penalties on the design, construction and supervision parties ranging from RMB50,000 to 200,000.

Climate mitigation policies in Hong Kong

The Hong Kong government has also recently incorporated climate change and low carbon development into its agenda. Instead of a standalone low carbon development plan, the basic policy framework in Hong Kong is shaped by the annual policy address, delivered by the governor until 1997 and by the chief executive thereafter. Prior to 2006, the environmental policies covered by policy addresses were mainly concerned with quality-of-life–related problems such as air pollution, urban hygiene and green coverage. Climate change issues were first mentioned in the 2006 address:

> As a world city, Hong Kong should not only attach importance to local pollution problems but also look at the wider picture. The crisis of global warming caused by the greenhouse effect is a major concern. Although the greenhouse gas emissions per capita in Hong Kong in 2004 were 7% below the 1990 level, we should maintain our efforts in this regard.
>
> (Appendix II: Chief Executive 2006: 22)

In 2008, low carbon development was first identified as a mitigation approach to climate change issues:

> We will make early preparations to meet the challenge of climate change. In particular, we will enhance energy efficiency, use clean fuels, rely less on fossil fuel, and promote a low carbon economy – an economy based on low energy consumption and low pollution
>
> (Appendix II: Chief Executive 2008: 20)

In 2008, the first capital injection into the ECF, the local grant earmarked for environmental issues, was made specifically for carbon-reduction activities: a total of HKD150 million was reserved to 'partially subsidise building owners to conduct comprehensive energy and carbon audits' (Appendix II: Chief Executive 2008: 21).

In 2010, a formal public consultation on the government-proposed climate action plan was carried out. Although the consultation results had not yet been released by the government at the time of writing, the document proposes concrete carbon-reduction targets to be achieved by 2020 (based on 2005 levels): i) a 19 to 33 percent absolute reduction in greenhouse gas emissions, ii) a 50 to 60 percent reduction in carbon intensity and iii) a 27 to 42 percent reduction in per-capita greenhouse gas emissions. The consultation paper contains several mitigation actions intended to achieve reduction targets, including maximising energy efficiency in buildings and reducing road transport emissions with cleaner fuel for motor vehicles. The paper reiterates the role of the Inter-departmental Working Group on Climate Change in coordinating the government's cross-sectoral actions. It also calls for societywide participation, market acceptance and community support of the low carbon transition process. Taking climate change issues seriously is frequently emphasised in Hong Kong as part of the responsibility of being an 'international' or 'world city'.

Hong Kong had started to introduce *sectoral mitigation actions* into its building and transport policies even before the launch of low carbon development, although they were not framed within the context of climate change. Preliminary sectoral actions can be found in the 1994 address by the governor (Appendix II: Patten 1994), when Hong Kong was still under British colonial rule. Table 3.4 summarises the sectoral actions proposed and taken up by the administration in Hong Kong between 1994 and 2013, according to the policy addresses of the period.

Between 1994 and 1996, the colonial administration started to explore the feasibility of establishing an institutional framework for green buildings and electric vehicles by drafting legislation, devising guidelines, developing a database and conducting research (Table 3.4).

After the handover of sovereignty to China in 1997, sectoral-mitigation actions slowed down due to the economic turmoil of the Asian financial crisis in 1997 and the shock of the Severe Acute Respiratory Syndrome epidemic in 2003, which seriously affected the city's economy and focused the new administration's attention towards environmental health. Until early 2005, there was a policy vacuum in the development of green buildings in Hong Kong, and progress stalled on the feasibility studies and trials of electric vehicles (Table 3.4).

In late 2005, sectoral actions were again launched in a self-governing manner, while in 2006 they were articulated as solutions to the problem of climate change. More decisive policy proposals on green buildings and electric vehicles came only after 2007 (Table 3.4).

There have been several vital institutional changes in policy development on green buildings and electric vehicles in Hong Kong. A government report published in 2010 identified that green buildings can easily turn into 'inflated buildings', meaning that their sellable size is artificially increased by including government concessions on gross floor areas (Appendix II: SDC, 2010). The public and business stakeholders started to call for more stringent policy requirements and information disclosure on the verification and certification of green buildings.

Table 3.4 Sectoral actions in Hong Kong (1994–2013 policy address / address by the governor)

Year	Green buildings	Low carbon transport (electric vehicles)
1994	– Devise comprehensive guidelines on energy-efficient design of buildings and building services for professionals (**enabling** initiatives)	– Explore the feasibility of using electricity vehicles (service **provisions**)
1995	– Introduce legislation to control maximum amount of heat transfer through the building envelope of new commercial and hotel buildings (**authority** rules) Conduct energy audits and energy management in government buildings (**self-governing** approach)	(*vacuum*)
1996	– Control asbestos in demolition and construction by code of practice and registration system (**authority** rules) Develop a comprehensive energy end-use database to locate energy users and identify the trend and pattern of energy usage (service **provisions**)	– Further study the feasibility of EV cars in the inter-departmental working group (service provisions)
Sovereignty handover		
1997		
1998		(*vacuum*)
1999	(*vacuum*)	– Conduct feasibility studies on an electric trolley-bus system and other electrically powered vehicles (service **provisions**)
2000		– Launch and complete a pilot scheme on the use of clean LPG and electricity by public light buses, to decide on measures for long-term implementation (service **provisions**)
2001		
2003		(*vacuum*)
2004		
2005 (1)		

(*Continued*)

Table 3.4 (Continued)

Year	Green buildings	Low carbon transport (electric vehicles)
2005 (2)	– Require all government office buildings to reduce annual electricity consumption by 1.5% (*self-governing* approach)	– Consider using hybrid vehicles in the government fleet (*self-governing* approach) – Reduce first vehicle registration tax by 30% (max. reduction $50,000 per vehicle) for purchase of vehicles with low emissions and high fuel efficiency (*enabling* initiatives)
2006	– Promote building energy codes: 1) full compliance in new government buildings; 2) energy retrofit in old government buildings; 3) green rooftop concept (*self-governing* approach)	*(vacuum)*
2007	– Consult the public on the proposed mandatory implementation of the Building Energy Codes as legislation (*authority* rules) – Conduct a carbon audit and implement an emissions reduction campaign in the new Central Government Complex at Tamar (*self-governing* approach) – Encourage different organisations via ECF to carry out greening on rooftops and podiums in suitable buildings (*enabling* initiatives)	
2008	– Legislate the mandatory compliance with Building Energy Codes (*authority* rules) – Reserve $150 million from the ECF to subsidise building owners to conduct comprehensive energy and carbon audits and another $300 million for energy efficiency projects in buildings (*enabling* initiatives)	
2009	– Introduce and enforce mandatory compliance with Building Energy Codes (*authority* rules) – Encourage more organisations to conduct carbon audits in buildings and reduce carbon emissions by the Buildings Energy Efficiency Funding Scheme (*enabling* initiatives)	– Cooperate with electric-vehicle manufacturers; expect a supply of around 200 electric vehicles for the local market (*enabling* initiatives) – Launch an electric-vehicle leasing scheme in collaboration with the two power companies (*enabling* initiatives)

2010

– Introduce measures requiring the incorporation of green building design elements (building separation, building permeability, building setback and greenery) (authority rules) Raise the building energy efficiency standards (service provisions)
– Require developers to provide environmental and energy consumption information on buildings for the reference of potential users (*authority* rules)
– Control 'inflated buildings' by altering GFA concessions conditions (*enabling* initiatives)
– Set up a $300 million Pilot Green Transport Fund to encourage the transport sector to test green and low carbon transport means and technology (*enabling* initiatives)
– Support low-emission bus testing: fund the full cost of procuring six hybrid buses for use by the franchised bus companies along busy corridors to test their operational efficiency and performance and collect operational data; provide the same financial support if the bus companies wish to test other greener buses such as electric buses (*enabling* initiatives)

2011

– Encourage owners to apply for green-building certification of converted buildings after revitalising industrial building projects (*enabling* initiatives)
– Bring about new business opportunities for the green building sector (*enabling* initiatives)
– Ask all franchised bus companies to test zero-emission electric buses (*authority* rules)
– Earmark $180 million for franchised bus companies to purchase 36 electric buses for trial runs on a number of routes to assess their performance in different conditions (*enabling initiatives*)

2013

– Develop Kowloon East into a low carbon community (*enabling* initiatives) Lead an interdepartmental steering committee to promote green building, strengthen the coordination among departments to formulate implementation strategies and action plans (service *provisions*) while maintaining close dialogue and cooperation with the relevant sectors and stakeholders (*enabling* initiatives)
– Take the lead by using more electric vehicles in the government fleet (*self-governing* approach)
– Solicit participation from public bodies and leading enterprises; provide funding for franchised bus companies to try out electric buses; subsidise the testing of electric taxis, coaches and goods vehicles by the Pilot Green Transport Fund (*enabling* initiatives)

Supporting policies in the area of electric vehicles were not fully developed in a consistent manner. However, after being absent from the 2007 and 2008 policy addresses, major policy proposals were put forward in 2009 to promote the wider use of electric vehicles in Hong Kong (Table 3.4).

More recently, new laws and regulations have indirectly encouraged the development of green buildings and the use of electric vehicles. Mandatory compliance with building energy codes has been in effect since the enactment of the Buildings Energy Efficiency Ordinance in September 2012. The Motor Vehicle Idling (Fixed Penalty) Ordinance, which prohibits drivers from idling vehicles on a road for more than 3 minutes in any 60-minute period, was fully implemented in December 2011. As electric and hybrid vehicles operating solely in electric mode do not emit pollutants, they are outside the scope of this prohibition (EPD 2013). Technically, neither legislation (on the energy efficiency of buildings or idling of motor vehicles) can be seen as a direct attempt to promote green buildings or electric vehicles, but they form part of Hong Kong's institutional foundation to prepare for more progressive integration of the two sectoral innovations.

In addition to the Inter-department Working Group on Climate Change set up in 2007 as part of the Environmental Protection Department, the sectoral focus of government leadership has also been institutionalised. Back in 1996, the interdepartmental working group conducted feasibility studies of electric vehicles (Table 3.4). The Transport Advisory Committee also engages expertise in its sector. Nonetheless, with its secretariat located in the Transport and Housing Bureau, the advisory committee does not provide cross-sectoral coordination; it only advises the chief executive on transport policies. In the building sector, an interdepartmental steering committee was established in early 2013 to strengthen *intra*governmental coordination in formulating implementation strategies and action plans to promote green buildings in a holistic manner in both public and private sectors in Hong Kong (Environment Bureau 2013). The steering committee is chaired by the newly appointed Secretary for the Environment, K.S. Wong, himself an architect specialising in green building design and technology.

Conclusion

The profiles of Guangzhou, Shenzhen and Hong Kong show that although the carbon intensity of these cities might be lower than the national average, they perform key roles in testing both nationally endorsed and locally developed low carbon experimental projects with a larger resource capacity and governance potential. Strong budgetary support for environmental projects is found in the local fiscal expenditures of each city, with enhanced endorsement of climate change and low carbon components through the provision of earmarked grants.

Politically, these cities are located in a special vertical *inter*governmental structure in relation to the central government. They have special administrative and constitutional status: Guangzhou as a nationally controlled city, Shenzhen as a Special Economic Zone and Hong Kong as a Special Administrative Region.

In terms of the national view on climate change problems, the central government has clearly specified a national target, and a centralised national climate regime has been established based on the existing organisational network of the China Meteorological Administration, the National Development and Reform Commission and other ministerial-level agencies. At the same time, both top-down and bottom-up decentralised actions are being carried out: top-down actions through centralised information sharing and nationally endorsed low carbon projects and bottom-up actions through locally devised low carbon practices. In this process, the three cities are engaged in central-local interactions.

The review of local climate mitigation policies suggests two integral components: first, the generation of incentives and their institutional foundation and second, local governments' perception of the importance of sectoral innovations in low carbon development. Two types of incentives are being implemented through local policies – direct economic benefits in monetary terms and preferential resources such as land-planning priority or development concessions. As for the perception of importance, the analysis shows an incremental development of the sectoral-mitigation approach in Hong Kong, relative to the rapid growth in Guangzhou and Shenzhen as part of their low carbon development plans.

In a nutshell, the review in this chapter of urban attributes, resource capacity, political structure and climate-policy development in the three cities under study contextualises CMNs in a loosely bounded sociopolitical system that connects the national regime and the global city networks. The special vertical *inter*governmental structure with the central government drives the formation of a network prototype in the form of *intra*governmental mechanisms for coordinating climate change mitigation and the transition to a low carbon environment (Chapter 4). The provision of a solid resource capacity and strong state support are incentives that initiate the dynamic interactions of innovation integration among policy sectors, thus accelerating the development of the networks in CMNs (Chapter 5 and 6). The unique political status and global functions of the three cities represent the start-up conditions for collaboration on climate initiatives that serve experimental and exemplary functions (Chapter 7).

Notes

1 The First National Assessment Report on Climate Change was launched in 2002 and published in 2006; the Second National Assessment Report on Climate Change was launched in 2008 and published in 2011. The Third National Assessment Report on Climate Change, launched in 2008, is under review at the time of this writing, to be published at the end of 2014.
2 Translation based on the Xinhua News on 19 March 2013 [Online]. Available: http://news.xinhuanet.com/english/china/2013-03/19/c_132245851.htm [Accessed 27 August 2013]
3 The other four National Central Cities are Beijing, Tianjin, Shanghai and Chongqing.
4 The other four 'specifically designated' cities are Dalian, Qingdao, Ningbo and Xiamen.
5 Energy intensity refers to the energy consumption per unit of gross domestic product (GDP).

References

Central Government, P. R. C. 2009. *State Council Executive Meeting Decided on Target of Controlling the Greenhouse Gas Emission in China.* Beijing, China [Online]. Available: www.gov.cn/ldhd/2009–11/26/content_1474016.htm [Accessed 15 June 2013] (*Guowu Yuan Changwu Weiyuanhui Yanjiu Jueding Woguo Kongzhi Wenshi Qiti Paifang Mubiao*)

Cheung, P. T. Y. 2011. Intergovernmental Relations between Mainland China and the Hong Kong S.A.R., in Berman, E. M. (ed.) *Public Administration in Southeast Asia: Thailand, Philippines, Malaysia, Hong Kong, and Macao,* Boca Raton, FL, CRC Press, Taylor & Francis Group.

CMA. 2011. *Brief Introduction of China Meteorological Administration.* Beijing, China: China Meteorological Administration [Online]. Available: http://2011.cma.gov.cn/bmgk/200805/t20080514_6824.html [Accessed 22 August 2013]

CMA. 2012a. *C.M.A. and Guangdong Provincial Government Signed Agreement, to Jointly Accelerate Meteorological Modernisation in Pilot Province Construction* [Online]. Available: www.cma.gov.cn/2011xwzx/2011xqxxw/2011xqxyw/201203/t20120312_164481.html [Accessed 22 August 2013] (*Zhongguo Qixiangju Yu Guangdong Sheng Zhengfu Qianshu Xieyi, Gongtong Jiakuai Qixiang Xiandai Hua Shidian Jianshe*)

CMA. 2012b. *Guangdong Province Take the Lead in Basic Achievement of Meteorological Modernisation Pilot Works* [Online]. Beijing. Available: www.cma.gov.cn/2011xz t/2011zhuant/20111214/2011121406/201112140602/201210/t20121024_188131.html [Accessed 22 August 2013] (*Guangdong Sheng Shuaixian Jiben Shixian Qixiang Xiandaihua Shidian Gongzuo Jinzhan*)

Dong, Y. 2013. *Consolidate Strength to the World: 30th Anniversary Remarks of Launching Guangdong–Hong Kong–Macau Cooperation on Meteorology* [Online]. China Meteorological Administration Newspapers Office. 27 January 2013. Available: www.cma.gov.cn/2011xwzx/2011xqxxw/2011xqxyw/201301/t20130125_204058.html [Accessed 22 August 2013] (*Ningju Liliang, Zouxiang Shijie: Ji Yuegang'ao Qidong Qixiang Hezuo 30 Zhounian*)

ECF. 2012. *Capital Injection to ECF* [Online]. Hong Kong: Secretariat of Environment and Conservation Fund. Available: www.ecf.gov.hk/en/about/introduction.html [Accessed 15 June 2013]

EIA. 2010. *Carbon Intensity Using Market Exchange Rates (Metric Tons of Carbon Dioxide Per Thousand Year 2005 U.S. Dollars)* [Online]. International Energy Statistics: U.S. Energy Information Administration. Available: www.eia.gov/cfapps/ipdbproject/IEDIndex3.cfm?tid = 91&pid = 46&aid = 31 [Accessed 15 June 2013]

Environment Bureau, H. K. 2013. *2013 Policy Address: Policy Initiatives of Environment Bureau – Environmental Protection* [Online]. Hong Kong: Legislative Council Panel on Environmental Affairs. Available: www.epd.gov.hk/epd/english/news_events/legco/files/EA_Panel_20130128a_eng.pdf [Accessed 24 October 2013]

EPD. 2013. *The Statutory Ban against Idling of Motor Vehicle Engines* [Online]. Hong Kong. Available: www.epd.gov.hk/epd/english/environmentinhk/air/prob_solutions/idling_prohibition.html [Accessed 8 September 2013]

Foreign & Commonwealth Office, U. K. 2013. *Guidance: China Prosperity Projects and Targeted Bidding Round 2013–2014* [Online]. Beijing, China. Available: https://www.gov.uk/government/publications/china-prosperity-projects-and-targeted-bidding-round-2013–2014 [Accessed 27 August 2013]

GaWC. 2011. *The World According to GaWC 2010* [Online]. Available: www.lboro.ac.uk/gawc/world2010t.html [Accessed 5 September 2013]

Government. 2011. *Chapter in National Five-Year Plan Dedicated to Hong Kong and Macao* [Online]. Hong Kong: Government Press Release. Available: www.info.gov.hk/gia/general/201103/16/P201103160315.htm [Accessed 28 August 2013]

GZ-EPD. 2011. *2011 Year-End Departmental Financial Settlement of Guangzhou Environmental Protection* [Online]. Guangzhou Environmental Protection Department Guangzhou Municipal Government. Guangzhou. Available: www.gzepb.gov.cn/zwgk/gs/cgyzb/201211/P020121122641497161308.pdf [Accessed 15 June 2013] (*Guangzhou Shi Huanjing Baohu Ju 2011 Nian Bumen Juesuan*)

HK-EPD. 2012. *Estimates of Government Resources Devoted to Environmental Protection and Conservation Work in 2011* [Online]. Environmental Protection Department. Hong Kong. Available: www.epd.gov.hk/epd/english/resources_pub/spending/spending_govt.html [Accessed 15 June 2013]

Hu, B. 2010. "Leader's Internal Instruction (*Pishi*)": Subtle Techniques and Circulation Logics. *Nanfang Weekly* [Online]. Published by Nanfang Newspaper Media Group. 1 July 2010. Available: www.infzm.com/content/47027 [Accessed 23 August 2013] ('*Lingdao Pishi*': *Weimiao Jiqiao Yu Chuanyue Luoji*)

Hu, J. 2012. Guangdong Provincial Government and C.M.A. Held Signing Ceremony of Memorandum of Cooperation. *Nanfang Daily* [Online]. Published by Central Government, P.R.C. 13 March 2012. [Accessed 22 August 2013] (*Guangdong Sheng Zhengfu Yu Zhongguo Qixiangju Juxing Hezuo Beiwanglu Qianshu Yishi*)

IPCC. 2013. *IPCC Focal Points* [Online]. Available: www.ipcc.ch/pdf/ipcc-principles/ipcc-focal-points.pdf [Accessed 28 November 2013]

Lin, G. C. S. 2004. The Chinese Globalizing Cities: National Centers of Globalization and Urban Transformation. *Progress in Planning,* 61, 143–157

Ministry of Water Resources. 2012. *M.W.R. and C.M.A. Signed MOU on Accelerating the Development of Water Conservancy and Meteorology* [Online]. Beijing, China: Central Government, P.R.C. Available: www.gov.cn/gzdt/2012–02/08/content_2061315.htm [Accessed 22 August 2013]. (*Shuilibu yu Qixiangju Qianding Jiakuai Shuili Qixiang Fazhan Hezuo Beiwanglu*)

NCC. 2012. *History and Development of the National Climate Center* [Online]. Beijing, China: National Climate Center. [Accessed 23 August 2013]. (*Guojia Qihou Zhongxin Lishi Yange*)

NCSC. 2012. *National Center for Climate Change Strategy and International Cooperation (NCSC).* Beijing, China: NCSC [Online]. Available: http://news.nost.org.cn/wp-content/uploads/2012/10/NCCS-presentation-oct2012.pdf [Accessed 15 August 2013]

NDRC. 2011. *Xie Zhenhua Deputy Director Attend Central-China Section Meeting on Planning for National Responses to Climate Change* [Online]. National Development and Reform Commission. Beijing, China. Available: www.sdpc.gov.cn/gzdt/t20110809_427842.htm [Accessed 15 June 2013]. (*Xie Zhenhua Fuzhuren Chuxi Guojia Yingdui Qihou Bianhua Guihua Bianzhi Gongzuo Zhongbu Pianhui*)

Wang, H. & Wu, C. 2005. Environmental Institutions in China, in *Urbanization, Energy, and Air Pollution in China: The Challenges Ahead – Proceedings of a Symposium,* Washington, DC, National Academies Press.

World Bank. 2009a. CO_2 Emissions (Metric Tons Per Capita) [Online]. *Climate Change, World Development Indicators.* Available: http://data.worldbank.org/indicator/EN.ATM.CO2E.PC/countries [Accessed 15 June 2013]

World Bank. 2009b. GDP Per Unit of Energy Use (Constant 2005 PPP $ per kg of Oil Equivalent) [Online]. *Energy & Mining, World Development Indicators.* Available: http://data.worldbank.org/indicator/EG.GDP.PUSE.KO.PP.KD/countries [Accessed 15 June 2013]

World Bank. 2011. *GDP Per Unit of Energy Use (Constant 2005 PPP $ Per Kg of Oil Equivalent)* [Online]. Energy & Mining, World Development Indicators. Available: http://data.worldbank.org/indicator/EG.GDP.PUSE.KO.PP.KD/countries [Accessed 15 June 2013]

Wübbeke, J. 2013. China's Climate Change Expert Community: Principles, Mechanisms and Influence. *Journal of Contemporary China,* 22, 712–731

Part II

Dynamics of climate collaborative municipal networks

4   *Intra*governmental coordination – the first step

The cross-sectoral nature of climate change–related issues has posed new governance challenges to municipal governments, such as the need to enable collaboration among various actors and to coordinate policy across different sectors (Betsill and Bulkeley 2006; Jordan *et al.* 2010; Wellstead and Stedman 2011). In this regard, various types of *intra*governmental coordination mechanisms have been established within Chinese municipal governments to address the different facets of climate change. However, no in depth research to date has explored how effective these mechanisms are in overcoming the governance dilemmas posed by climate change. This fact raises two questions: how do these *intra*governmental structures collaborate horizontally with external network actors to realise climate-change governance? And do their internal organisation and operations address the pervading functional fragmentation among agencies?

In this chapter, we analyse both the internal and external configurations of *intra*governmental coordination mechanisms as a network prototype stage of CMNs (see Figure 2.2) and as tools of institutional solutions for climate change. We present a comparative analysis of the established *intra*governmental mechanisms in Guangzhou, Shenzhen and Hong Kong: the Low Carbon Economy Development Leadership Group organised in 2011 in Guangzhou; the Leadership Group on Responding to Climate Change, Energy Efficiency and Emissions Reduction Works formed in 2010 in Shenzhen; and the Interdepartmental Working Group on Climate Change established in Hong Kong in 2007. We also examine the effects that these groups have had on low carbon transitions in the building and transport sectors.

However, we argue that these *intra*governmental coordination mechanisms experience levels of institutional deficit and demonstrate varying levels of effectiveness in overcoming horizontal fragmentation. We find that the *political* and *institutional* characteristics of the *intra*governmental mechanisms determine the extent of *intra*governmental coordination effectiveness in terms of both processes and outcomes. The data used in this chapter were collected through semistructured interviews with (a) programme specialists in core government departments participating in the coordination mechanisms and (b) activists working in nongovernmental organisations in collaboration with city governments (Appendix I).

Analytical rationale and perspectives: *intra*governmental structure, the coordination prototype

As mentioned in Chapter 2, horizontal coordination in an *intra*governmental setting can be defined and analysed from an *outcome-process* perspective and/or from a holistic view of its *internal capacity* and *external linkages* (6 2004; Dunsire 1978; Peters 1998; Wollmann 2003).

Gulick and Urwick (1937) propose two means of coordination: first, *by organising* the authoritative structure and orders; and second, *by the dominance of an idea.* Gulick and Urwick's arguments can be extended to the means of achieving *intra*governmental coordination, as outlined in the matrix presented in Table 4.1. The matrix shows that coordination can be achieved by *institutional support* from authority and resource structures and also *by political priority*, such as positioning policy goals and enhancing political will in the governing system. Hypothetically, high political priority results in coordinated *outcomes* in the end-state of specified policies, while strong institutional support leads to coordinated *processes* in programme development. In Table 4.1, Scenario A characterises the optimal situation, while Scenario D represents fragmentation in both process and outcome, with Scenarios B and C in between. These conceptual scenarios are tested with empirical cases.

To coordinate policies and actions in response to climate change and to implement low carbon development projects, the three cities have initiated institutional innovations in climate governance – a prominent one is through building up *intra*governmental networking groups (coordination mechanisms). Table 4.2 summarises their constitution (including leaders, secretariats, involved agencies and supporting units). These coordination mechanisms institutionalised in *intra*governmental settings invite comparative analysis through empirical and theoretical questions:

- How can issue-driven climate governance arrangements achieve effective horizontal coordination, circumventing the inherent functional fragmentation (i.e. cross-sectoral silo effects) of agencies?

Table 4.1 Analytical *intra*governmental coordination scenarios

		Political priority	
		Low	*High*
Institutional support	**Weak**	D. Fragmentation in *intra*governmental networking process and outcome	C. Coordinated outcome but fragmented implementation process
	Strong	B. Outcome fragmentation but effective coordination in *intra*governmental governance process	A. Effective coordination in both process and outcome

Table 4.2 Intragovernmental coordination mechanisms in Guangzhou, Shenzhen and Hong Kong

Name	Year	Chair	Agencies included	Secretariat		Expert base [Y/N]	External staff@ [Y/N]	Supporting institute(s)
				Location	Staff			
Leadership Group on the Development of a Low Carbon Economy in Guangzhou	2010	Mayor	30 agencies (Mayor's Office; 5 commissions; 19 departments)	GZ-DRC	3+	Y	N	Guangzhou Energy Saving Supervision Centre
Leadership Group on Building Energy Efficiency in Guangzhou	2011	Deputy Mayor	31 agencies (Mayor's Office; 3 commissions; 14 departments)	GZ-URCC	3+	Y	N	Management Office of Building Energy Efficiency & Wall Materials Innovation
Joint Session System on Promoting Building Energy Efficiency and Developing Green Building in Shenzhen	2008	Deputy Mayor	20 agencies (Mayor's Office; 4 commissions; 5 departments)	SZ-HCD	max. 8	N	Y	Construction Science & Technology Promotion Centre
Leadership Group of Environmental Protection, Energy Efficiency and Emission Reduction in Shenzhen Transport Industry	2010	Director (SZ-TC)	Units internal to and agencies subordinate to SZ-TC	SZ-TC	1	N	N	N
Leadership Group on Responding to Climate Change, Energy Efficiency, and Emission Reduction Works in Shenzhen	2010	Mayor	30 agencies (Mayor's Office; 7 commissions; 10 departments)	SZ-DRC	max.13	Y	N	Shenzhen Research Centre for Urban Development
Joint Session System on Constructing National Low Carbon Ecological Demonstration City in Shenzhen	2011	Deputy Mayor & Ministry Director	16 municipal agencies and MOHURD	SZ-UPLRC	max. 15	N	N	Shenzhen Urban Planning and Land Development Research Centre
Inter-department Working Group on Climate Change	2007	Deputy Director	21 agencies (5 bureaus; 16 departments)	HK-EPD	7+	N	N	Council for Sustainable Development
Steering Committee on the Promotion of Electric Vehicles	2009	Financial Secretary	5 agencies & non-governmental members	HK-EPD	1	Y	N	N
Steering Committee on the Promotion of Green Buildings	2013	Environment Secretary	11 agencies (3 bureaus, 8 departments)	Environment Bureau	1	N	N	N

Note: @The ability of the secretariat to deploy staff from other departments

Analysing *intra*governmental structures is also theoretically relevant to the broader discussion of this volume:

- How are coordination (and potentially collaboration) dynamics configured in the network prototype stage of the collaborative municipal networks (CMNs) for climate mitigation?

Referring to the characteristics of the *intra*governmental management networks identified in Chapter 2, Figure 4.1 depicts the scope of engagement and the pattern of resource-authority-information distribution inherent in an *intra*governmental

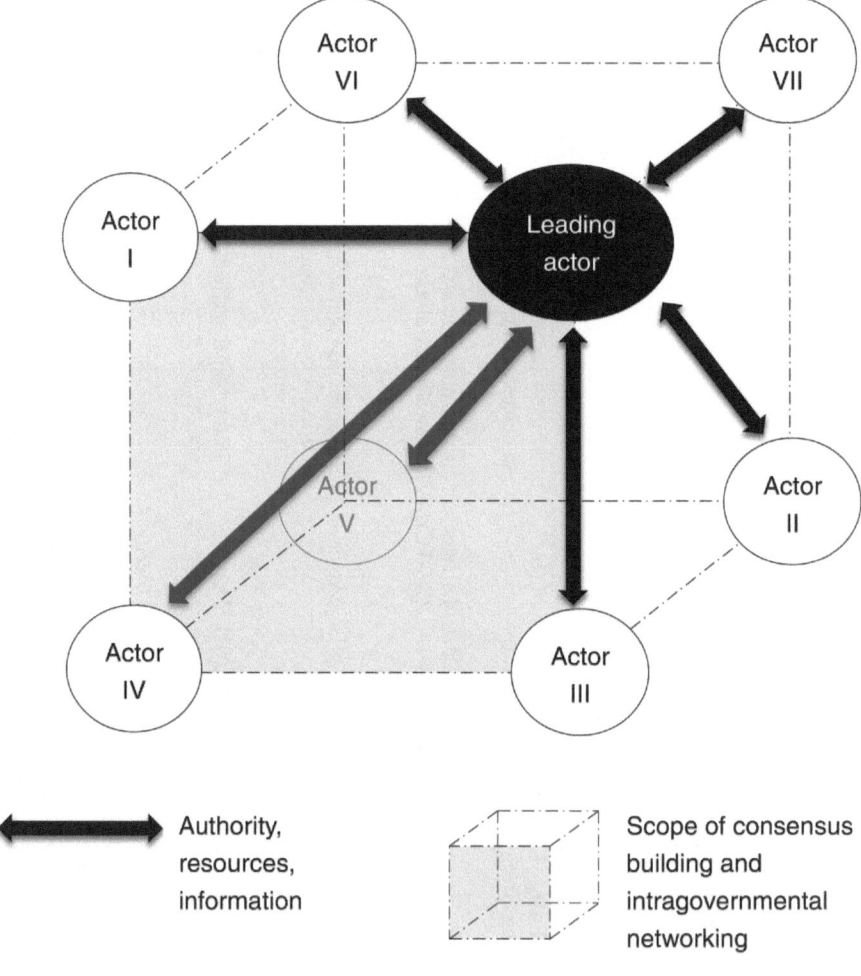

*Figure 4.1 Intra*governmental structure: coordination prototype
Source: Compiled by authors

setting. The leading actor in the *intra*governmental structure typically initiates and manages the coordination mechanism. Other actors are the participating agencies at the core of engagement. All of these actors hold different sources of authority, resources and/or information with respect to policy issues.

The dominant values in the scope of engagement (dashed-line box) are often shaped by the administrative culture, the policy objectives and the nature of the targeted issues. Given the existing dominant values, the established mechanisms aim to mediate conflicts and build consensus among the actors. Resources, authority and policy information are exchanged (double-ended arrows) through the *intra*governmental structure in this process.

*Intra*governmental coordination mechanisms

Following the proposed coordination prototype (Figure 4.1) and the analytical matrix (Table 4.1), this section identifies the core actors involved in the *intra*governmental structures of the three cities, given their engagement in network activities and in the authority-resource-information relationship of the established coordination mechanisms. The process and end-state effectiveness of coordination are gauged by analysing the resulting interaction dynamics (Tables 4.2, 4.3).

Guangzhou, unified forces of **intra*governmental coordination***

The objective of the Guangzhou Leadership Group on the Development of a Low Carbon Economy is to devise strategies for energy conservation and low carbon development in accordance with local conditions, based on the policy framework set out at the provincial and national levels (Appendix II: Guangzhou Municipal Government 2010).

As low carbon development relates to various issues of urban governance, 30 offices, bureaus and commissions in the municipal government are convened within the leadership group (Table 4.2). However, only five core actors in this group hold key decision-making power and institutional responsibilities. Figure 4.2 provides a visual representation of the coordination structure and internal-external linkages between the core actors. The leading actor (the black rounded rectangle in Figure 4.2) is the municipal Development and Reform Commission (GZ-DRC), which, as convener and secretariat, plays the role of strategic leader for low carbon development. Through the *intra*governmental platform of the leadership group, the GZ-DRC engages with three functionally relevant departments (white rounded rectangles in Figure 4.2): the Economy and Trade Commission (GZ-ETC), the Transport Commission (GZ-TC) and the Urban–Rural Construction Commission (GZ-URCC). To strengthen its financial capacity, the GZ-DRC also works closely with the Finance Department (GZ-FD) in evaluating and distributing funds for innovative low carbon development projects. These authority relationships and ties between information and resource exchange are depicted by double-ended arrows in Figure 4.2. The *external links* between core *intra*governmental actors and the *nongovernmental space* are represented by the grey rounded rectangles.

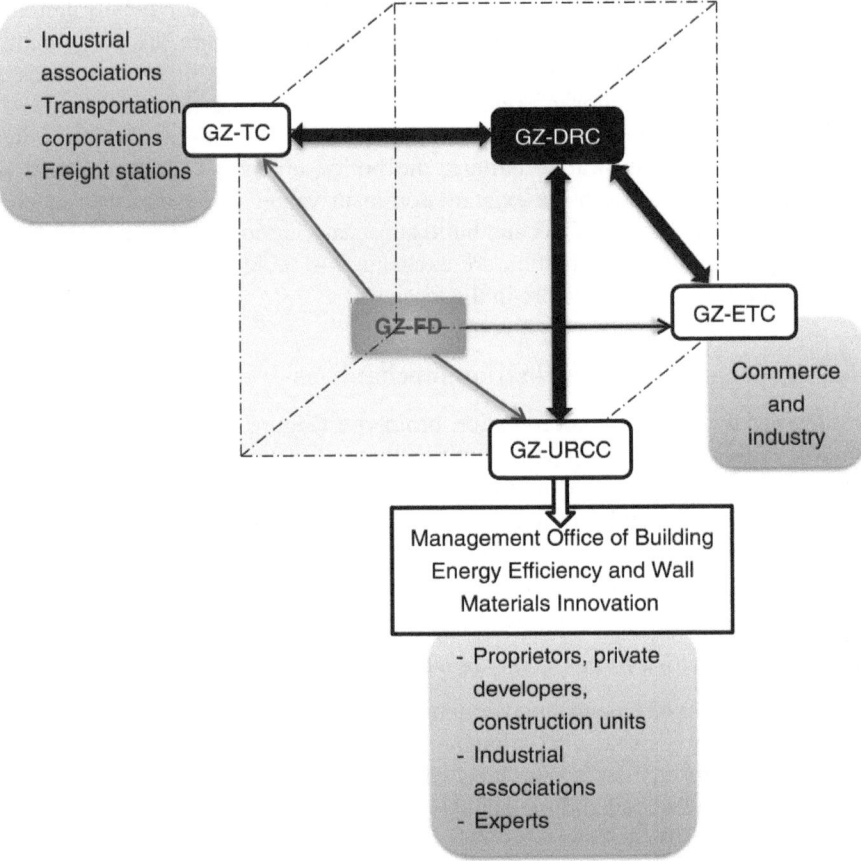

*Figure 4.2 Intra*governmental coordination mechanisms in Guangzhou
Source: Compiled by authors

Leading through change

The GZ-DRC currently leads the *intra*governmental coordination on the city's low carbon transition. It provides an *intra*governmental platform within the municipal government to devise policies and development plans. An official working in the secretariat of the leadership group reflected on its primary tasks:

> Actually, our department, the GZ-DRC, is a comprehensive economic managerial department [. . .] in Guangzhou's low-carbon development, our most important tasks are, firstly, formulating rules, secondly, building up platforms internal to the municipal government, and then mobilising the initiatives of each department.

(Appendix I: 9/GZ/08–02–2012)

Therefore, while the administrative ranks designated to the five core departments (GZ-TC, GZ-DRC, GZ-FD, GZ-ETC, GZ-URCC) are the same, the GZ-DRC's responsibility is to coordinate the various policy functions and to strategically lead the governance process with overall management and planning (Appendix I: 2/GZ/17–08–2011). The GZ-DRC, indirectly subordinated to the NDRC, is a member of the vertical policy portfolio on development and reform. The GZ-DRC has conventionally enjoyed a higher level of authority than its counterparts in other policy portfolios (Yang 2008) because the development and reform portfolio devises macroeconomic plans for its respective jurisdictions, with the *authority* and autonomy to allocate an earmarked grant set up specifically for its own operation. With sufficient capacity and interdependent interests, the GZ-DRC has established *internal legitimacy* in leading the *intra*governmental leadership group, which may help to overcome the inherent silo effect of horizontal fragmentation (Emerson *et al.* 2012).

The official further described the rule-setting function of the secretariat and how this task is achieved in the leadership group:

> Generally, [our task] is planning and drawing up the work of the implementation program mainly in three forms: first of all, there is a meeting; then the initial outline of the plan is drawn up, which are the two major jobs; the third that follows is to produce detailed policies on low carbon industry, technological innovation, manufacturing, industrial parks and so forth, based on the planning policies that have gone through official approval.
>
> (Appendix I: 9/GZ/08–02–2012)

The GZ-DRC frequently emphasises its role as an intra*governmental platform*. First, it serves as a *formal communication platform* connecting the departments involved in low carbon development. As part of its coordination efforts, the leadership group hosts cross-departmental meetings such as the Leadership Conference on Building a Low Carbon City, held in February 2011. Second, the commission attempts to stand as a *supervision and measurement platform* for the public sector in the municipality. As one official reflected:

> We have another job – monitoring the energy-saving of public agencies – to become a monitoring platform. We review, summarise and supervise the energy-conservation policies proposed by each public institution, as well as monitoring of public funding [. . .] each department in the meeting introduces its own sectoral work and we can see how much emissions each can reduce.
>
> (Appendix I: 9/GZ/08–02–2012)

The leadership group is also a *platform for strategic resource distribution*. An Energy Efficiency Special Fund has been established in Guangzhou, which sponsors the research and industrial development of innovative energy-saving projects. Administered by the GZ-DRC, the fund has been incorporated with low carbon development components (Appendix II: GZ-DRC, 2011b). Approximately

60 to 70 units successfully solicited the fund in 2011, around two thirds of which targeted carbon reduction (Appendix I: 9/GZ/08–02–2012). The GZ-DRC steers the leadership group to manage funding allocations across sectors, from funding approval to in-project assessment and postproject evaluation.

The leading role in the *intra*governmental coordination of climate mitigation–related matters has not always been performed by the same actor. The establishment of the leadership group on low carbon development altered the power structure among departments in the Guangzhou government. In the past, governmental attempts to maintain energy efficiency focused mainly on coercive practices in the commercial and industrial sectors. The leading authority at that time was the Guangzhou Economic and Trade Commission (GZ-ETC). However, the situation has become more complex and the functional areas have been broadened since the climate change component was formally added to the national agenda in late 2009 and was incorporated in the areas related to development and reform (Chapter 3; Appendix I: 2/GZ/17–08–2011, 9/GZ/08–02–2012). Thus, the designated administrative roles among municipal departments. have changed.

Although the leading role of *intra*governmental coordination has shifted away from the GZ-ETC, its *external links* in the existing *organisation* are still strong due to decades of involvement in steering the municipal agenda on energy conservation. The activities of the GZ-ETC rely upon various societal forces, as identified by the commission's director:

> There are probably several forces. On the one hand, it is the institutional units (*'shiye danwei'*), with managerial responsibilities and channels subject to the governmental mandates. On the other hand, there are some nongovernmental organisations working on some other tasks. We can also plan some similar activities through enterprises [. . .]. Our government plays the role of planning and initiating then lets social groups such as the Guangzhou Energy Academy operate specific activities.
>
> (Appendix I: 2/GZ/17–08–2011)

In this sense, as the former administrative leader, the GZ-ETC maintains consistent institutional power in the coordination of low carbon solutions at the municipal level, which the current leadership group should have been able to make good use of.

Sectoral coordination of low carbon transport

The governing activities around low carbon development are dispersed across different sectors. The GZ-TC is the sectoral leader in coordinating low carbon transport.

> We are only a managerial department, only an executive department, one of the component departments in the municipal government. Our function is

more about organising [government-led activities]. Regarding the requirements of low-carbon transport or low-carbon economic development led by
the government, we are commissioned to coordinate the work and carry it out.
(Appendix I: 11/GZ/17–02–2012)

As the sectoral leader, the GZ-TC is committed to enhancing its *internal institutional capacity* and sustaining *external organisational links* to carry out its tasks
effectively. One informant described the rationale for coordination as follows:

We coordinate various enterprises with their relevant associations, as well
as research and development institutes, to bring them into a joint force, and
carry out the policy work collaboratively.
(Appendix I: 11/GZ/17–02–2012)

Internal institutional capacity is reflected in the coordination between the commission and industrial participants, associations and technology innovators. The
GZ-TC tries to manage governmental and industrial actors in the transport sector
to fulfil the emission-reduction tasks assigned by the municipal government. It
keeps close track of progress in energy saving by conducting monthly checks on
subordinate units and providing seasonal reports to higher levels and by communicating and publicising industrial good practices and new technological innovations (Appendix I: 11/GZ/17–02–2012).

Furthermore, intermediary associations in the transport sector strengthen the
external links of the GZ-TC. They assist the intrasectoral coordination of the commission in four different respects: training, synergetic promotion, public–private
communication and government-led policy transfer.

The sectoral coordination role of the GZ-TC is enhanced through these activities. An illustrative example is the Green Trucks Pilot, initiated in 2009. The
pilot involved the GZ-TC, with external financial sponsorship from the World
Bank through Clean Air Asia[1] and technical guidelines provided by the American
SmartWay[2] programme (a detailed case study is provided in Chapter 8). While
promoting a green supply chain and energy-efficient freight stations for the project, the GZ-TC has been active and open in absorbing external experiences, which
has facilitated the building of cross-sector partnerships.

Concurrent and delegated coordination in the building sector

Subsequent to the establishment of the Low Carbon Leadership Group in November 2010, the GZ-URCC established a *Leadership Group on Building Energy Efficiency* in April 2011. The secretariat is located in the office of the building energy
efficiency unit in the GZ-URCC and the group is led by the deputy mayor, who
supervises construction-related matters. It thus has a lower administrative status
than the Low Carbon Leadership Group chaired by the mayor (Table 4.2). The
two leadership groups are engaging in *concurrent coordination* in the low carbon
transitions of the building sector. Compared with other sectors such as transport

and commerce, the construction sector has instituted more intensive intrasectoral interactions within the *intra*governmental structure.

To facilitate policy implementation, the Management Office on Building Energy Efficiency and Wall Materials Innovation (Figure 5.2) was established as a subordinate (although independent) organisation to the GZ-URCC. Offering organisational and policy support, the GZ-URCC *delegates authority* to this Management Office to monitor whether construction projects comply with green building standards. The Management Office is a public institution ('*shiye danwei*') that hires employees with civil-servant provisions and receives regular state financial support. It has also been engaged in the discussions of the Building Energy Efficiency Leadership Group. As an entity with discretionary power, the Management Office strengthens the *institutional capacity* of the GZ-URCC by promoting new guidelines on green buildings, providing training to different social groups and cooperating with nongovernmental associations on innovative projects (Appendix I: 12/GZ/21–02–2012).

The Management Office and GZ-URCC *together* maintain *external links* with nongovernmental associations and the professional community. Two such organisations are the Surveying and Design Association and the Construction Inno-Tech Promotion Station (CITPS) in Guangzhou. A new organisation was established in 2013 and was named the Guangzhou Association of Energy Efficiency and Technology in Buildings. This is steered by requirements specified in the Special Plan for Building Energy Efficiency in the 12th Five-Year Plan of Guangzhou (Appendix II: GZ-URCC 2011). This plan, drafted by the GZ-URCC and implemented in collaboration with the Management Office, is a fundamental policy document for the building sector in response to climate change.

Intra*governmental interactions*

The leadership group is commissioned to serve as an *institutionalised channel of* intra*governmental communication, conflict mediation and resource distribution.* Departments involved in the leadership group often encounter conflicts due to their different organisational orientations in the process of low carbon transition, which cannot be easily reconciled without cross-sectoral consensus. These *intra*governmental problems are submitted to discussion panels at regular meetings of the leadership group, which then becomes a conflict-mediation mechanism located at a higher administration level (Appendix I: 11/GZ/17–02–2012).

In addition to settling conflicts, the leadership group enables *boundary expansion and engagement* and *resource mobilisation* by maintaining the sustaining *external links* of each participating government agency. Resource mobilisation is vital when trying out institutional innovations such as the carbon-emissions trading market in Guangzhou. Support from different stakeholders and social actors needs to be solicited for initiatives at the experimental stage. Thus, the GZ-DRC held 'mobilisation meetings' through the platform of the leadership group, inviting all relevant actors in an effort to gain their endorsement for the emissions trade market. Both the leadership of the GZ-DRC and the institutional support of the leadership group are prerequisites for organising these activities.

Another important activity of the leadership group is *information sharing and resource allocation.* The Guangzhou energy-efficiency special fund and the provincial low carbon special fund cover industrial and technological innovations with a low carbon component (Appendix II: GZ-DRC 2011b). These special funds are allocated by means of earmarked grants, the dominant means of fiscal transfer[3] for specific purposes in China (Hoffman and Guerra 2004; Ma 1997). The administrative role of municipal departments in allocating funding mainly involves making a fair assessment of the potential contribution of the application proposal. In the assessment process, the GZ-DRC, working with the GZ-FD, holds the authority for allocating resources among the applications submitted by each sector of the entire municipality.

Nonetheless, the capacity for allotting the received grants is shared amongst the sectoral coordinators – such as the GZ-TC and GZ-URCC – which are charged with making suggestions and decisions concerning project selection, based on the rationale that they hold more *information and expertise in their own field* and can thus provide funding recommendations (Appendix I: 2/ GZ/17–08–2011, 9/GZ/08–02–2012, 11/GZ/17–02–2012). In addition, sectoral commissions are responsible for the management and supervision of whether the fund is being deployed appropriately in each sector (Appendix I: 9/GZ/08– 02–2012). If the fund consumption is not closely monitored and effectively distributed by the sectoral coordinator, the amount of funding allocated to the entire sector from the municipal budget will be reduced in the next fiscal year. Therefore, both the GZ-TC and the GZ-URCC devote attention to managing the awarded projects. Core actors are thus mobilised by the *intra*governmental funding mechanism.

Shenzhen: diversified coordination of intragovernmental contestation

The *intra*governmental structures in Shenzhen differ from those in Guangzhou. The arrangements are much more dispersed in Shenzhen compared to the united core group in Guangzhou. Almost every Shenzhen counterpart of the core actors in Guangzhou has established its own *intra*governmental coordination group (Table 4.2). A Joint Session System on Promoting Building Energy Efficiency and Developing Green Buildings in Shenzhen was established in 2008, with its secretariat placed in the Shenzhen Housing and Construction Department (SZ-HCD) (Appendix II: Shenzhen Municipal Government 2008b). In early 2010, the Shenzhen Transport Commission (SZ-TC) set up a Leadership Group on Environmental Protection, Energy Efficiency and Emission Reduction in the Shenzhen Transport Industry. The Shenzhen Development and Reform Commission (SZ-DRC) coordinated a Leadership Group on Responding to Climate Change, Energy Efficiency and Emission Reduction Works in Shenzhen at the end of 2010 (Appendix II: Shenzhen Municipal Government 2010a). In 2011, the Shenzhen Urban Planning, Land and Resources Commission (SZ-UPLRC) set up another

Joint Session System on Constructing National Low Carbon Ecological Demonstration City (Appendix II: Shenzhen Municipal Government 2011a).

The operation of the joint-session system is similar to that of the leadership group, based on a convention protocol of meeting schedules and administrative structure. However, the joint-session system places more emphasis on regular formal meetings to engage key actors in an equal manner, whereas the leadership group is often convened by a higher-ranking organisational actor (Appendix I: 34/BJ/06–08–2013).

These various *intra*governmental coordination mechanisms collectively govern the decision-making and implementation of climate-change policy in Shenzhen. Their structural configurations are illustrated in Figure 4.3, and their institutional specifications are summarised in Table 4.2. Unlike the *intra*governmental structure institutionalised in Guangzhou (Figure 4.2), there are multiple leading actors in Shenzhen (black rounded rectangles in Figure 4.3): not only the city's development planner (SZ-DRC), but also its urban spatial planning authority (SZ-UPLRC) and its public housing provider (SZ-HCD). Participating agencies are SZ-TC and the Shenzhen Finance Department (SZ-FD). Information and resource exchanges have been activated among the core actors (black double-ended arrows in Figure 4.3)

As Shenzhen is a special economic zone and a specifically designated city in the state plan (Chapter 3), the city-level administration maintains direct interactions with the national government, bypassing certain provincial controls. These direct interactions embedded in the vertical *inter*governmental hierarchy (grey downward arrows in Figure 4.3) influence the coordination capacity of and authority relationships among the core actors engaged in Shenzhen's *intra*governmental coordination for low carbon transition.

Leading through complexity

To coordinate citywide structural reform to mitigate climate change, the SZ-DRC accommodates the secretariat of the Leadership Group on Responding to Climate Change, Energy Efficiency and Emission Reduction Works (Table 4.2). The SZ-DRC's major tasks are to *mobilise* citywide low carbon development and to *implement* the low carbon tasks initiated within the city or delegated from the national government (Appendix I: 16/SZ/07–05–2012).

The SZ-DRC coordinated the preparation and promulgation of the recently published *Medium and Long Term Plan of Shenzhen Low-Carbon Development (2011–2020)* (Appendix II: SZ-DRC 2012). The SZ-DRC collected opinions from the departments involved in the leadership group before issuing the plan to ensure that the proposed tasks could be implemented across sectors with full consent (Appendix I: 17/SZ/11–05–2012). The SZ-DRC also organises annual and seasonal meetings with the leader and participants of the leadership group to analyse city-wide energy usage patterns.

The SZ-DRC routinely seeks *external links*. Maintaining external connections with the nongovernmental space serves the functions of exploring innovative

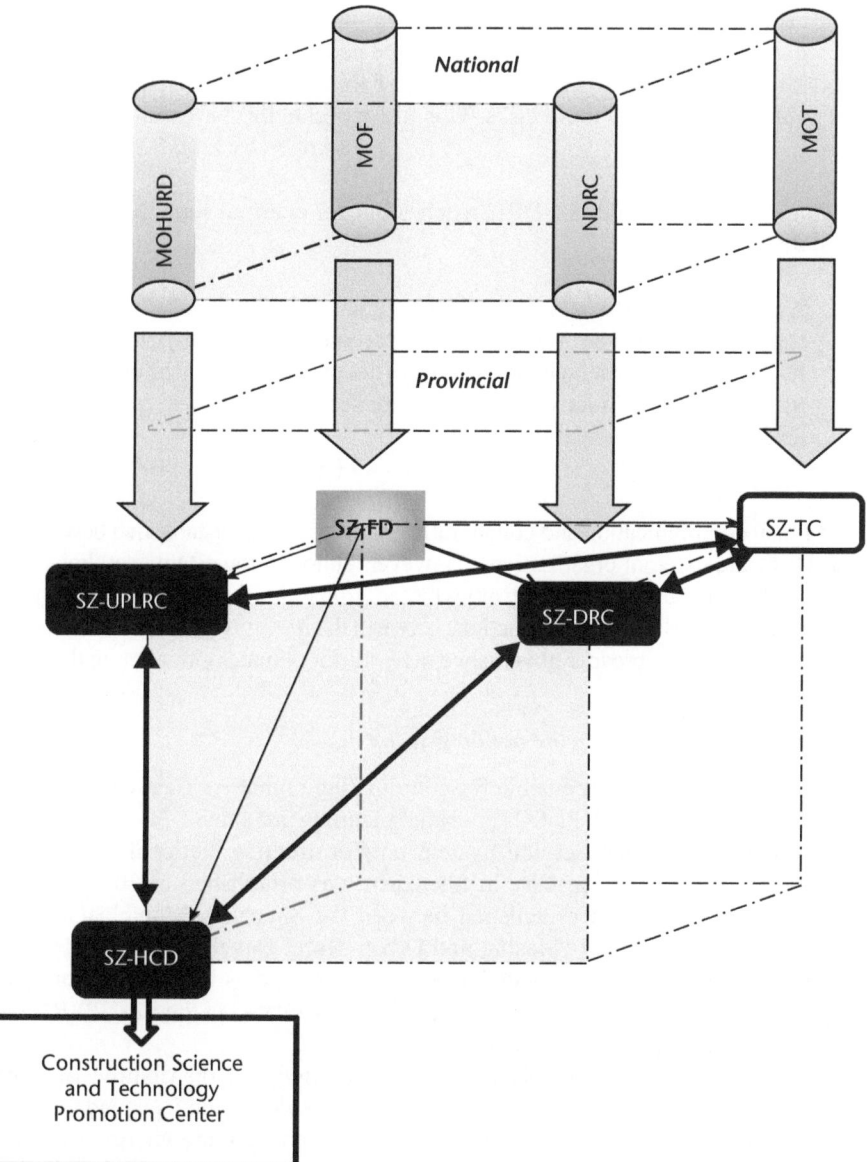

*Figure 4.3 Intra*governmental coordination mechanisms in Shenzhen
Source: Compiled by authors

research and expanding the impact of government-led, low carbon transition policies. The manager of the SZ-DRC reflected on this process:

> For example, for the [government-led] energy-efficiency public promotion week, detailed undertakings are carried out by [external] associations. We

also need to conduct certain research, to make some explorations – after all, low carbon is quite a recent thing – so we need to get help from these research institutes. It is due to the strength of these third-party organisations that we are able to accomplish these tasks. This [the external link] is certainly necessary.

(Appendix I: 16/SZ/07–05–2012)

In describing how the SZ-DRC reaches out for external support, the manager said that

As long as the external organisations are based in Shenzhen, and their organisational missions are related [to low carbon development], we try to establish feasible contacts with them. They [the external organisations] also reach us on their own initiatives, while we are searching for them given the available resources in society.

(Appendix I: 16/SZ/07–05–2012)

Mutual communication and collaborative links have been established between the SZ-DRC and external organisations. However, although the established links serve as a supplementary resource for citywide coordination, they are not fully institutionalised to initiate regular interactions between the *intra*governmental coordination mechanism and the broader governance network for climate mitigation in the city.

Expanded coordination in the building sector

Two departments are responsible for coordinating climate mitigation tasks in the building sector: the SZ-UPLRC for spatial planning and the SZ-HCD for building construction. The Joint Session System on Constructing National Low Carbon Ecological Demonstration City in Shenzhen was established in 2011 based on a framework agreement established between the Shenzhen Municipal Government and the Ministry of Housing and Urban–Rural Development (MOHURD), to coordinate the tasks subdivided among the participants involved in constructing a low carbon ecological city. The secretariat is located in the SZ-UPLRC with a staff of up to 15 people (Table 4.2).

The SZ-HCD is responsible for the construction and distribution of public housing in Shenzhen. It first put energy-efficient buildings on its agenda in 2006, with the promulgation of a municipal regulation on building energy efficiency. The SZ-DRC subsequently launched the implementation of green buildings in 2008 with the issue of a formal policy, the *Action Plan for Constructing a Green Building Metropolis* (Appendix II: Shenzhen Municipal Government 2008a). In the same year, to coordinate government activities in the implementation of the targets required by the regulation and the action plan, the SZ-HCD called for the establishment of the Joint Session System on Promoting Building Energy Efficiency and Developing Green Building in Shenzhen.

Fewer government agencies have been involved in these two joint-session systems led by the SZ-UPLRC and SZ-HCD than in the leadership group steered by the SZ-DRC.[4] The exchange of professional opinions and the discussion of implementation issues are confined to the building and construction sectors, particularly regarding green and energy-efficient buildings. Moreover, these two joint sessions are headed by the respective deputy mayors rather than the city mayor, indicating lower political significance.

External links are weak and marginalised in the building sector, while most of the government-led mitigation activities are internalised among the government agencies. For the SZ-UPLRC, the external organisations are merely entities that collect public opinion on proposed development plans (Appendix I: 17/SZ/11–05–2012). The SZ-HCD, as the governmental implementation agency, regards external organisations such as the Shenzhen Green Building Association as '*non-technical outsiders*' that are not able to offer professional assistance (Appendix I: 20/SZ/23–05–2012). The specific technical support tasks of green-building verification are delegated to the Construction Science and Technology Promotion Centre, a public institution (*shiye danwei*) that is directly subordinate to the SZ-HCD.

Connecting nodes in the transport sector

The leadership group on environmental protection, energy efficiency and emission reduction in the Shenzhen transport industry was initiated in response to the recent requirements on energy efficiency that were vertically communicated and delegated directly from the Ministry of Transport (MOT) across the transport-governing system. Although energy efficiency has been formally institutionalised as a task in the transport sector, the SZ-TC coordinator is not fully focused on it: only one manager in the SZ-TC works as the communicator on energy-efficiency tasks and is at the same time responsible for two other functions – transport *safety* and transport *environmental protection and energy efficiency*. The chair of the transport leadership group is merely a director in the SZ-TC, and thus has a lower administrative rank than a mayor or deputy mayor, which diminishes the SZ-TC's ability to integrate low carbon initiatives in the transport sector with other sectors (Appendix I: 21/SZ/24–05–2012). Given the lack of administrative capacity, the scope of the leadership group is limited to the *internal* transport sector (Appendix I: 21/SZ/24–05–2012). Municipal departments external to the SZ-TC are not engaged in this leadership group. Therefore, the SZ-TC is only a participating rather than a leading actor (Figure 4.3).

In terms of external links with the nongovernmental space, the existing networks are not active in either information provision or resource mobilisation. Instead, based on the *intra*governmental structure, the SZ-TC seeks to maintain external coordination given its limited internal capacity: horizontally, with other departments working on energy efficiency and emission reduction such as the SZ-DRC, and vertically, with the Guangdong Transport Bureau and the MOT (Appendix I: 21/SZ/24–05–2012). Its role is to acquire resources that are mandated from the transport bureau and MOT, first to 'expand the current staff establishment by reinforcing the intention of deploying external assistance', and second to 'strengthen the existing self-administered institutions by putting in proposals to establish an expert committee to help out the decision making process of energy efficiency and emission reduction' in the transport sector (Appendix I: 21/SZ/24–05–2012).

Intragovernmental interactions

The *intra*governmental structures in Shenzhen carry certain distinct features. First, the independent legislative power conferred on Shenzhen (Chapter 3) motivates government agencies to initiate law-making pursuits favourable to their sectoral activities and vested interests in the process of low carbon transition. City-based legislators have been active, with the support of *intra*governmental structures. The SZ-DRC coordinated the drafting of the *Shenzhen Special Economic Zone Circular Economy Promotion Regulations* in 2006, claiming that it was 'the first of its kind across the nation' (Appendix I: 16/SZ/07–05–2012). The SZ-HCD drafted the *Shenzhen Measures of Green Building Promotion*, which was recently promulgated in 2013 after several rounds of negotiations and conflict meditation. The drafting and consultation processes were coordinated solely by the SZ-HCD to maintain 'consistencies of legislative ideas' (Appendix I: 20/SZ/23–05–2012). *Intra*governmental conflicts arose during the legislation's final consultation process: after seeing the draft legislative proposal, the implicated governmental agencies responded with 'a lot of disruptive comments', mainly because the proposed legislation 'disrupted the established interests of many departments' and 'a lot of them do not want to take up new duties' (Appendix I: 20/SZ/23–05–2012). The involved departments negotiated for more potential benefits and fewer administrative duties during the *intra*governmental communications by demanding revisions to the legislative draft in exchange for their endorsement and participation.

Second, Shenzhen's economic management authority, which is independent from the provincial administration, has intensified the city's *intra*governmental communication and stimulated its attempts at *inter*governmental cooperation. Shenzhen receives direct fiscal transfers from the central government, thus bypassing the province. The public revenues and expenditures of Shenzhen are directly connected to the central budget. Technological innovations based in Shenzhen are not eligible for grants incorporated at the provincial level, such as the Guangdong low carbon special fund. Individual government departments are active in building cooperative networks between national ministries and municipal commissions, which allow them to receive central-government grants and establish corresponding matched funding at the municipal level.

In this process, establishing joint-session systems has become a necessary means to securing central grants through the application and implementation of national demonstration projects and pilot city experiments (a detailed discussion is provided in Chapter 7). These projects, which are vertically connected to the ministerial level, aim to disseminate locally developed experiments and replicable experiences. Participation allows the local governments to receive direct fiscal support from the central government to implement the pilot projects and to deliver the assigned outputs based on general instructions. This also spurs fierce contests among localities.

From 2007 to 2012, the SZ-HCD successfully secured five demonstration building projects, such as energy retrofitting and renewable energy applications in public buildings, from the Ministry of Housing and Urban–Rural Development (MOHURD). Each of these projects came with central grants upon successful application, in conjunction with the provision of municipal matching funds during implementation. The Joint Session System on Promoting Building Energy Efficiency and Developing Green Buildings was first established due to the backlog of work that had piled up during the application process for an early demonstration project. An informant from the SZ-HCD explained how this joint session system operated:

> The focus of the joint session system is actually on the coordination department itself, on the matters it needs to deal with, [. . .] and on certain tasks involving many departments. Applications for national demonstration projects must be lodged with the central office of municipal governments. The specific department that wants to coordinate the project must first prepare the application report, forward it to the municipal government office and get it approved in a joint session.
>
> (Appendix I: 20/SZ/23–05–2012)

As an *intra*governmental coordination mechanism, the joint session is required to gain consent from various departments with different interests in one complex issue before the city can take action. Consent is also necessary during the implementation process. Prescribed by the cooperative framework agreement signed between the MOHURD and the SZ-UPLRC in 2010, the Joint Session System on Constructing a National Low Carbon Ecological Demonstration City was established to enable coordination among departments (Appendix I: 17/SZ/11–05–2012). It compels actions and subdivides tasks among the departments involved (Appendix I: 20/SZ/23–05–2012).

Third, the *intra*governmental structures of the multiple leading actors in Shenzhen reflect hidden departmentalism, which has led to governmental neglect of external links with the nongovernmental space. There has been departmental competition between the SZ-DRC and SZ-UPLRC over the role of the leading actor in governing climate-change problems and low carbon transitions, with the former focusing on energy and economy dimensions and the latter targeting spatial planning. Both departments have solicited national endorsement to create a low carbon model in Shenzhen through different pilot projects. However, they have failed to engage with each other in their core *intra*governmental coordination mechanisms (i.e. dashed-line box in Figure 4.3); instead, they are competing for greater authority in the coordination of a complex policy issue, stronger political endorsement from the municipal government and larger public attention for their projects. Departmentalism intertwined with issue complexity diminishes *intra*governmental coordination.

The expansion of *intra*governmental groups has further diverted governmental actors from the nongovernmental space. Numerous environmental nongovernmental organisations established in Shenzhen in recent years – such as the Shenzhen Green Building Association and the Shenzhen Energy Conservation Association – are eager to build collaborative networks with the government. Yet strong inertia can be found in the overstretched *intra*governmental groups. Governmental actors in Shenzhen are overwhelmed by the bureaucratic procedures involved in proceeding the numerous leadership groups and joint sessions and have not been active in building external links with these organisations.

Hong Kong: internal coordination demanded by external support

In 2007, the Hong Kong government established an Interdepartmental Working Group on Climate Change (IWGCC) to deal with the governance challenges resulting from climate change (Appendix II: Environment Bureau 2008). Five bureaus and 16 departments are included in the group, with the Environmental Protection Department (EPD) as the prime convener (Table 4.2).

The configuration of the IWGCC is illustrated in Figure 4.4. The *intragovernmental* structure configured by the IWGCC is contextualised in the two-layer administrative structure of Hong Kong, that is, policy formulation bureaus heading their subordinate implementation departments and agencies. These hierarchical relationships are indicated by the grey downward arrows in Figure 4.4. As the leading actor in the IWGCC, the EPD is represented by a black rounded rectangle. Participating agencies in the transport sector include the Transport and Housing Bureau and the Transport Department. The Transport and Housing Bureau is also involved in the building sector, together with the Development Bureau (Works Branch), the Electrical and Mechanical Services Department, the Housing Department, the Planning Department and the Buildings Department, which are connected to the IWGCC. All of the participating agencies are depicted by the white rounded rectangles in Figure 4.4.

The key activities initiated by the IWGCC – represented by double-ended arrows in Figure 4.4 – include a public consultation study on climate change, regular reviews of tasks undertaken by individual departments in relation to adapting and mitigating initiatives for climate change and capacity-building workshops on adaptation to climate change impacts (ERM 2010; Appendix I: 15/HK/30–04–2012). Before the outputs from the public consultation were available, the EPD proceeded to implement mitigation activities with sectoral focuses on building and transport. Mitigation activities carried out in the building sector included enacting the Buildings Energy Efficiency Ordinance, constructing a district cooling system at an experimental site and implementing the mandatory Energy Efficiency Labelling Scheme. In the transport sector, the government set up a Pilot Green Transport Fund to encourage the testing of green innovative transport technologies, released a preferential taxation policy to businesses purchasing environmentally friendly vehicles and encouraged the wider use of electric vehicles (EVs) by expanding the network of EV charging stations and procuring EVs for the government's own fleet.

Stagnation

As mentioned, a three-month public consultation was launched in 2010 over the policy proposal developed by the IWGCC's coordination platform regarding Hong Kong's climate-change strategy and action agenda (Appendix II: Environment Bureau 2010a). By the end of 2010, around 30 responses had been submitted

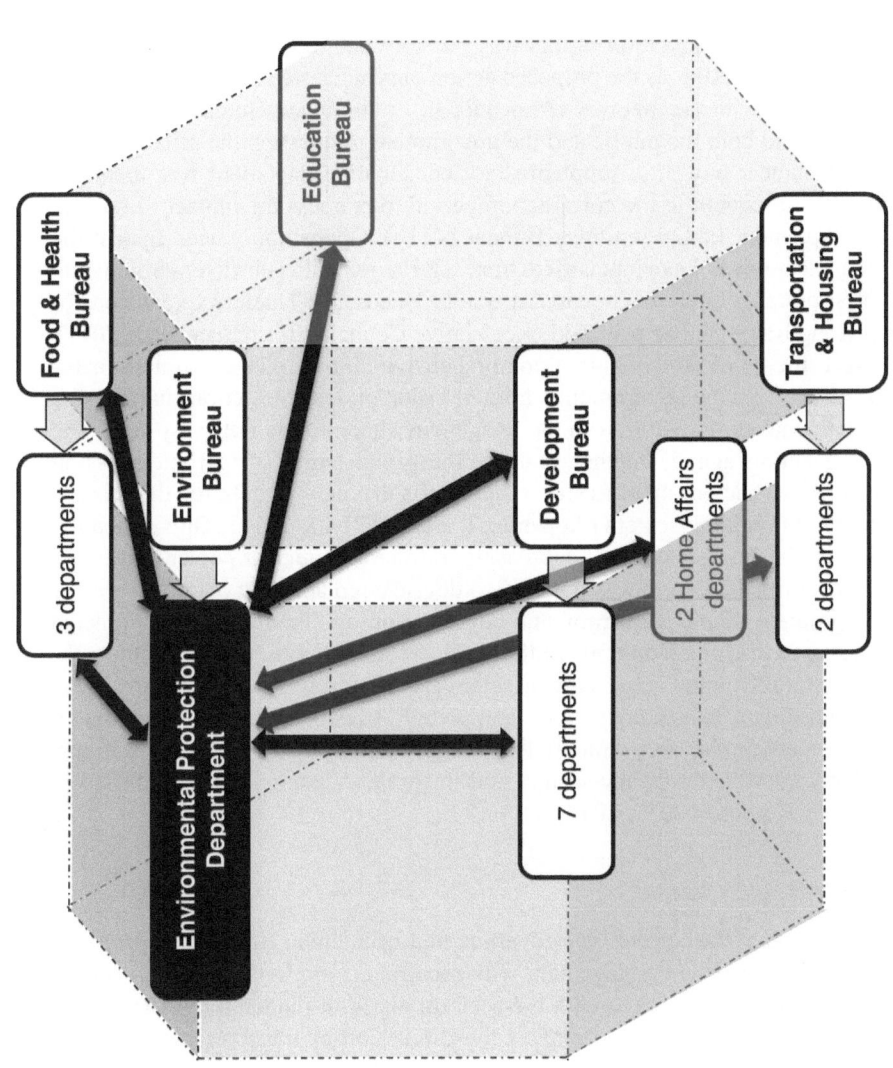

Figure 4.4 Intragovernmental coordinating mechanisms for climate change in Hong Kong

Source: Compiled by authors

to the EPD by power companies, environmental nongovernmental organisations, professional institutes, industrial groups and individual academics. The Environment Bureau planned to analyse the feedback received during the consultation period and compile a consultation report by April 2011 (Appendix II: Environment Bureau 2011a).

However, the result of the consultation exercise has been postponed to date, and Hong Kong has not yet seen a comprehensive climate change policy. The postponement of the consultation results was possibly due to the Fukushima nuclear disaster in March 2011, as the proposed action plan suggested increasing the share of nuclear power in the fuel mix (Appendix II: Environment Bureau 2010b). The disaster caused both the public and the government to question the reliability and safety of nuclear power. A supplementary consultation document was therefore circulated subsequent to the consultation period to reassess the impact of nuclear power (Appendix II: Environment Bureau 2011b). Opposition voices against the use of nuclear power have intensified, from legislators' criticisms to environmental activists' protests, such as Greenpeace's Join Hands: No Nuclear Expansion for Hong Kong Rally on 24 April 2011 (Legislative Council 2011; Greenpeace 2011).

In fact, the delay in devising a comprehensive climate change strategy actually resulted from the government's limited vision on feasible actions in response to climate change, in addition to the exogenous shock to the policy system. The proposed action agenda neither reduced the grand target of 'carbon-emission reduction' to concrete measures nor specified subsidiary targets for the energy, transport and building sectors (Appendix I: 27/HK/20–08–2012). Thus, when the one important possible means – increasing the share of nuclear power in the fuel mix – was turned down, it was not easy to devise a replacement plan.

The progress of *intra*governmental coordination on climate change problems has stagnated in Hong Kong, given the leading actors' failure to finalise the public consultation output and settle a relatively comprehensive citywide action plan. With only a consultancy study completed in 2010, the IWGCC has failed to render a comprehensive strategy that is supported by dedicated organisations, effectively coordinates policy sectors and integrates scattered climate initiatives (Francesch-Huidobro 2011, 2012).

Fragmented policy sectors

The stagnation of the current *intra*governmental structure in Hong Kong probably results from insufficient engagement with sectoral actors. In the transport sector, the IWGCC does not engage with the sectoral platform rendered by a Transport Advisory Committee that coordinates the discussion of transport planning and infrastructure. There is no evidence that the IWGCC is communicating with the Transport Advisory Committee on the mitigation of climate change. For example, encouraging the wider use of EVs is one of the proposed sectoral-mitigation activities. A Steering Committee on the Promotion of Electric Vehicles was established for this purpose in 2009. The committee was chaired by the city's financial secretary and engaged members from the relevant policy bureaus, academia and the private sector (mainly power companies and real estate/infrastructural developers). Again, the IWGCC failed to engage actively with the sectoral steering committee (Appendix I: 37/HK/07–01–2014). The discussion output of the steering committee was neither translated nor incorporated in the climate-change action plan proposed by the IWGCC, due to the limited capacity of the IWGCC and its leading actor (the EPD) to initiate *intra*governmental coordination and absorb input regarding policy information and resources.

In the building sector, the 2013 Policy Address declared the launch of an interdepartmental steering committee to promote green buildings (Appendix II: Chief Executive 2013). With the appointment of Wong Kam-sing – an architect by profession and expert in green buildings – as Secretary for the Environment, it is anticipated that sectoral transitions will become more closely integrated into the broader issue of climate change. However, it is still too early to assess whether this newly established *intra*governmental coordination mechanism will actually *facilitate* the coordination of the IWGCC and enhance the leadership's commitment to the issue or whether it will simply *migrate* the broader policy discussion of climate change problems into a specific sectoral issue of green buildings by abandoning the former.

Without effective sectoral coordination, the IWGCC has been unable to achieve horizontal cross-sectoral synergy in an *intra*governmental setting. The function of the IWGCC (and the EPD) is merely to feed top-down policy directives regarding low carbon transitions from the national government to the issue-related bureaus of the city in a *hierarchical* way (Francesch-Huidobro 2011, 2012). Knowledge transfer is not synergetic among different sectoral or functionally specialised policy departments. Despite the 'flourishing' of sectoral steering committees, the policy sectors remain detached from the core discussion on climate change problems. As a member of the steering committee on electric vehicles lamented, 'over the years, I have served on a lot of government committees here [in Hong Kong]; one very striking thing is that unless there is some overlapping membership, the interaction is minimal' (Appendix I: 37/HK/07–01–2014).

Even with certain members concurrently serving on different committees, the transfer of expertise is impeded by administrative barriers in the *intra*governmental groups. 'I did point out that there had been similar discussions in another committee; but one person is not going to swing the whole committee. These committees are still subject to protocols and processes' (Appendix I: 37/HK/07–01–2014). The discussion in different steering committees is often confined by the stipulated agenda. It is often the case that different departmental representatives are assigned to different steering committees, thus hampering the exchange of discussion output across policy functions. With minimal interactions across established *intra*governmental groups, the *intra*governmental coordination structure has failed to integrate cross-sector policy actions around the climate change problem.

Powerful external resources

Despite the weak capacity of the existing *intra*governmental structure to integrate sectoral approaches, nongovernmental actors on climate change in Hong Kong are relatively powerful in terms of resources and potential. Equipped with professional expertise and external funding to develop climate change initiatives, a group of longstanding environmental nongovernmental organisations, think tanks and sectoral institutes has carried out advocacy on climate change, demanding more proactive action from the city government. In 2009, local branches of Greenpeace China, Oxfam Hong Kong and the World Wildlife Fund formed the Combating Climate Change Coalition, a nongovernmental group that also brings together other local nongovernmental organisations. These are all potential platforms that the EPD and IWGCC can use to gain stakeholder support and legitimacy 'through informal involvement of sectors external to government horizontally' (Francesch-Huidobro 2012: 5).

The IWGCC is therefore embedded in an extensive external nongovernmental space. The nongovernmental space generates climate knowledge for the development of a local climate change action plan. The IWGCC operation is also shaped by the professional reviews and public engagement exercises concerned with energy saving and carbon reduction in buildings conducted by the Council of Sustainable Development,[5] a body external to the *intra*governmental group. It is feasible for the leading actors and core sectoral participating agencies in the *intra*governmental structure to establish operational contacts with a broader range of organisations in the nongovernmental space. Currently, only resourceful green groups are marginally involved in the discussion of two steering committees on electric vehicles and green buildings, whereas the IWGCC – which has received only a passive reception to its opinions through public consultation – has not actively engaged with the nongovernmental space.

Intra*governmental interactions*

Various institutional factors have led to the IWGCC's ineffective coordination. Previous analyses have identified possible causes, including a lack of sources of legitimation in the political system, no mandate or authority conferred to the group, deficits in the group's persuasion ability and political indifference to the issue of climate change, especially from government leadership (Francesch-Huidobro 2011; Francesch-Huidobro 2012).

These analyses, particularly those concerning authority and political leadership, were confirmed by practitioners working on government liaison for climate change policy advocacy:

> Generally, the IWGCC is not working so effective[ly] . . . That's one of our problems, one of our complaints. We have waited to see that elevated and get a lot more authority. And part of that problem is we are not going to get things done, until the IWGCC has people with actual political power sitting in its chairperson seats. People who sit at these meetings are not the ones who have the ability to make political decisions, so they need to have a supportive bureau behind them. And until they do, they are not going to be able to do anything.
>
> (Appendix I: 26/HK/29–08–2012)

In Hong Kong, departments stand at a lower administrative rank than bureaus. It is apparent that the EPD, which clearly has insufficient authority, has not taken a strategic leadership role relative to other bureaus and departments in the IWGCC. The EPD, as the leading actor in the IWGCC, has failed to establish *internal legitimacy* among the core participants in the process of *intra*governmental coordination. It lacks sufficient authority to either compel actions or press negotiations in the process of *intra*governmental coordination.

On the other side of the coordination equilibrium, the EPD is in a quandary about initiating cross-sector cooperation and obtaining consent for its actions. Another specialist who has worked on climate change policy in Hong Kong said:

> I haven't heard anything about it (IWGCC). I mean, I know it exists, but I don't know what they do. There are a lot of people who would like that to be a Climate Change Authority or something; so at the moment, climate change would sit under the Environment Bureau. Actually because it crosses transport, waste, buildings and electricity, it needs to be sitting somewhere where it can delve into these issues. So I think that's a problem. Again it is the [silo effect] also in the environment [sector].
>
> (Appendix I: 27/HK/20–08–2012)

The recently established Steering Committee on the Promotion of Green Buildings has certainly been positioned at a more strategic administrative rank within the decision-making body. It is chaired by the Secretary for the Environment, the political appointee with the highest administrative ranking in the environmental policy portfolio. The political significance attached to the sectoral-mitigation option of the green-building approach has certainly been enhanced (Appendix I: 28/HK/19–11–2013). This newly established Steering Committee has held regular meetings, campaign presentations and focus groups in an attempt to engage nongovernmental stakeholders from the construction sector and green-building assessment organisations (Appendix I: 28/HK/19–11–2013; 30/HK/19–11–2013; 36/HK/16–12–2013). The presumed leading role and conferred authority of the EPD in climate-change issues is currently being elevated by the presence of a more committed leader in the Environment Bureau. The building sector in the CMNs of climate mitigation in Hong Kong has instituted intensive intrasectoral coordination, similar to the situation in Guangzhou.

However, the fragmentation of outputs remains a persistent problem. There is a missing link between the IWGCC and the Steering Committee on the Promotion of Green Buildings, with the former instituted under the then chief executive (Tsang Yam-kuen) and his successor Leung Chun-yin. The governing emphasis has shifted across the government's two terms. The supposed attention to the cross-sectoral issue of climate change has been diverted back to segregated sectoral approaches. Together with the steering committee on EVs, the adoption of a sectoral approach appears to be a pragmatic way to move forward with mitigation options for climate change in the city. Nevertheless, focusing only on either the built environment or on road transport does not directly lead to effective *intra*governmental coordination on the broader problem of mitigating climate change, which requires cross-sectoral synergic action. Placing the two steering committees of sectoral responses in a position far above that of the coordination group of integrative responses to climate change has further diminished previous efforts at putting together a policy programme of coherent plans and objectives. The city government still lacks a strategic vision for improving coordination outcomes.

Conclusion

We have identified four key elements underpinning effective horizontal coordination from the empirical analysis. First, *political significance* needs to be attached to climate change problems and solutions, in order to assimilate useful policy information and gather resources for relevant sectoral policy initiatives such as green building and low carbon transport. Second, given the long-term and complex nature of climate-change problems, it is essential that the local administrative elites show *political will* and determination. The successful horizontal coordination of different municipal government agencies depends on the endorsement of the Chief Executive in Hong Kong and of the mayors in Guangzhou and Shenzhen, who confer political authority on the *intra*governmental coordination mechanism and ensure the specification of a clear timetable on how climate-change action will go forward (Appendix I: 11/GZ/17–02–2012; 26/HK/29–08–2012). Such endorsement is reflected in the administrative ranks of the assigned chairpersons of the *intra*governmental coordination mechanisms (Table 4.2).

Third, it is necessary to grant *authority* and a political mandate to the leading agencies of the *intra*governmental mechanisms so that they can enhance their ability to persuade other departments into joint actions, to reformulate the agendas of the responsible departments and to initiate an integrative rather than a sectoral approach to climate mitigation. The primary agency is most likely to be the department where the secretariat of the *intra*governmental coordination mechanism is located. In Guangzhou, within the Leadership Group on Development of a Low Carbon Economy, the GZ-DRC is the strategic leader, and it has more authority than other departments, although they have the same administrative rank. The coordinating departments of leadership groups and joint session systems in Shenzhen are competing for more authority, resources and attention. In contrast, Hong Kong's IWGCC is convened by the EPD and led by the EPD's deputy director and thus has much less authority.

Finally, the internal *manpower* of the secretariat and its *external links* with supporting units and nongovernmental actors are important for leveraging the influence of the specific *intra*governmental coordination mechanism. The effectiveness of *intra*governmental coordination mechanisms is proportional to the number of staff employed in the secretariat and its ability to deploy additional personnel from other departments (Table 4.2). *Intra*governmental coordination mechanisms with larger staff – such as the Leadership Group on Responding to Climate Change, Energy Efficiency and Emission Reduction Works (in Shenzhen) and the Joint Session System on Constructing a National Low-Carbon Ecological

Demonstration City (also in Shenzhen) – are more capable of manoeuvring the behaviour of other departments than are those with limited human resources. External links with other supporting units such as research centres and sectoral associations are additional sources of human resources (Table 4.2). These units provide research support and advice, including training, promotion of best practices and publicising new policies.

In light of these empirical findings, in Table 4.3 we revise the analytical matrix proposed earlier in this chapter (Table 4.1) with scenarios ensuing from the analysis of the empirical findings in the three cities.

The discussion in this chapter has considered whether the mechanisms established as organisational attempts (summarised in Table 4.2) are effective (or merely cosmetic) in improving the outcomes and processes of *intra*governmental coordination. The process of coordination involves maintaining a *dynamic stability* that is contingent on *inter*organisational *equilibrium* – a balance between arousing conflict and achieving cooperation and between imposing coercion and obtaining consent (Dunsire 1978). Conversely, the coordination outcomes examine end-state problems – task redundancies, core-task lacunae, overlapping clientele groups and incoherence between coordination goals and imposed organisational requirements (Peters 1998).

The complexity of governing climate change has given rise to various *intra*governmental coordination mechanisms in the three city governments. To gain consent to initiate climate policy innovations and achieve cooperation to mobilise resource support, the leadership group in Guangzhou has mediated conflicts among the participating government agencies. These *intra*governmental interactions maintain dynamic stability among the participants, resulting in a coordinated governing process. Guangzhou's coordination outcomes are also assessed. Core tasks are performed or allotted by a specified leading actor with authority leverage. The sectoral coordinators in the transport and building sectors interact

*Table 4.3 Intra*governmental mechanisms for mitigating climate change in practice

| | | *Political priority (dominant issues, will of political elites)* | |
		Low	*High*
Institutional support (authority of leading agency, manpower)	**Weak**	D. Bureaucratic inertia, with no action plan or overall strategy on climate change and piecemeal sectoral actions	C. Rapid introduction of climate change policies, yet sectoral segregation in the design of action plans and horizontal fragmentation in their implementation
	Strong	B. Effective communication within *intra*governmental mechanism, but slow in goal setting and implementation	A. Effective coordination with clear sectoral emission reduction goals and efficient actions

with their respective and mutually independent clientele groups. Guangzhou does not exhibit end-state problems such as lacunae and overlapping clientele groups. However, the city's administration does not attach political significance to climate-change issues, which has held back the delivery of clear policy goals and reduction targets. Scenario B in Table 4.3 matches the status quo in Guangzhou.

In Shenzhen, the making of local laws has involved compelling sectoral actions and negotiating with conflicting established interests. The *intra*governmental structure has been overstretched, with leadership groups and joint sessions merely fulfilling administrative requirements rather than gaining consent and aligning departmental goals. The equilibrium is in constant crisis, which has reduced the overall effectiveness of *intra*governmental coordination. End-state problems are also identified. The hidden departmentalism and fragmented leadership in Shenzhen result in overlapping clientele groups and task redundancies between the low carbon policy plans devised by different leading departments. The existing development in Shenzhen fits Scenario C in Table 4.3.

There has been a transition of *intra*governmental coordination in Hong Kong. With weak authority in the leading agency, limited manpower, low issue significance in the political system and limited political will, the stand-alone IWGCC has experienced fragmented coordination in both processes and outcomes (Scenario D in Table 4.3). The newly instituted steering committees have elevated the authority of the leading actor. The city has departed from its previous coordination quandary. The process of *intra*governmental coordination has become more streamlined, with improved capacity to engage the external nongovernmental space. Yet, in terms of end-state coordination, the elevated authority has not solved the problems of core-task lacunae and overlapping clientele groups. The process of issuing an integrative climate-change action plan remains stagnated, even though sectoral actions are active. Although some nongovernmental actors are repeatedly engaged across the two steering committees – the private real-estate developers in the promotion of both EVs and green buildings – other external voices are barely heard. In this regard, Hong Kong is undergoing a gradual shift from Scenario D to Scenario B, as represented in Table 4.3.

Given the discussion on *intra*governmental coordination in this chapter, we shall develop further analytical elements in the next two chapters to study the process and network dynamics of innovation integration for low carbon transitions in the building (Chapter 5) and transport (Chapter 6) sectors. As essential connections exist between the environmental organisations in the nongovernmental space and the core nodes that form the coordination mechanisms within the city governments, the next chapter focuses on the role of the expanding nongovernmental space in climate governance (Chapter 5). More importantly, with *intra*governmental coordination processes and outcomes remaining functionally fragmented in these three southern Chinese cities, Chapters 5 and 6 investigate how the legitimation process and collaborative functions of the CMNs for climate mitigation may cope with these coordination deficits. Three dimensions are examined: policy diffusion and innovation transfer, resource mobilisation in an incentive structure and the communication of sectoral innovations internal and external to the network.

Notes

1 Clean Air Asia was established in 2001 by the Asian Development Bank, the World Bank and the United States Agency for International Development, with a mission to 'promote better air quality and livable cities by translating knowledge to policies and actions that reduce air pollution and greenhouse gas emissions from transport, energy and other sectors' (Clean Air Asia 2013). Its China office was set up in 2004 in Beijing.

2 SmartWay® was launched in 2004 by the United States Environmental Protection Agency in collaboration with the freight industry. The programme aims to reduce transport-related emissions by creating supply-chain incentives. It also seeks to provide guidance and resources for other countries on sustainable freight projects.

3 Fiscal transfer refers to the mechanism by which national or upper-level governments coerce the collection of local revenues and redistribute them across localities according to local needs and capacities. This mechanism has been implemented in China since its 1994 tax-sharing reform. The current fiscal transfer system in China is stratified into three tiers: central-provincial transfer (including transfers to cities with independent planning status such as Shenzhen), provincial-municipal/county transfer (which applies to Guangzhou) and municipal-district/county transfer (Guangzhou and Shenzhen). Fiscal transfers are generally one of two types: general-purpose (unconditional) and specific-purpose grants.

4 As summarised in Table 4.2, 20 agencies have been engaged in the joint session system led by the SZ-HCD, including the SZ-DRC, SZ-UPLRC and district governments. The session coordinated by the SZ-UPLRC involved 16 municipal agencies on urban infrastructural planning and a national ministry. In contrast, more than 30 municipal departments are coordinated in the SZ-DRC's leadership group, including the directors of the SZ-TC, SZ-UPLRC and SZ-HCD.

5 The Council for Sustainable Development is a forum for exchanging ideas on sustainability in Hong Kong. It was established by the chief executive with secretariat support from the Environment Bureau.

References

6, P. 2004. Joined-up Government in the Western World in Comparative Perspective: A Preliminary Literature Review and Exploration. *Journal of Public Administration Research and Theory,* 14, 103–138

6, P., Goodwin, N., Peck, E. & Freeman, T. 2006. *Managing Networks of Twenty-First Century Organisations,* Basingstoke, UK, Palgrave Macmillan.

Betsill, M. M. & Bulkeley, H. 2006. Cities and the Multilevel Governance of Global Climate Change. *Global Governance: A Review of Multilateralism and International Organizations,* 12, 141–159

Clean Air Asia. 2013. *About Us: Our History and Mission* [Online]. Available: http://cleanairinitiative.org/portal/aboutus [Accessed 13 December 2013]

Dunsire, A. 1978. *The Execution Process, Vol. 2: Control in a Bureaucracy,* London, Martin Robertson.

Emerson, K., Nabatchi, T. & Balogh, S. 2012. An Integrative Framework for Collaborative Governance. *Journal of Public Administration Research and Theory,* 22, 1–29

ERM 2010. *A Study of Climate Change in Hong Kong – Feasibility Study,* Hong Kong, Environmental Resources Management.

Francesch-Huidobro, M. 2011. Governance of Climate Change in Coastal Cities: The Example of Hong Kong, in Aerts, J., Botzen, W., Bowman, M., Ward, P. & Dircke, P. (eds.) *Climate Adaptation and Food Risk in Coastal Cities,* Oxford, Earthscan Climate from Routledge.

Francesch-Huidobro, M. 2012. Institutional Deficit and Lack of Legitimacy: The Challenges of Climate Change Governance in Hong Kong. *Environmental Politics,* 22, 1–20

Greenpeace. 2011. *Over 460 Citizens Joined Greenpeace Rally to Demand the Government Drop Nuclear Expansion Plan* [Online]. Hong Kong: Greenpeace Press Release. Available: www.greenpeace.org/eastasia/press/releases/climate-energy/2011/hk-no-nuke-rally/ [Accessed 14 May 2013]

Gulick, L. & Urwick, L. F. (eds.) 1937. *Papers on the Science of Administration,* New York, Institution of Public Administration, Columbia University.

Hoffman, B. & Guerra, S. C. 2004. Fiscal Disparities in East Asia: How Large and Do They Matter? [Online]. *East Asia Decentralizes: Making Local Government Work,* Washington, DC: World Bank. Available: http://siteresources.worldbank.org/INTEAPDECEN/Resources/Chapter-4.pdf [Accessed 1 December 2011]

Jordan, A., Huitema, D. & Van Asselt, H. 2010. *Climate Change Policy in the European Union: Confronting the Dilemmas of Mitigation and Adaptation?* Cambridge, Cambridge University Press.

Legislative Council. 2011. *LCQ3 – Climate Change Consultation and Fuel Mix.* Hong Kong: Government Press Release [Online]. Available: www.info.gov.hk/gia/general/201111/02/P201111020243.htm [Accessed 14 May 2013]

Ma, J. 1997. *Intergovernmental Fiscal Transfers in Nine Countries: Lessons for Developing Countries.* World Bank Policy Research Working Paper, 1822.

Peters, B. G. 1998. Managing Horizontal Government: The Politics of Co-Ordination. *Public Administration,* 76, 295–311

Wellstead, A. M. & Stedman, R. C. 2011. Climate Change Policy Capacity at the Sub-National Government Level. *Journal of Comparative Policy Analysis: Research and Practice,* 13, 461–478

Wollmann, H. 2003. Coordination in the Intergovernmental Setting, in Peters, B. G. & Pierre, J. (eds.) *Handbook of Public Administration,* London, Sage.

Yang, T. 2008. The Implementation Challenge of Mitigating Greenhouse Gas Emissions in the Developing World: The Case of China. *Georgetown International Environmental Law Review,* 20, 1–26

5 Networking for green buildings – why is it so attractive?

We have argued that collaborative municipal networks are configured to integrate sectoral innovations in the broader institutional setup of the environmental state. Regarding green buildings, a pertinent question is how the emerging modes of interaction and collaboration among the different types of actors involved may have influenced the existing environmental governance structure. This chapter analyses the dynamic interactions of driving forces across societal sectors and the patterns of nongovernmental space expansion for the sectoral integration of green buildings to examine:

- how network actors *interact* to integrate green-building innovation into urban-mitigation responses and the environmental state through collaborative municipal networks (CMNs); and
- how the pattern of interactions *evolves*, with changing forms of resource provision and a varying scope of network engagement.

The first part of this chapter is a discussion of the process of integrating green-building innovations into the broader governance context. It starts with an overview of the current state of green-building practices in China and in each of the three cities. To generate more verifiable and valid results, a quantitative analysis of data from certified green buildings is triangulated with a qualitative analysis of semistructured interviews with informants from key organisations participating in the three cities' green-building development. This triangulation analysis sheds light on the roles of different driving forces and their interactive effects supported by the emerging CMNs during the process of integrating green buildings.

Institutionalising green buildings

The progress of green-building practices in China has experienced different paradigmatic stages. Theoretically, the configuration of technical niches precedes the opening up of market niches for the integration of sociotechnical innovations (Lovell 2007; Schot and Geels 2008). With the *technical niches* remaining open but not yet completely filled, the institutionalisation of supportive policies and organisations has created *market niches* for the development of green buildings.

The government has continued to provide 'protected space that allows nurturing and experimentation' with the vanguard technologies and their user practices in green buildings (Lovell 2007: 35). Such technical niches cover architectural initiatives such as applying passive design,[1] innovative cooling and dehumidification systems[2] and 'energy-plus' operations[3] in new government buildings or exemplary construction projects. Meanwhile, green-building practices have gradually become the norm in the architectural engineering industry, which opens up market niches to achieve broader engagement and integration across societal sectors through already stabilised technologies and designs.

In 2005, China hosted an international conference on green buildings, followed by the publication of national green-building standards (Evaluation Standard for Green Building) in 2006. The top-down promotion (i.e. by the national government) of green-building practices started in 2007 with a series of accreditation measures based on a green building labelling (GBL) system. In 2008, the promotion of green buildings was institutionalised through the establishment of the China Green Building Council (CGBC). The CGBC is an affiliate organisation of the Chinese Society for Urban Studies: the former does not have an independent legal personality but is an integral part of the latter.[4] The Chinese Society for Urban Studies is a ministerial-level government-organised nongovernmental organisation (GONGO), and since its incorporation, it has been chaired by the then vice ministers with a housing and construction policy portfolio. The CGBC was initiated under the current leadership of Qiu Baoxin, an incumbent Vice Minister of Housing and Urban–Rural Development who is a professional urban planner. In addition to the quasigovernmental support rendered by the CGBC, a government agency, the Management Office of Green Building Evaluation and Labelling, was also set up to streamline government-led green-building development at the national level.

In Hong Kong, the institutionalisation of green-building practices has seen incremental progress since 1995, a decade earlier than the rapid launch of national policy advocacy. The design of a building environmental assessment method (BEAM) was initially steered by the government (the then Planning, Environment and Lands Bureau, and the Housing Authority) and supported by academics and resourceful private corporations who collectively named themselves the 'BEAM Steering Group'. A BEAM Society was later formally founded in 2002, based on the steering group, to institutionalise professional support for local green-building practices. In addition, the Hong Kong Green Building Council (HKGBC) in 2009 and the China Green Building (Hong Kong) Council (CGBC[HK]) in 2010 broadened the organisational network for mainstreaming green buildings in the city. Subsequent to its initial release in 1996, the BEAM as a local evaluation system for green buildings has been revised a few times, the most recent in 2012. Also in effect from 2012, all newly built public housing development projects are subject to mandatory compliance with green-building standards (Appendix I: 28/HK/27–08–2012).

The action plan for constructing a green building metropolis in Shenzhen was launched in 2008, but it consisted only of noncompulsory action guidelines (Appendix II: Shenzhen Municipal Government 2008a). During the same year,

Shenzhen Green Building Association (SGBA) was formed as a local branch of the national CGBC, and a Joint Session System on Promoting Building Energy Efficiency and Developing Green Buildings was also established, strengthening *intra*governmental coordination for institutionalising green buildings (discussed in Chapter 4). After the 2009 release of green-building evaluation methods tailor-made for Shenzhen, recognised green-building accreditations became mandatory for all of the city's new public housing (mainly the city's indemnificatory housing scheme),[5] effective from 2010 (Appendix I: 20/SZ/23–05–2012; Appendix II: SZ-HCD 2010). Closely following this mandatory requirement, the city's attempt to integrate green buildings was formally institutionalised into local legislation. The Shenzhen Green Building Promotion Measures, drafted by the city government's Housing and Construction Department (SZ-HCD) under a prioritised legislation arrangement, went through public consultation in 2012 and were eventually promulgated the following year.

In Guangzhou, an interdepartmental leadership group on residential building energy efficiency was formed in 2008 and, three years later, was converted to a leadership group on building energy efficiency with the mission of mainstreaming green-building practices in all building types – not only those that are residential. Policy guidelines on green buildings were published by the city government's Urban–Rural Construction Commission (GZ-URCC), with an update specifying the timeframe for municipality-wide implementation (Appendix II: GZ-URCC 2010). In 2012, these voluntary guidelines became mandatory for all new buildings in Guangzhou (Appendix II: Guangzhou Municipal Government 2012). Although this mandatory requirement was a swift governmental announcement, its enforcement feasibility has remained largely in doubt given the lack of established organisational support in the city (Appendix I: 14/GZ/22–02–2012). Institutionalising actions were subsequently requested by the municipal government to pull together local expertise by forming new GONGOs and spurring support for green buildings. In the same year, the Design Guide for Green Buildings in the Guangzhou Area was released as the city's initial attempt to fill the vacuum of area-specific and locally applicable evaluation standards.

Multilevel convergence of green building concepts and standards

The practice of local green building is shaped by adopted concepts and established assessment systems. The following section briefly reviews and compares the green-building concepts and assessment systems adopted and applied in Guangzhou, Shenzhen and Hong Kong. At the moment, the rating systems that have originated globally, nationally and locally are to different extents officially recognised and applicable in the three cities.

Globally transferred practices

The U.S. Environmental Protection Agency defines the practice of green building as: 'creating structures and using processes that are environmentally responsible and resource-efficient throughout a building's life-cycle from siting to design, construction, operation, maintenance, renovation and deconstruction' by reducing the 'overall impact of the built environment on human health and the natural environment' through 'efficiently using energy, water, and other resources' while 'protecting occupant health and improving employee productivity' and 'reducing waste, pollution and environmental degradation' (EPA 2010: n.p.). Based on this concept, a rating system – Leadership in Energy and Environmental Design (LEED) – has been developed in the United States. LEED exceeds resource efficiency and indoor environmental quality, as emphasised in Chinese standards (i.e. the Evaluation Standard for Green Buildings in China and Chinese Green Building Labeling – see next section), by adopting a whole-building approach to green buildings and integrating additional elements including sustainable sites, transportation links, awareness education and design innovation (USGBC 2011).

In Singapore, the Building and Construction Authority (BCA) has issued another rating system for green buildings – the BCA Green Mark, which has five criteria: energy efficiency, water efficiency, environmental protection, indoor environmental quality and other green features/innovation (BCA 2012). Strong emphasis is placed on energy efficiency by giving it significantly more weight in the rating system's point distribution (Liu *et al.* 2010). The BCA Green Mark generally appears to be less sophisticated than LEED because unlike what has been prescribed in LEED, the BCA Green Mark does not consider pollution and emissions, sustainable sites or design innovation as detailed indicators or as broader topics worth discussing (Liu *et al.* 2010). Nevertheless, because the BCA Green Mark is based on a tropical/subtropical building environment, it is regarded by the assessment authorities of Hong Kong, Shenzhen and Guangzhou as compatible and applicable to their local physical conditions.

Locally devised and nationally endorsed practices

As mentioned, green-building practices in Hong Kong were initiated earlier than their national counterparts. The Building Environmental Assessment Method (BEAM) of Hong Kong was first introduced for new and existing office buildings in 1996, and then for new residential buildings in 1999. These early BEAMs were revised in 2004 to extend to all building types at all life stages by integrating the previously separated standards for office and residential buildings into respective manuals for new and existing buildings. Given that these pre-2004 BEAM standards were developed mainly by the private sector and the professional community, the city government's Buildings Department launched an official Comprehensive Environmental Performance Assessment Scheme (CEPAS) for buildings in 2006, which not only adopted a holistic life-cycle approach (including demolition) to building assessment but also incorporated credit weighting for different aspects of assessment to calculate an overall grade – the first such attempt in Hong Kong. The CEPAS covers different environmental concerns in eight categories of building-assessment criteria, including indoor environmental quality, building amenities (management and operational), resources use (materials used, water, waste and energy efficiency), loadings (emission and pollution), site amenities, neighbourhood amenities, site impacts and neighbourhood impacts. The resources use category carries a slightly heavier weighting than the other categories in the overall calculation. This credit-weighting system is critical in differentiating the significance of environmental impacts. The BEAM, which incorporates the credit-weighting system introduced in the CEPAS, was revised as BEAM-Plus in 2010 and updated again in 2012. Six categories are considered in the current version of the BEAM-Plus: site aspects, materials aspects, energy use, water use, indoor environmental quality and innovations/additions. Despite omitting the category of education and awareness, the existing BEAM-Plus categories do converge with LEED specifications. The energy use category is weighted more heavily than the other categories in the assessment of both existing and new buildings.

In 2006, the first Evaluation Standard for Green Buildings in China was launched in Beijing. A rating system for this evaluation standard, the Chinese Green Building Labeling (GBL), was released in 2007. This national standard and rating system emphasises the core concept of 'four savings (energy, land, water and materials) and one environmental protection' (Appendix II: Ministry of Housing and Urban–Rural Development [MOHURD] 2006). However, it has limited scope regarding building type in that it merely distinguishes between residential and public buildings but neglects the drastically different needs of new and existing buildings. The life-cycle and whole-building approaches are overlooked, given the heavy focus on resource efficiency (i.e. the 'four savings'). The GBL rating system does not incorporate credit weighting, and it is therefore viewed as relatively undeveloped by practitioners (architects, structural engineers and local government officials in charge of green-building enforcement) and academic researchers (Appendix I: 20/SZ/23–05–2012; Zhang and Lau 2010). Nevertheless, based on the still-elementary national standard and rating system, localised versions have been released in various Chinese provinces and cities. In 2010, the Green Building Evaluation Standard (Hong Kong version) was compiled, the Evaluation Standard for Green Buildings in Guangdong was published in 2011 and the Design Guide for Green Buildings in the Guangzhou area was introduced in 2012. A Hong Kong edition of the Chinese GBL was also launched in March 2012.

Although there has been a large pool of green-building assessment methods in Hong Kong, the BEAM-Plus as updated in 2012 is the most commonly recognised by the local government. The Chinese GBL (HK) is not officially endorsed by the Hong Kong government due to the 'One Country, Two Systems' constitutional principle of the special administrative region (Appendix I: 28/HK/27–08–2012). Nevertheless, new public-housing projects in Hong Kong under the mandatory green-building requirement can be accredited by either BEAM-Plus or the Chinese GBL (HK). Given the Closer Economic Partnership Arrangement (CEPA) between Hong Kong and China, BEAM-Plus and the Chinese GBL (HK) also apply to the construction projects of Hong Kong developers in mainland cities, which expands the applicability of Hong Kong's locally devised standards.

As an alternative to the national evaluation standard, Shenzhen has compiled its own municipal Evaluation Specifications for Green Buildings, published in 2009 (Appendix II: Shenzhen Department of Quality and Technical Supervision 2009). Although the publication of the Shenzhen standard came three years later than that of its national counterpart, Shenzhen's construction authority actually started compiling its standard earlier than the national government because MOHURD accelerated its compilation so that the national standard could be completed within a short time and released in 2006. However, Shenzhen's slower compilation had one clear advantage – that is, it allowed more time for overseas research visits, which has enabled the Shenzhen standard to follow international benchmarks (mainly LEED and BCA Green Mark) more closely than its national counterpart in terms of evaluation items and assessment criteria (Appendix I: 20/SZ/23–05–2012). Yet in practice, the Shenzhen standard as a set of locally devised practices covers a limited geographical area confined within the Shenzhen municipality, and thus it has not been widely accredited in construction projects.

We have seen that green-building practices in Hong Kong, Shenzhen and Guangzhou are influenced by locally devised methods, a nationally endorsed system and internationally accredited standards (LEED and BCA Green Mark). The convergence of these green-building concepts and accreditation standards leads to the empirical basis for analysis in the next section of this chapter.

Innovation and governance in expanding nongovernmental space

Mainstreaming green buildings: the key actors

Core governmental and newly emerging nongovernmental actors are the prime participants in the process of mainstreaming green-building practices in the three cities under discussion.

The core governmental actors are the major steering agents in the *intra*governmental management of sectoral mitigation options. As discussed in Chapter 4, the Joint Session System on Promoting Building Energy Efficiency and Developing Green Buildings in Shenzhen was incorporated internally (within the municipal government) in 2008, whereas in Guangzhou, the Leadership Group on Building Energy Efficiency was reformed in 2011 to include green buildings on its agenda.

In Hong Kong, several governmental departments[6] have policy portfolios on and direct involvement with green buildings. Given the Inter-departmental Working Group on Climate Change's ineffective coordination (Chapter 4), policy functions for green buildings have been fragmentally spread across these departments without an institutionalised interdepartmental steering and coordination body. The Council for Sustainable Development (SDC) once planned to set up an interdepartmental group encompassing all of the relevant departments and bureaus in green-building development, but the attempt failed due to insufficient attention from the political elite (Appendix I: 26/HK/29–08–2012, 28/HK/27–08–2012). Eventually, in 2013, an interdepartmental Steering Committee on the Promotion of Green Buildings was proposed in the Chief Executive's annual policy address and chaired by the Secretary for the Environment. This committee is intended to coordinate relevant government departments and engage with nongovernmental stakeholders in the relevant sectors (Chapter 4).

The actors newly emerged in the nongovernmental space are valuable resources with whom the established *intra*governmental coordination mechanisms within the city governments may seek to collaborate to enhance capacity in the green-building integration process. The nongovernmental actors are connected with the issue at varying levels of intensity, from being core actors to staying on the periphery. The core (of more intensive connectivity) and peripheral (of less intensive connectivity) actors in green-building development in Hong Kong, Shenzhen and Guangzhou are graphically depicted in Figure 5.1.

Tier-1 actors are the core nongovernmental actors in the green-building mainstreaming process;[7]

Tier-2 actors are supportive professional associations in the construction sector, including associations that promote the transfer and integration of sectoral innovation,[8] organisations concerned with land and real estate development[9] and professional building consultant bodies;[10]

Tier-3 actors are local environmental nongovernment organisations (ENGOs) campaigning for awareness of the broader issues of environment and climate change.[11]

Figure 5.1 tries to capture the dynamic process of nongovernmental space expansion, in which key nongovernmental actors are actively networking for larger roles and for broadened engagement in the governance of sectoral innovations. Core nongovernmental actors (Tier 1) directly interact with the *intra*governmental coordination structures in the respective city governments and take up boundary-spanning roles across the three tiers to expand the cross-sector protected space of the sectoral mitigation options (6 *et al.* 2006; McGuire and Silvia 2010; Schot and Geels 2008). In this process, the core actors seek to reach out to the peripheral actors in Tier 3, who are capable of generating wider societal support. Collaboratively, actors in multiple tiers and inside the government may ultimately achieve sociotechnical transitions (Hodson and Marvin 2010; Zeemering and Romero 2012). Meanwhile, boundary-spanning activities are taking place horizontally among the three cities to maintain broader communication and resource-exchange networks to bargain for stronger influence on their respective locally devised innovation practices.

Driving the expansion of nongovernmental space

The driving forces of nongovernmental space expansion are found in the public sector (government agencies), the private sector (businesses) and the professional community (academia and professional associations), giving rise to government-, business- and professional-led nongovernmental space expansion.

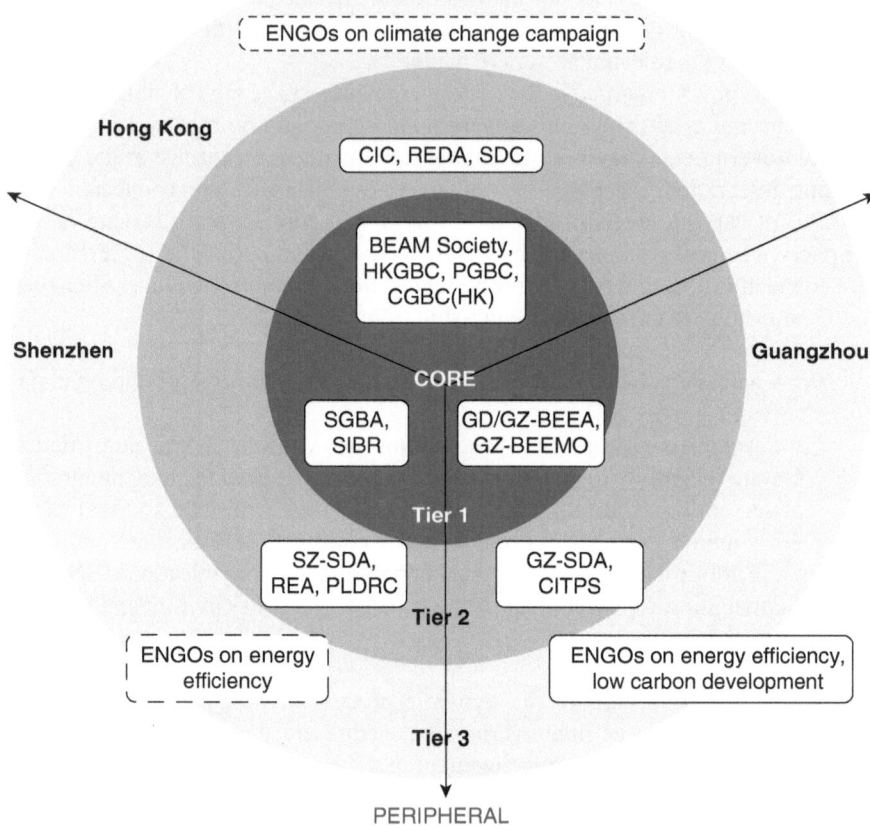

Figure 5.1 Multitier actors in the nongovernmental space for green building
development

Source: Compiled by authors

Government-led nongovernmental space expansion

Through a January 2012 policy directive from the municipal government, Guangzhou has put in place a mandatory green-building requirement for all new building projects in the municipality (Appendix II: Guangzhou Municipal Government 2012). To speed up the implementation of green buildings in Guangzhou, the policy directive sought organised support from the nongovernmental space. The organising of the Guangzhou Association of Energy Efficiency and Building Technology (GZ-BEEA in Figure 5.1) began in March 2012, only 2 months after the policy directive, and it was officially formed in June of the same year. Also in March 2012, the first Expert Committee on Buildings Energy Efficiency was incorporated by the GZ-URCC to provide government-appointed consulting for green-building development.

Due to the impetus provided by the Guangzhou municipal government, there has been swift action in the expansion of nongovernmental space in Guangzhou. The output legitimacy of collaborative arrangements, when referring to effectiveness in problem-solving capacity, consists of two dimensions: i) whether the expected environmental and development outcomes are reached and ii) whether there is an adequate and effective institutional design to reach those outcomes (Bäckstrand 2006). With the upfront regulatory approach imposed on the construction sector, it appears that high *output legitimacy* is feasible in green building mainstreaming, given that the core participating agencies (such as the GZ-BEEMO in Figure 5.1) are granted mandated enforcement authority, which has improved their ability to reach the expected environmental outcomes. In addition, the institutional feasibility is inherent in the newly established GONGOs (the GZ-BEEA and the expert committee) formed by a group of like-minded experts working closely with the municipal party-state system – the GZ-URCC, the mayor's office and the city's party leadership. However, because these nongovernmental actions are called on and indirectly controlled by the local government, the *discursive* and *input legitimacy* have not been widely established among the citizens and the professional communities of the building and construction sectors. In this sense, the newly established GONGOs must behave proactively to broaden the organisational support and mainstream green buildings in the city through the CMNs.

Business-led nongovernmental space expansion

The Business Environment Council (BEC) has driven the BEAM standards, the BEAM Society and the HKGBC in Hong Kong. The BEC is a nongovernmental organisation set up by the private sector in 1989 to promote corporate sustainable development and environmental responsibility. It has pushed forward various business-led environmental initiatives. In 1995, it drew together the private sector, the government (the then Planning, Environment and Lands Bureau), the nongovernmental space (the REDA in Figure 5.1) and academia (Hong Kong Polytechnic University) to form a steering group on the BEAM standards, which launched the first BEAM green-building standards in 1996. As an informant who was involved in the initial set-up stage of the BEAM standard recalled:

> It was purely an initiative driven by the private sector: only a few people of the Housing Authority participated in the initiating process representing the government side, while several major property developers had teamed up – including Swire Properties, Hong Kong Land, and Sun Hung Kei Development – with the vision to establish a standard belonging to ourselves, belonging to Hong Kong.
>
> (Appendix I: 29/HK/22–10–2012)

In 2002, the steering group was eventually formalised into the current BEAM Society (Figure 5.1), organised on a membership basis. To increase public awareness and knowledge of green buildings, the BEC again joined the BEAM Society and two other sectoral associations – the Construction Industry Council (CIC in Figure 5.1) and the Professional Green Building Council (PGBC in Figure 5.1) – to establish a new nongovernmental organisation, the Hong Kong Green Building Council (HKGBC in Figure 5.1). The BEAM Society is currently the owner of the BEAM standards and the HKGBC is the training and accreditation body of BEAM professionals. Even after these new nongovernmental organisations related to green building were officially incorporated, the BEC has continued to provide resource support and indirectly advocate for integrating the BEAM standards into local green-building assessment (Appendix I: 26/HK/29–08–2012).

In this regard, the development of green buildings in Hong Kong is closely intertwined with raising awareness and taking pioneering steps in the private sector, which has led to the open and still-expanding nongovernmental space for green-building sectoral integration. Thus, the *discursive legitimacy* of integrating green-building innovations in CMNs is strong, with the expansion of nongovernmental space driven by the private sector. Although the *output legitimacy* of innovation transfers has been incremental, the expanded nongovernmental space in Hong Kong has ensured the diverse representation of organised interests in the relevant sectors and among the stakeholders, which has generated *input legitimacy* in the coordination and collaboration processes of the CMNs to integrate the green-building sectoral innovation into the broader governing of climate change mitigation.

Professional-led nongovernmental space expansion

Shenzhen Green Building Association (SGBA in Figure 5.1) was first collaboratively initiated in the academic community and then formalised with the final endorsement of the municipal government. It was institutionalised initially through a proposal compiled by the Shenzhen Institute of Building Research (SIBR in Figure 5.1), a think tank with expertise in urban planning, construction engineering, architectural design and environmental protection. The proposal was then submitted by the SIBR and co-endorsed by business entrepreneurs, experts and scholars from more than 50 organisations who, together, formed an initiating group to reach the eligibility requirement for setting up a new social organisation in China. Taking the opportunity to fill the vacuum of locally based green-building associations and perform as a local extension of the newly established China Green Building Council (CGBC), the proposal soon gained the endorsement of the heads of the Shenzhen Housing and Construction Department (SZ-HCD) who were committed to carrying the initiative through the tedious application process and complicated licensing procedures for new social organisations. The committed participation of SZ-HCD made a real difference in overcoming major obstacles in the application process, such as negotiating with the local Civil Affairs Department – which holds licensing authority – over the questions of whether green building had become an independent industry and whether the proposed organisation could represent this industry given the existing associations in architecture, surveying and design (Appendix I: 25/SZ/18–06–2012).

Through the process of professional-led nongovernmental space expansion in Shenzhen, a moderate level of *input legitimacy* has accumulated in the city's sectoral mitigation responses to climate change, with established forums in the SGBA for multisectoral collaboration and information sharing. The SGBA, as a core network participant in the CMNs, has obtained high *discursive legitimacy* within the city's epistemic community comprised of concerned practitioners and like-minded experts closely connected to the local government.

To promote and implement the Green Building Labelling (GBL) certification standards in Hong Kong, the China Green Building (Hong Kong) Council (CGBC(HK) in Figure 5.1) was formed by three academics affiliated with three different local universities and specialising in architecture, structural engineering and town planning, respectively. The organisation is professionally oriented towards building and construction assessment methods and technologies. It maintains a close relationship with the CGBC national headquarters in Beijing endorsed by the Vice Minister of Housing and Urban–Rural Development, Qiu Baoxing. Qiu is a well-known technocrat among China's administrative elites. With expertise in economy and urban planning, he has supported various initiatives to mainstream sustainable development practices in China (Appendix I: 25/ SZ/18–06–2012, 28/HK/27–08–2012).

One key aspect of green-building assessment under BEAM-Plus or the GBL in Hong Kong is the *supplementary human resources* support provided by accredited BEAM professionals or GBL managers. BEAM professionals are trained and accredited by the Hong Kong Green Building Council (HKGBC), and their role is to join and advise construction project teams on how to meet BEAM-Plus standards and eventually facilitate the submission of green-building assessment data to BEAM assessors. The training of BEAM professionals was introduced to the industry in April 2011 with the launch of the BEAM-Plus standards. Experienced professionals in the area of built environment are eligible to apply for BEAM professional accreditation (training and examination). Currently more than 2,000 registered BEAM professionals are based in Hong Kong serving as practicing engineers, surveyors, architects, construction managers and environmental consultants in government and private firms. Conversely, GBL managers are trained and accredited by the CGBC(HK). With a similar role to that of BEAM professionals, GBL managers are consultants in the GBL assessment process. The accreditation scheme for GBL managers was introduced in September 2011, with the first batch of 100 professional practitioners qualifying in 2012 – a smaller group than the BEAM professionals.

The training of BEAM professionals and GBL managers is tailored to the specific circumstances of Hong Kong. In Guangzhou and Shenzhen, similar functions are performed by engineers and architects employed as authorised personnel within public institutions ('*shiye danwei*'), typically the government-funded architectural practice and research institutes (Appendix I: 28/HK/27–08–2012). Hong Kong, however, lacks the institutional support to provide human resources for conducting consultations and documentary preparation for green-building assessment (Appendix I: 28/HK/27–08–2012). To fill this gap, specialised personnel are recruited from different governmental departments and from relevant firms in the private sector. They go through technical training provided by the HKGBC, based on BEAM or the CGBC(HK), to be accredited to support the green-building evaluation in each underassessed project, which connects the professionals and practitioners in the field by providing them with an ongoing networking platform during the integration of green building practices. Thus, the political space for nongovernmental participation has been expanded.

Contextualised driving forces

Having identified and distinguished these three types of driving forces, one finding is pertinent: different driving forces tend to agglomerate in different types of municipal authorities, leading to different levels of publicness in the policy environment of sectoral innovations. The analysis of nongovernmental space captures the notion of *environmental publicness* measured by the aggregate publicness of organisations in the multilevel governance environment, that is, the relative proportions of public agencies, nonprofit organisations and privately owned organisations in the policy area of climate change (Miller and Moulton 2013). Given the different compositions of driving forces in the three cities, their interactive processes for green-building integration are contextualised with various levels of environmental publicness. The public sector is mostly at work in Guangzhou to drive nongovernmental space expansion (high environmental publicness), the private sector is strong in Hong Kong (low environmental publicness) and the professional community has a strong effect in Shenzhen (medium environmental publicness) and a somewhat weaker effect in Hong Kong. If Guangzhou has the highest level of authoritarianism, Shenzhen is more of a mixture of entrepreneurship and pluralism and Hong Kong is the most liberal of the three cities. The variations in these three types of municipal authority provoke a question: How do these different extents of publicness in the CMNs' system contexts and networking drivers shape the sectoral integration process of green building? The quantitative analysis in the third section of this chapter is thus necessary to further inquiry.

Networking in the expanded nongovernmental space

The previous section identifies the nongovernmental space as providing a legitimation environment and generating additional resources for CMNs' innovation transfers for climate mitigation. This section discusses how diversity in nongovernmental space facilitates interactions integrating different CMNs' segments (Gilchrist 2009). The interactive dynamics of governance networks are presented in the boundary-spanning activities among the core and peripheral network actors (listed in Figure 5.1) within the expanding nongovernmental space. These interactions meet the ends of driving policy development, making public impact and encouraging knowledge exchange.

First, for the *policy end*, interactions are activated between the *government* and *policy entrepreneurs* in the nongovernmental space. The HK3030 Campaign is an initiative proposed and advocated by the HKGBC, with a set of strategies to be adopted by both the public and private sectors to reach the vision of a low carbon built environment in Hong Kong; for example, a 30 percent absolute reduction in buildings' electricity consumption by 2030 (with 2005 as the baseline). After collecting stakeholder opinions, the HKGBC finalised the HK3030 proposal towards the end of 2012 with an attempt to make actual policy changes (Appendix I: 30/ HK/16–11–2012). Soon after HK3030 was finalised, the Steering Committee on the Promotion of Green Buildings was set up by the government in January 2013. One of this committee's integral missions is to 'formulate implementation strategies and action plans' to mainstream green buildings 'in both public and private sectors' (Appendix II: Environment Bureau 2013). At that point, the policy window (Kingdon 1995) opened for nongovernmental input from policy entrepreneurs trying to drive policy change for mainstream green-building practices. The HKGBC has been engaged in this *intra*governmental coordination. The organisation representatives were invited to present their HK3030 vision and implementation plan to the first meeting of the Steering Committee (Appendix I: 30/ HK/19–11–2013). Today's policymakers in the Hong Kong government are 'well aware' of this nongovernmental initiative (Appendix I: 30/HK/19–11–2013). Although real policy change has not yet emerged, interactive dynamics have been observed in the networking between the *intra*governmental coordination mechanism (Chapter 4) and the nongovernmental space.

Second, for the *impact end*, constant interactions appear between the construction and building *sectors* and green building *advocates* in the nongovernmental space. One of the primary missions of most newly formed nongovernmental organisations promoting green buildings is to attain wider industry recognition for the green-building standards they advocate. The basic task of CGBC(HK) is to promote the application of GBL in construction projects. Because the GBL is not an officially endorsed green-building assessment standard in Hong Kong, the CGBC(HK) invests its limited resources in training GBL managers to expand the reach of the GBL method across the industrial sector in Hong Kong.

This *impact end* of interactions is also found in *organisational ties* with the general *public*. The core nongovernmental actors are engaged in awareness-raising activities to mainstream green building. The SGBA has been proactive in organising various activities for both the general public and its own members in Shenzhen. One of its founding visions is to operate independent of government control, with self-sufficient resource regeneration and nonprofit service provision in mainstreaming and publicising green-building practices (Appendix I: 25/ SZ/18–06–2012). A major activity hosted annually by SGBA is the Green Buildings Summer Camp for University Students, which combines lectures on green-building concepts with study tours to green-building projects in the city.

With close networking in the local business sector and the official endorsement of the Buildings Department on the BEAM standards, the HKGBC is capable of gaining access to media attention, resources and projects awarded green-building certificates. Through a committee dedicated to public education, the HKGBC is involved in building Hong Kong citizens' awareness of how their actions may affect the building environment and even their quality of life. HKGBC has built its reputation in the community through a school competition (innovative ideas about green space) for primary- and secondary-school students, training work-shops (on green-building concepts and methods) for teachers in the schools par-ticipating in the competition, summer workshops for secondary-school students, talks and lectures to students in tertiary institutions and tours of development proj-ects in Hong Kong and the PRD that have been granted green status. All of these activities are conducted with the support of the Education Bureau, the Buildings Department and the Construction Industry Council (CIC), who provide funding, venues and accreditation.

Third, interactions are also configured in a *diversified platform of knowledge management* embedded in the expanding nongovernmental space. The knowl-edge platform serves to encourage dynamic learning among engineers and archi-tects through site visits planned by the core nongovernmental organisations. In Hong Kong, sponsored by the CIC and the Professional Green Building Council (PGBC), the HKGBC regularly calls on members to visit local green building projects such as the Zero Carbon Building and projects that have received the Green Building Award[12] to share the implementation experiences of local green building standards across the industry. In contrast, the CGBC(HK) board mem-bers as professional assessors pay regular visits to construction sites in other Chi-nese cities in the company of the CGBC(HK) secretariat staff and accredited GBL managers to create opportunities for the on-site exchange of innovative ideas and mutual learning (Appendix I: 28/HK/27–08–2012).

SGBA is active in providing industrial support to Shenzhen's small to medium-sized enterprises in the green-building sector by organising study tours with vanguard enterprises, which enables learning and interaction between emerging enterprises and their established counterparts on how to survive in this burgeoning industry (Appendix I: 25/SZ/18–06–2012). Visits to local green-building demon-stration projects in newly developed districts in Shenzhen are organised by the SGBA for its own members, for members of the Real Estate Association in Shen-zhen and for experts travelling from other places to enhance communication with local officials responsible for land planning (Appendix I: 25/SZ/18–06–2012).

Comparatively, the nongovernmental space is less robust in Guangzhou, but visits to local demonstration projects still serve as a means for government offi-cials and property developers to exchange ideas and engage a broader range of interested nongovernmental actors (Appendix I: 14/GZ/22–02–2012).

Dynamic interactions of green building drivers

The qualitative analysis of the forces driving the expanded integration of green building invites questions regarding how the innovation integration actually works. Various factors shape the process of transferring green-building practices. To probe the actual interactive dynamics of integrating green-building innovation with an interconnected array of actors requires quantitative analysis of data on certified green buildings, which reveal the causal path of green-building development.

Measuring dynamics: data collection and rationale

Quantitative statistical analysis is triangulated with the case study in this chapter's analysis. The data for statistical analysis were retrieved from various databases and government documents related to green building certification. The major assessment tools applicable to the three cities[13] are covered in the dataset. Some assessment tools – such as LEED, BEAM and BEAM-Plus – have published detailed information about previously certified green building projects on their own websites, which is easily accessed and collected.[14] However, other assessment tools applicable to these three cities – including the Chinese GBL, the Chinese GBL (HK), the Evaluation Standard for Green Buildings in Guangdong and the Shenzhen Evaluation Specification of Green Buildings – do not provide consolidated information on all certified projects. Documents have been published by the assessment authority, disclosing information about buildings certified during the specified time range. For example, the Chinese GBL assessment authority published 21 separate documents in 2011 (and 23 in 2012) informing the general public which buildings had been certified. Given this situation, all related documents were collected and the information on each certified project in the three cities was screened and entered manually into the dataset. The consolidated data provided by LEED and BEAM had some missing values in essential variables, for which information was supplemented from alternative sources. The finalised dataset covered green-building certification information from 1999 to February 2013. It contained comprehensive information including project name, address, city location, developer and owner, project end-use type,[15] assessment tool adopted, rating (provisional and final), years of certification, newly built or existing buildings and gross floor area (with units converted to square meters) assessed.

Two variables derived from basic information about building projects were introduced into the dataset to facilitate the analysis: the *urbanisation scale* of project location and the *publicness score* of driving organisations. The urbanisation scale of project location is classified by urban center, 'rurban' space and peri-urban space. 'Rurban' space is an area that 'generally lies between a city's business district and the surrounding residential districts' and has been marginalised in the urbanisation scale due to its relative inaccessibility to some central transportation point (Firey 1946; Land 1967). Peri-urban space – newly developed zones designed to further expand the city's infrastructural and residential development given the decline of agricultural and rural employment opportunities (Allen 2003) – is more distant from the urban center than rurban space. The urbanisation scale of project siting presumably affects the green-building assessment process and outcomes. There is more potential space for green building construction in more remote urban areas. New development districts (such as Pingshan and Guangming New Districts in Shenzhen, and Nansha New District in Guangzhou, shown in map, Figure 1.1) are often located in rurban or peri-urban spaces to release land and population pressures in the urban center and experiment with green-building technologies and designs. In the dataset, different values are given to projects sited in locations with different perceived urbanisation scales; that is, a value of 3 is given to an urban center, 2 to a rurban space and 1 to a peri-urban space. Thus, the resulting variable is *ordinal.*

The publicness score is assigned based on the classifications of driving organisations. Organisational publicness is determined by the concepts of political and economic authority and shaped by the multilevel context of policy environment (Bozeman and Bretschneider 1994; Miller and Moulton 2013). Thus, the driving organisations are classified into seven types: government, government-led body/ corporate (including statutory bodies, semipublic organisations, state-owned enterprises, public institutions and shareholding companies with the government as the major or sole shareholder), public–private partnerships, publicly held private corporates (publicly listed markets), privately held companies, nongovernmental organisations and tertiary-education institutions. Each type is given a different numeric value to calculate the publicness score,[16] which can be measured as an *interval scale* variable. The publicness score variable is a key independent variable in this analysis because it reflects how the different patterns of organised driving forces identified in the second section affect the integration level of green buildings (i.e. the drivers of sectoral integration).

The data collected from different assessment authorities need to be effectively combined into a mutually compatible dataset. Essentially, ratings conferred by different assessment tools are in different forms, but they can be made equivalent and convertible. For example, 'excellent' in older versions of BEAM is equated with 'platinum' in BEAM-Plus and LEED, and 3 stars in Chinese GBL is equivalent with Level 3 in Guangdong's Evaluation Standard for Green Building, which is stipulated in relevant policy documents. Therefore, after converting the rating into a value with a unified benchmark, the 'rating' variable is one of the key dependent variables in this dataset.

The variables were identified and the hypotheses were established during data preparation. Table 5.1 summarises both of the variables analysed and the hypotheses tested in this dataset.

Analysis and results of dynamic interactions

There is a mix of drivers in green-building development across the three cities. Figure 5.2 illustrates the composition of drivers in the three cities' certified green-building projects. The private sector alone takes up 54 percent of all of the certified green-building projects, while the public sector (including government and government-related body corporates) assumes another 37 percent. However, the professional and nongovernmental sectors account for only 5 percent of the green-building projects, perhaps because the actors in these two sectors do not possess the political power and financial resources needed to launch major construction projects. The last 4 percent of the green-building projects are jointly undertaken by the public and private sectors.

The three cities also have varied green-building development paces, as shown in Figure 5.3. Beginning in 1999, even though Hong Kong had the largest cumulative number of certified green buildings, Shenzhen experienced steep upward

Table 5.1 Summary of variables and hypotheses

	Independent variables		Dependent variables
	Explanatory variable	*Intervening variables*	
Variables	Project Drivers: 1) Publicness score of driving organisations;	Project Location: 2) City; 3) Urbanisation scale; Project Type: 4) New/existing building; 5) Project end-use type; Project History: 6) Years of certification	Policy Choice: 7) Assessment tools adopted; Quantitative Output: 8) Gross floor area assessed; Qualitative Outcome: 9) Assessment rating.

Hypotheses

H1: If the publicness of project drivers directly or indirectly affects the project owners' selection of assessment tools, then the pattern of organisational involvement shapes the decision making in the green-building assessment process.

H2: If higher publicness of project drivers directly or indirectly leads to larger gross floor area assessed, then the public sector's stronger involvement improves the quantitative outputs of green building mainstreaming.

H3: If the green building assessment rating increases due to higher publicness of project drivers, then the public sector's stronger involvement in the green building project development processes enhances the qualitative outcomes of green-building integration.

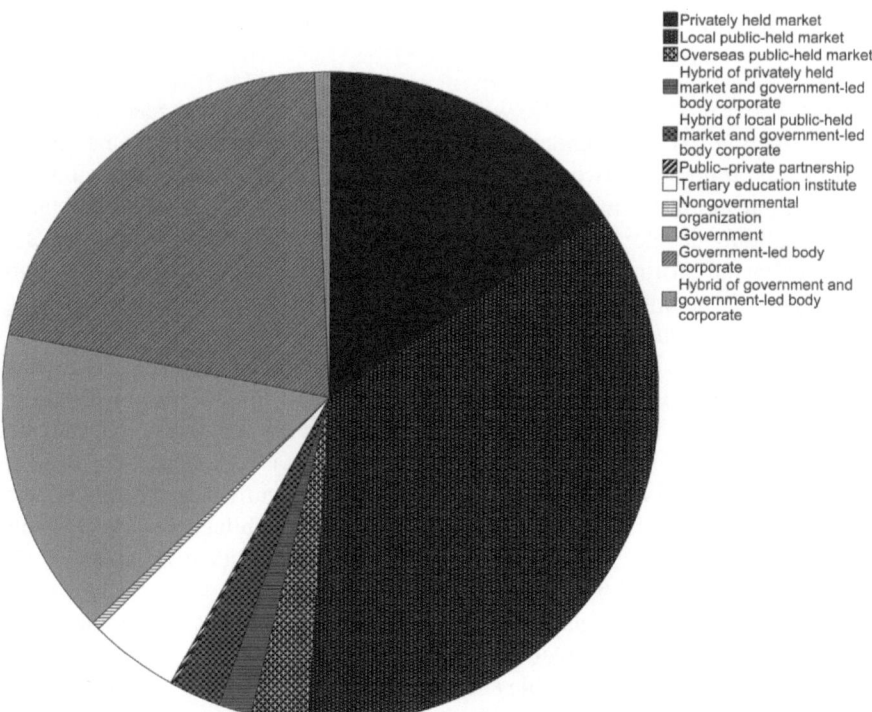

Privately held market
Local public-held market
Overseas public-held market
Hybrid of privately held market and government-led body corporate
Hybrid of local public-held market and government-led body corporate
Public–private partnership
Tertiary education institute
Nongovernmental organization
Government
Government-led body corporate
Hybrid of government and government-led body corporate

Figure 5.2 Composition of cross-sector drivers in green building projects
Source: Compiled by authors

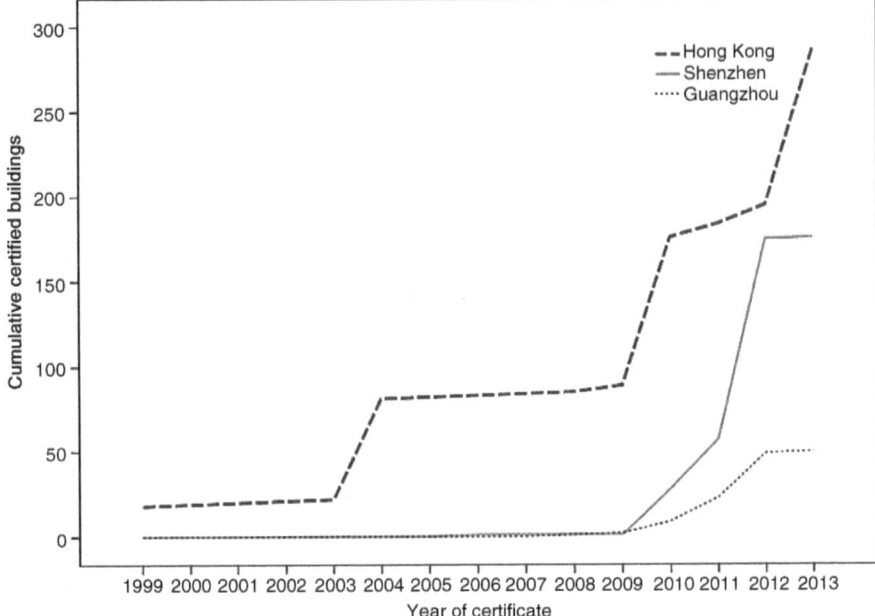

Figure 5.3 Cumulative certified green buildings in Guangzhou, Shenzhen and Hong
      Kong

Source: Compiled by authors

growth between 2009 and 2012, with a total cumulative number reaching that of
Hong Kong in 2012. The impact of publicness is possibly affected by the municipal factors, in terms of two loci: first, the *political* locus of municipal authority,
considering the different policy interventions for and political commitments to
green-building development in the three cities; and second, the *physical* locus of
cities given the existing development landscape and trajectories of green buildings.

    Judging from this initial analysis, it is pertinent to analyse the following:
whether *level of publicness* is the key explanatory factor in the three cities' different levels of green-building development, whether it has a direct or indirect effect
on the dependent variables and how the intermediate causal mechanism affects
these effects. Given these questions, Table 5.2 illustrates the descriptive statistics
and correlations of variables to be analysed in this dataset. The 'city' variable, as
a controlled condition of the dependent variables, is significantly correlated with
most of the other tested variables in the dataset. The 'publicness score' variable,
however, is *insignificantly* correlated with two of the dependent variables yet *significantly* correlated with most of the intermediate variables.

    To probe how the correlated intermediate variables change depending on the
publicness of the project drivers, a flipped-over one-way ANOVA was conducted
on the 'publicness score' variable and the intermediate variables were found to be
significantly correlated with it (Table 5.3).

Table 5.2 Descriptive statistics and pearson correlation matrix (for the entire set of tested variables)

Variables	Mean	Standard deviation	1	2	3	4	5	6	7	8	9
1. Publicness Score of Driving Organisations	2.876	1.3718	1.00								
2. City	1.55	0.670	−0.002	1.00							
3. Urbanisation Scale	2.36	0.733	−0.287***	−0.401***	1.00						
4. New/Existing Building	0.82	0.387	0.306***	0.159***	−0.356***	1.00					
5. End-use Type	3.97	2.717	0.766***	−0.083†	−0.364***	0.414***	1.00				
6. Years of Certification	2.82	3.373	−0.001	−0.307***	0.235***	−0.320***	−0.142	1.00			
7. Assessment Tool Adopted	2.33	0.876	−0.300***	0.313***	−0.535***	0.360***	0.333***	−0.169***	1.00		
8. Gross Floor Area Assessed	74069.08	94147.28	−0.020	0.278***	−0.166***	0.154***	−0.037	−0.097*	0.245***	1.00	
9. Assessment Rating	2.862	1.1215	−0.041	−0.531***	0.481***	−0.191***	−0.144***	0.269***	−0.442***	−0.233***	1.00

Note:

$N = 498$

† $p \leq 0.10$

* $p \leq 0.05$

** $p \leq 0.01$

*** $p \leq 0.001$ (2-tailed)

Table 5.3 One-way ANOVA of key variables (by city)

		2. City	3. Urbanisation scale	4. New/existing building	5. Project end-use type	7. Assessment tool adopted
				F Statistics		
1. Publicness score of driving organisations	Hong Kong	n/a	27.608***	38.331***	113.531***	11.477***
	Shenzhen	n/a	19.488***	3.511†	42.810***	9.383***
	Guangzhou	n/a	5.406**	11.254**	2.347†	8.763**
	All 3 cities	4.470*	22.381***	51.195***	124.427***	22.152***
6. Years of certification	Hong Kong	n/a	3.005†	25.061***	5.297***	3.182*
	Shenzhen	n/a	1.34	0.015	2.155*	31.583***
	Guangzhou	n/a	0.202	5.957*	0.361	0.992
	All 3 cities	33.511***	14.463***	56.759***	6.449***	26.693***

Note:

$N = 498$

† $p \leq 0.10$

* $p \leq 0.05$

** $p \leq 0.01$

*** $p \leq 0.001$

To further investigate the flow of causal impact, various types of regression analyses were conducted on the intermediate and dependent variables, given their measurement levels. To test the causal relationship of the publicness of project drivers with the urbanisation scale of the project location, ordinal regressions were conducted regarding the ordinal nature of the variables (urban center = 3; rurban space = 2; peri-urban space = 1). Table 5.4 indicates the results of these ordinal regressions, in which the partial variance (26.3 percent) of the project location's urbanisation scale can be explained by the publicness of the project drivers, given the respective municipal location and whether the building is newly built or already existing.

A multinomial logistic regression was conducted to see whether the publicness of driving organisations determines the end-use type of the green-building projects. The analysis results are presented in Table 5.5. The publicness of project drivers – together with the project location and whether the building is newly built or already exists – produces a valid explanation on 47.4 percent of the variance in the project end-use type.

Indirect effect of publicness on assessment tool selection (H1)

With respect to the nominal nature of the variable regarding the selected assessment tool, a multinomial logistic regression was conducted for this specific dependent variable as the indicator of the political influence of each assessment authority. The analysis results are demonstrated in Table 5.6, in which the two project location variables are causally significant in assessment tool selection, with 53.5 percent of the variance explained. Although the publicness of project drivers does not directly affect the choice of assessment tool, it has an indirect effect on it through an intermediate variable of the urbanisation scale of project location, with the testified direct causal effect of publicness on urbanisation scale in project location (see Table 5.4). The combined results, presented in Tables 5.6 and 5.4, indicate that the publicness of project drivers indirectly affects the selection of the assessment tool(s). This suggests that the pattern of organisational involvement in project development shapes the decision-making process in green-building assessment through the siting of construction projects.

Limited effect of publicness on gross floor area assessed (H2)

To establish the causal path leading to the quantitative output and physical landscape of green-building development indicated by the gross floor area assessed, a multiple-regression analysis was run and the results summarised in Table 5.7. The volume of gross floor area assessed was causally related to the city location and the project type. However, the publicness had only a minimal and indirect effect on the assessed gross floor area, through the two project type features (see Tables 5.3 and 5.5). This argues against the idea that the public sector's stronger involvement may improve the quantitative output of mainstreaming green building as measured by the assessed gross floor area.

Table 5.4 Ordinal regression on the urbanisation scale of the green-building project location

Variables	Parameter estimates	
	Exploratory model (all independent variables)	Final model (after manual backward elimination)
DV: Urbanisation Scale		
Peri-urban Space	−17.951	−0.404
Rurban Space	−15.448	2.07***
IV: Publicness Score of Driving Organisations		
[Publicness = 1.0]	1.711***	1.794***
[Publicness = 2.0]	1.474***	1.51***
[Publicness = 2.5]	1.704*	1.842*
[Publicness = 3.0]	2.061*	2.056**
[Publicness = 4.0]	1.156***	1.201***
[Publicness = 4.5]	−18.238	−20.139
[Publicness = 5.0]	0[a]	0[a]
IV: City		
Hong Kong	1.092*	1.515***
Shenzhen	−1.011*	−0.959**
Guangzhou	0[a]	0[a]
IV: New/Existing Building		
Existing Building	2.429***	2.386***
New Building	0[a]	0[a]
IV: Years of Certification		
[Years = 0]	−17.031	-
[Years = 1]	−17.396	-
[Years = 2]	−18.083	-
[Years = 3]	−17.026	-
[Years = 4]	−1.228	-
[Years = 5]	−0.31	-
[Years = 7]	−17.162	-
[Years = 9]	−17.24	-
[Years = 14]	0[a]	-
Goodness-of-Fit: Pearson chi-square	179.25**	97.349***
Pseudo R-square (McFadden)	0.275	0.263

Note:

$N = 498$

[†] $p \leq 0.10$

[*] $p \leq 0.05$

[**] $p \leq 0.01$

[***] $p \leq 0.001$ (2-tailed)

[a] Parameter is set to zero

Table 5.5 Multinomial logistic regression on project end-use types

Independent variables	Coefficients: end-use type[a]							Log likelihood
	Commercial	Hotel	Private residential	Mixed use	Industrial	Infrastructure	Public residential	
Intercept	-6.563	-11.063	-1.233	-8.577	-16.510	-19.151	-7.573	283.250***
Publicness Score of Driving Organisations								681.502***
[= 1.0]	23.193	24.047	16.165	21.276	21.921	20.312	1.891	
[= 2.0]	17.898	17.841	10.913***	16.656	15.643	15.485	-3.055	
[= 2.5]	4.234	4.348	7.232***	12.786	2.849	1.921	-7.971	
[= 3.0]	4.033	3.615	7.029***	13.448	14.117	1.690	-8.462	
[= 4.0]	11.569	3.146	4.422***	11.993	11.839	10.892	1.400**	
[= 4.5]	13.436	10.316	2.978	20.409	8.161	18.849	17.151	
[= 5.0]	0[b]	0[b]	0[b]	0[b]	0[b]	0[b]	0[b]	
City								343.161***
Hong Kong	-6.423***	-3.760	-4.927**	-4.571**	-12.227	5.611	5.374	
Shenzhen	-4.417*	-4.064†	-3.571*	-3.585*	-3.887*	4.674	6.158	
Guangzhou	0[b]	0[b]	0[b]	0[b]	0[b]	0[b]	0[b]	
New/Existing Building								362.945***
Existing	0.060	-2.576	-4.498**	-10.918	-0.243	-1.867	-7.660	
New	0[b]	0[b]	0[b]	0[b]	0[b]	0[b]	0[b]	
Urbanisation Scale of Project Location								394.158***
Peri-urban	-2.068†	2.574	2.028*	-9.299	10.086	-6.705	1.876*	
Rurban	-2.847***	-1.612	-0.501	-0.615	6.459	0.746	-0.704	
Urban Center	0[b]	0[b]	0[b]	0[b]	0[b]	0[b]	0[b]	
Pseudo R-square (McFadden): 0.477								

Note:

$N = 498$

† $p \leq 0.10$

* $p \leq 0.05$

** $p \leq 0.01$

*** $p \leq 0.001$ (2-tailed)

[a] The reference category is: Government, Institution, Community

[b] This parameter is set to zero

Table 5.6 Multinomial logistic regression analysis of project location on assessment tool
selection

Independent variables	Coefficient: assessment tool(s) selected[a]			Log likelihood
	BEAM/ BEAM-Plus	Chinese GBL	Shenzhen green building evaluation specification	
Intercept	−18.404	−0.886*	−19.293***	53.287***
City				
Hong Kong	20.628	−0.789	3.434	488.016***
Shenzhen	13.026	0.281	18.444	
Guangzhou	0[b]	0[b]	0[b]	
Urbanisation Scale of Project Location				
Peri-urban	−9.250	2.814***	2.831***	109.295***
Rurban	3.042**	1.933***	1.862***	
Urban Center	0[b]	0[b]	0[b]	
Pseudo R-square (McFadden): 0.535				

Note:

$N = 498$

[†] $p \leq 0.10$

[*] $p \leq 0.05$

[**] $p \leq 0.01$

[***] $p \leq 0.001$ (2-tailed)

[a] The reference category is: LEED

[b] This parameter is set to zero

Direct causal effect of publicness on assessment rating (H3)

The causality between the drivers' publicness and relevant assessment outcomes
was also tested in the multiple-regression analysis, with the results listed in the
right-hand column of Table 5.7. A positive coefficient was recorded between the
publicness score and the assessment rating, which means that the higher the public involvement, the higher the qualitative outcome after certification. Hypothesis
3 is thus statistically confirmed. This implies that although the private sector takes
up a larger proportion quantitatively as the driver of green-building construction,
projects with greater public involvement demonstrate larger compliance with and
are more capable of fulfilling the green-building standards on final certification.

Driving the integration of sectoral innovations

A causal pathway – graphically depicted in Figure 5.4 – was constructed to theo-
rise how organisational publicness as a driver of sectoral integration affects the
decision-making processes and outcomes of green-building assessment. Based
on the quantitative analysis, the extent of project drivers' publicness has only an

Table 5.7 Multiple-regression analysis on indicators of green-building development

	Standardised coefficient	
	Gross floor area assessed	Assessment rating
Independent Variable:		
Publicness of Driving Organisations	–	0.230***
Control Variables: Project Location		
City	0.228***	−0.432***
Urbanisation Scale	−0.067	0.279***
Control Variables: Project Type		
New/Existing Building	0.141**	0.027
Project End-Use Type	−0.099†	−0.26***
Control Variable: Project History		
Years from now	0.021	0.043
Constant	–	–
Valid Observations (N)	493	497
F Statistics	10.558***	54.053***
R-square	0.098	0.398
Adjusted R-square	0.089	0.391

Note:

†$p \leq 0.10$

*$p \leq 0.05$

**$p \leq 0.01$

***$p \leq 0.001$ (2-tailed)

indirect causal effect on the choice of assessment authority and a limited effect on the actual quantitative outputs in terms of assessed gross floor area. In this case, the cross-sector drivers rely on the intervening mechanism of project specifications to impose influence. This intervening mechanism is composed of three essential dimensions – project location, type and history – which are not separate components but rather are organically integrated and mutually influencing. However, given the causal impact of the factors in the intervening mechanism, the publicness of project drivers still poses a direct and positive causal effect on the qualitative outcomes of green-building development as measured by assessment rating. This implies that the public sector's stronger involvement in the initial development stage of green-building projects results in higher-quality completed projects, assessed either at the end of the construction or during the initial years the building is in use.

Structuring incentives in sectoral integration

As the results of the quantitative analysis indicate, the publicness of project drivers and the locus of municipal authorities (see 'city' variable in the results presented in Tables 5.6 and 5.7) have statistically significant causal impacts on the

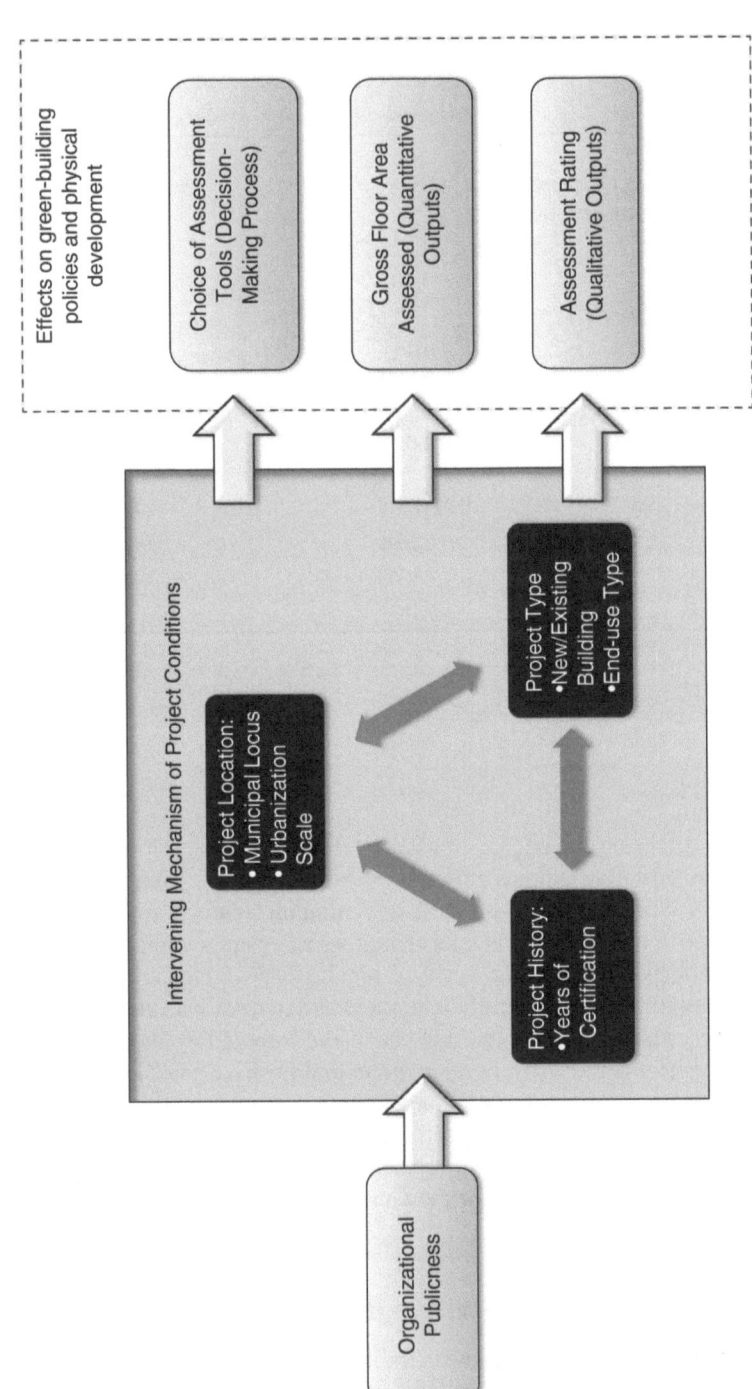

Figure 5.4 Organisational publicness and green building development

Source: Compiled by authors

decision-making process and mainstreaming outcomes of green-building develop-ment in the three cities. However, the quantitative dataset only reveals which fac-tors affect individual projects; it does not automatically tell us why this is so and how such individual effects aggregate to generate sectoral influence. To explain why i) the varying levels of the drivers' publicness and ii) the municipal locus of these green-building projects affect the integration process, further analyses must be conducted with qualitative analytical tools that recontextualise the quantita-tive analysis results. The qualitative analysis in this section thereby identifies the *interactive dynamics* of different collaborative functions and network participants in the CMNs for sectoral integration in the mitigation of climate change.

One observable characteristic of the city locus that is differentiated in the three cities is the type of municipal authority – that is, the different *governing modes* and *policy tools* used by the municipalities. Policy instruments governing envi-ronmental issues often comprise direct and mandatory regulations (governing by authority), incentive mechanisms (governing by provision of resources) and vol-untary approaches (governing through enabling; Bulkeley and Schroeder 2008; Li 2006). During the study period (1999–2013) in the three cities, green-building technologies and verification methods as sectoral innovations mainly relied on incentive mechanisms and the voluntary approach. It was only towards the end of the study period[17] when policy interventions placed administrative orders mandat-ing full or partial compliance with green-building standards in new construction.

The quantitative results showed that the organisational publicness of project drivers shapes the processes and outcomes of green-building assessments. The underlying factors differentiating organisational publicness are likely to be related to the different contextual constraints and opportunities in the public and pri-vate sectors regarding the governing and policy approaches to the mainstream-ing of green building (Miller and Moulton 2013). Governing sectoral innovation through authoritative, direct and mandatory regulations on green-building con-struction induces high administrative costs, that is, monitoring and supervising the entire construction process involving multiple parties.

Voluntary verification is constrained by the regulatory and marketing environ-ment in the private sector. For publicly held corporations, green certification is a useful marketing tool to brand their corporate image and increase consumer confidence. This was confirmed by the private property developer of an assessed Platinum green building in Hong Kong (Appendix I: 31/HK/25–02–2013). Green-building verification can also satisfy the information disclosure requirement of corporate social responsibility for companies listed in overseas stock markets. However, for local privately held companies, voluntary verification makes no dif-ference to or even reduces their profit share. Even for certain publicly listed major developers in Hong Kong, developing green buildings has never been a priority. Therefore, governing through the voluntary approach only enables limited groups of private-sector participants.

Governing by provision of resources through incentive mechanisms brings substantial opportunities for green-building participants to overcome the contex-tual constraints in the public and private sectors. Incentives constitute one of the collaborative functions in the CMNs. There are political and economic incentives for the sectoral integration of green buildings. Political incentives 'target government officials and seek to induce them to consider the environment in making regional and local development decisions', requiring 'continuous improvement in the environmental behaviour of government at all levels', whereas economic incentives 'harness market force and stimulate action' in the relevant private industries 'to benefit both the environment and the individual enterprises' (Li 2006: 10520). The political and economic incentive mechanisms shape the interactions of driving organisations in the public and private sectors.

As summarised in Table 5.8, more than 60 percent of green-building projects are driven by the private sector in Guangzhou and Hong Kong, whereas more than half of the projects in Shenzhen are led by the public sector. It is therefore pertinent to focus on Guangzhou and Hong Kong in the following discussion studying interactions with economic incentives in the private sector and on Shenzhen to investigate the dynamics of political incentives in the public sector.

Economic incentive mechanism in the private sector

The economic incentive mechanism is often more effective in altering the behaviour of actors in the private sector. To promote green-building development, the incentives are transferred explicitly as cash rewards or implicitly as the provision of gross floor area (GFA) concessions, shaping interaction patterns across the private, public and professional sectors with the presence of an expanded nongovernmental space for cross-sector innovation integration.

Cash or concessions?

To rapidly integrate green-building innovations, multiple governmental levels in China recently institutionalised a cash-incentive system (Appendix II: Guangdong Provincial Government 2013; Guangzhou Municipal Government 2013; MOHURD and MOF 2012). Table 5.9 summarises the financial rewards available for certified green buildings in Guangzhou administered by three tiers of authority on construction and finance.

The policy environment in Hong Kong, however, is different. There are strong contextual constraints on subsidising green-building certification with cash

Table 5.8 Sectoral drivers of green-building development in cities

	Public sector		Private sector		Nonprofit professional sector		Total green-building certifications (1999–Feb. 2013)
Guangzhou	16.0	32.0%	33.5	67.0%	0.5	1.0%	50
Shenzhen	93.5	53.1%	81.5	46.3%	1.0	0.6%	176
Hong Kong	83.5	30.7%	165.5	60.9%	23.0	8.5%	272

Table 5.9 Cash reward policy in Guangzhou

Multilevel assessment policy	Rating	Financial reward (Unit: ¥/m²)
National Standard (Chinese GBL)	3 stars	80
	2 stars	45
Provincial Standard (Evaluation Standard for Green Buildings in Guangdong)	Level 3	45 (max. 2 million/project)
	Level 2	25 (max. 1.5 million/project)
Guangzhou Matching Rewards		For projects obtaining the above financial rewards, additional rewards are appropriated from the municipal finance/Energy Efficiency Special Fund (effective from June 2013)

rewards in Hong Kong. As the green building practitioner and HKGBC director explained:

> It is certainly much easier [there] to have top-down implementation in green building. In the [China] mainland, you can directly pay [the developer] in cash per meter square after green building assessment. In Hong Kong, if you dare to do so, [people will definitely accuse you of] 'government-business collusion'! We are comparatively in a dilemma. It is difficult to implement a lot of globally applied market drivers in Hong Kong.
>
> (Appendix I: 30/HK/16–11–2012)

One economic incentive mechanism that provides an impetus to the private sector in Hong Kong is the GFA concession policy, institutionalised in 2001, which partially exempts the imposed GFA restrictions on private developments if green and innovative features are integrated in the building design and construction. To impose population and residential density control, the Planning Department has undertaken stringent administrative measures to regulate the permissible GFA relative to each site area.[18] Every private developer in Hong Kong must comply with the GFA controls, which limits profit maximisation.

To promote green and innovative buildings, a former director of the Building Department in Hong Kong in 2001 and 2002 initiated a revision of the Joint Practice Notes, exempting certain green features in the building from the ratio plot calculation of gross floor area (Appendix II: Buildings Department *et al.* 2001, 2002). This has become an important profit driver for many private real estate developers in Hong Kong. They are concerned about the consumers' mentality, such that 'the saleable areas are actually unchanged, but if you "bluff" on larger GFAs, potential buyers are more content' (Appendix I: 29/HK/22–10–2012).

However, the GFA concession policy at this stage has not effectively driven the positive development of green building. As a green-building professional commented, 'at that time, the private developers requested bonus GFA [and] in the end, everyone had obtained the GFA benefits, yet no one actually worked on the green features' (Appendix I: 28/HK/27–08–2012). Rather, it inflated the property prices in Hong Kong even further, as some developers took advantage of the GFA concession and turned it into public-use area saleable to the buyers in addition to the private-use interior areas.

The inflated property prices incited public discontent that exploded in June 2010, when the SDC published a report on the sustainable-built environment in Hong Kong (Appendix II: SDC 2010). The report – colloquially called 'inflated flat report' – revealed potential abuses of the then GFA concession policy, with some buildings having been granted 20 percent of GFA with controversial green features (Appendix I: 29/HK/22–10–2012). After a year of cross-sector discussion on the nongovernmental space, which ensured a sufficient level of *input legitimacy*, the Buildings Department issued a new practice note imposing a 10 percent cap on GFA concessions for all green and amenity features, effective for development proposals submitted after April 2011, to prevent further abuse (Appendix II: Buildings Department 2011).

The SDC report also recommended qualitative controls on green-building assessment in granting GFA concessions, by engaging with the local certification authority in the nongovernmental space – the BEAM Society (HKGBC had not yet been formally established when the report was released). With HKGBC established thereafter and a refined assessment standard (BEAM-Plus), the government reported that 'it is timely to give this green building movement a bigger push in the private sector' (Appendix II: Development Bureau 2010: 11). In the reformed GFA concession policy, registration for BEAM-Plus assessment conducted by HKGBC is a prerequisite for all new buildings seeking GFA concessions for green and amenity features (Appendix II: Buildings Department 2011; Development Bureau 2010). In this regard, *nongovernmental space* as a legitimating environment is broadly engaged in the implementation of economic-incentive policies. However, the quality control of assessment ratings is not mandated in this policy and private developments that have registered for the assessment but produced poor ratings can still claim the GFA concessions. This partly explains why higher public involvement in the construction project may secure better assessment ratings, as indicated in the quantitative analysis.

Despite the cap, GFA concessions remain a significant economic incentive for private developers to consider incorporating green-building innovations in their development projects. On Hong Kong Island alone, the most densely populated built landscape in the existing developed zone of the city, six newly proposed private development projects – submitted after April 2011 and approved with construction occupation permits before December 2012 – have solicited GFA concessions on green and amenity features (Buildings Department 2013). Even while subject to the 10 percent cap, projects incorporating innovative green-building technologies can be considered for GFA concessions above that limit. Using this rationale, one development project obtained 25 percent GFA concessions.

Interaction barriers

The economic-incentive mechanism, often initiated by the public sector, is targeted to stimulate private-sector action. It relies on effective interactions across both sectors, facilitated by the legitimation environment provided by the nongovernmental space. However, the established mechanisms in Guangzhou and Hong Kong have intrinsic and extrinsic barriers fragmenting the interaction of sectoral driving forces to integrate green-building innovations.

The intrinsic barrier is related to the power and authority needed to enforce the economic-incentive mechanisms. After the implementation of the reformed GFA concessions policy, the engaged nongovernmental organisation (HKGBC) has limited enforcement capacity. Although the Practice Note clearly states that obtaining the 'official letter issued by the HKGBC acknowledging the satisfactory completion of project registration for BEAM-Plus certification' is a mandatory prerequisite for the granting of GFA concessions, only two of the six new private development projects on Hong Kong Island (mentioned earlier) registered with the HKGBC before getting concessions (Appendix II: Buildings Department 2011: 2). Three projects registered instead with the BEAM Society, to be assessed using the more obsolete 2004 method. A fourth did not register with the HKGBC or the BEAM Society, let alone agree to provisional assessment with any versions of BEAM. Although the accreditation authority of the HKGBC has been formally endorsed by the public sector, the Practice Note that prescribes the delegated authority has low regulatory power. As one practitioner commented, 'We call it [the Practice Note] "indirect" regulation because it is not a formal ordinance, it is only a "note" supplementary to certain "practice", subordinate to a regulation, while the regulation is subordinate to the [Buildings] Ordinance' (Appendix I: 29/ HK/22–10–2012).

Similarly, the cash-reward system enacted in Guangzhou is institutionalised in a regulatory guideline issued by the mayor's office that is of lower regulatory authority to the private developers than is legislation. Policy guidelines can be revised annually, increasing the uncertainties of the cash-reward system, which is still in its experimental stage. Nongovernmental actors were included in studying the feasibility of the reward system, and as one reflected on his involvement in the policy process, 'The incentive policy certainly requires a longer time for research, because it is not only the business of the construction system, but also of other functional systems of the government; and the budgets are stratified across the province, the cities, and the districts, altering sources of the Special Fund that generates the cash rewards' (Appendix I: 14/GZ/22–02–2012). Unless the cash-reward provisions in the guideline are stated in concrete terms or ratified in relevant legal documents, as in the Green Building Promotion Measures of Shenzhen (Chapter 3), the private developers are not assured of receiving any actual economic benefits. The contextual uncertainties hinder further private investment and broader participation across the private sector in the integration of green-building innovations.

The extrinsic barrier in Hong Kong is the result of the limited demand-side awareness of the green-building concept in the property market and among citizens. This reduces the profitability of green building investment as perceived by marginal private developers (not the core ones who have long been in the green-building field and have established 'green' profiles). In Hong Kong, 'the professional knowledge on green building is substantial, yet the market acceptance is rather insufficient', while 'the average citizens are "summoned" by the property speculation in that they look at the property views and prospects, not the green features', according to two local green-building practitioners (Appendix I: 28/HK/27–08–2012, 29/HK/22–10–2012). This limited societal awareness confines the *output legitimacy* of collaborative networks for green-building integration in Hong Kong, as the peripheral actors cannot foresee any significant distributed benefits or regulatory advantages that they could acquire by participating in the network (Héritier 1999; Skogstad 2003).

Conversely, in Guangzhou, the extrinsic barrier is found in the limited discursive legitimacy of green-building innovation. According to a local government official working on green-building promotion, 'we have to establish a facilitative atmosphere in our society first; it needs more promotion, which requires actions of social organisations, such as the Surveying and Design Association [GZ-SDA in Figure 5.1] and the Construction Inno-Tech Promotion Station [CITPS in Figure 5.1]' (Appendix I: 12/GZ/21–02–2012). Without strong representation of *discursive legitimacy* in the nongovernmental space supporting Guangzhou's nascent green-building sector, it is difficult – despite the institutionalisation of an economic-incentive mechanism, that is, cash awards – to start engaging a broader scope of cross-sector interactions to enhance public awareness of sectoral innovation.

The intrinsic and extrinsic barriers to the existing economic-incentive mechanisms diminish the engagement clout of the networking interactions for green-building integration. It is ineffective to create shelter from immediate market pressures for potential (peripheral) actors to become engaged as (core) participants in the innovation process because there are no extended opportunities for potential participants to interact and learn about green-building innovation and about their own preferences and attitudes in relation to the development of green buildings (Caniëls and Romijn 2008: 615).

Political incentive mechanism in the public sector

As the results of the quantitative analysis indicate, publicness in the driving force exerts a significant effect on the qualitative outcome of green-building assessment as measured by assessment rating, and an indirect effect on the quantitative output as measured by assessed area. Instead of economic incentive as with the GFA concession or cash awards, local officials may rely on political incentives to successfully assist and encourage green-building development in their own jurisdiction. This is a possible explanation for why higher publicness leads to higher rating results. If, hypothetically, private developers care only about economic benefits, they do not need to invest much to improve their assessment rating; as long as they get the minimum rating, they are able to secure the stipulated amounts of GFA concession, not to mention the already-identified barriers in the existing economic-incentive mechanism (i.e. inferior authority and low predictability of the changing policies, with limited demand-side awareness and discursive legitimacy). In contrast, for public developers, extraordinary green-building rating outcomes are a symbolic success that embellishes their personal performance and polishes the image of their jurisdiction. Thus, high green-building ratings bring political incentives for the political elite and policy entrepreneurs, a stronger motive than the economic incentive for the private sector.

To be specific, the political incentive serves various ends: it secures or expands the functional jurisdictions of certain governmental entities, it polishes the performance evaluations of political leaders, it impresses citizens and it brand cities or subnational regions in the context of vertical *inter*governmental relations. Thus, one major way to generate political incentives in China is by building up pilot eco-cities, eco-towns or eco-projects to demonstrate local performance (Li 2006; to be further discussed in Chapter 7).

The effect of the political incentive would seem to be more prominent in Shenzhen than in the other two cities, with a larger presence of public-sector driving forces in green building projects (Table 5.8). Green building projects driven by the public sector are most densely located in the city's pilot ecological and low carbon districts. Two new districts of Shenzhen – Pingshan and Guangming – in the peri-urban area have been designated by the municipal and national governments as new development zones for meeting the needs of urban expansion. Anticipating extensive new construction in these new districts, the district management authorities and the relevant municipal departments are actively seeking green-building solutions in the development process. Out of the 176 green-building certifications between 2010 and 2012 in Shenzhen, 53 (30.11 percent) have been located in these two new districts – 35 in Guangming and 18 in Pingshan. All of the projects in the new districts are double certified – the Shenzhen Green Building Evaluation Specification and the Chinese National GBL standards. Figure 5.5 shows that most of the projects (84.9 percent) in these two new districts are being developed by the government or by government-related bodies. In contrast, less than half of the other 123 certifications (41.0 percent) in the city are driven by the public sector.

Guangming New District has been developed as a Green-Building Demonstration District, an initiative started by the SZ-HCD with the subsequent endorsement of the central government (detailed case study in Chapter 7). In this process, the SZ-HCD secured independent resources and administrative powers as political incentives. Due to Shenzhen's 2009 structural reform, the management of the SZ-HCD – together with that of other relevant agencies – was centralised to the coordination of the Shenzhen Urban Planning, Land and Resources Commission (SZ-UPLRC); specifically, centralised contacts (*'guikou lianxi'*) (Ye and Hua 2009). The SZ-HCD and SZ-UPLRC are of equal administrative rank, and SZ-HCD administrators were certainly against the centralised arrangement. This administrative resistance was reflected in an SZ-HCD officer's comments on the structural reform:

> The thing as to who is managing whom doesn't exist. It was never implemented before – it was only due to the structural reform. But what was written in the municipal government [document] was only to centralize the contacts

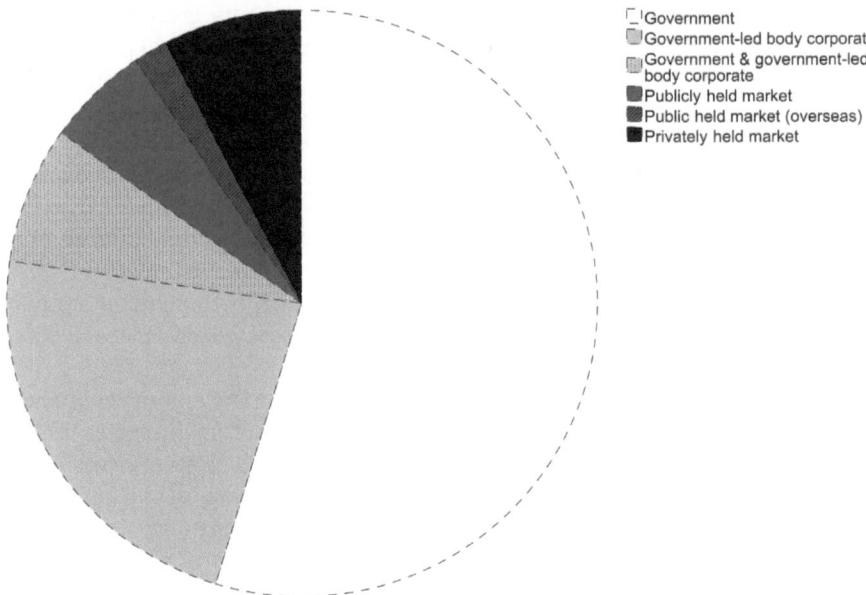

Figure 5.5 Organisational driving forces in new districts
Source: Compiled by authors

with us. It is only when the municipal government doesn't need that many departments, it can ask SZ-UPLRC to coordinate the data; by then we may give SZ-UPLRC something relevant to report to the municipal government. Compared to the past, we are still working on our own things.

(Appendix I: 20/SZ/23–05–2012)

Ever since the reform, SZ-HCD has been more cautious about securing independence and equality in terms of administrative power with SZ-UPLRC. The Green Building Demonstration Project in Guangming New District is thus an important tactic for generating political resources and reinforcing SZ-HCD's supremacy in the governing portfolio of public-housing construction and management.

Conversely, the Pingshan New District has been mainly initiated and managed by SZ-UPLRC as a trial scheme in the broader project of Shenzhen's National Low Carbon Ecological Demonstration City, which has been endorsed by the central government (detailed case study in Chapter 7). It serves slightly different political ends for the SZ-UPLRC. Through this initiative, the SZ-UPLRC is attempting to occupy the political space in the *intra*governmental network for climate-change mitigation that has not been occupied by the Shenzhen Development and Economy Commission (SZ-DRC). SZ-DRC, as the city's economic planning authority, is the current coordinator of the horizontal *intra*governmental management of low carbon development (Chapter 4). Highlighting the performance of the Ecological Demonstration City with intermediate project outcomes generated from the Pingshan New District, SZ-UPLRC seeks to share the role of coordination leader in the *intra*governmental mechanism with SZ-DRC. In SZ-UPLRC's perspective, SZ-DRC and SZ-UPLRC are responsible for two independent policy areas, the former for economic planning and the latter for spatial planning; accordingly, SZ-UPLRC should have its own discursive legitimacy in the CMNs for climate change mitigation. The political incentive for the SZ-UPLRC emerges in *intra*governmental interactions for legitimate leadership.

Political incentives have also been presented in Hong Kong and Guangzhou to reach the ends of city image branding and political marketing by creating positive impressions in the community. In Hong Kong, the Construction Industry Council (CIC), sponsored by and collaborating with the city government, has developed a Zero Carbon Building that integrates and experiments with various advanced environmental building technologies. The Zero Carbon Building is explicitly a building only for demonstration purposes rather than actual use. It has been rated as a Platinum green-building project in BEAM-Plus provisional assessment. The Zero Carbon Building, as a communal demonstration, is located within an urban gentrification area that features a mix of industrial, commercial and residential use. Comprising urban native woodland and an outdoor activity area, it provides the community with a cozy public space – a rare public asset in densely populated Hong Kong that could potentially improve citizens' impression of their government, that is, it carries political incentives.

Symbolic buildings have also been developed in Guangzhou. A prominent one is the Pearl River Tower, situated in the newly planned and developed central business district. The Pearl River Tower is developed and owned by a major state-owned enterprise in China (i.e. a government-led body/corporate). It is a LEED-assessed green building with a provisional Platinum rating. Although the building has actually incorporated many technological innovations imported and borrowed from other countries, it has been used as an important tool to brand Guangzhou's image, covered in foreign media as an indicator of the city's changing external image (Watts 2009). Meanwhile, promotional materials internal to the city government have also published supportive articles on its technological specifications (e.g. Li 2011).

Conclusion

This chapter focused on the dynamics of integrating green-building innovations through the collaborative municipal networks (CMNs). It analysed the actors and drivers in different societal sectors during the governance and integration processes of green-building practices, specifically how these actors and drivers interacted and how they altered existing governance structures and capacity. The analytical results provide insights into the core research question in this study: How is the climate mitigation process in urban settings institutionally transforming the Chinese environmental state?

The analytical method used in this chapter involved a triangulation of qualitative and quantitative analyses. A comprehensive empirical review of China's current green-building practices in the three cities was provided in the first section, covering the administrative regulations and sectoral standards used.

The qualitative analysis of semistructured interviews and documentary reviews found that the nongovernmental space has been expanded to support green-building development in the three cities. As an integral component of the CMNs, the nongovernmental space has integrated sectoral innovations. To sort through the *inter*governmental complexity in the mitigation of climate change, interorganisational and intersectoral interactions occurred in the nongovernmental space. The second section identified three different forces driving the expansion of nongovernmental space: the government, associated businesses in the private sector and concerned professionals. The analysis also found that nongovernmental space provides a legitimation environment that generated additional resources for innovation transfers in the CMNs' climate mitigation. In addition, the nongovernmental space's value for diversity fostered interactions for driving policy development, generating social impacts and streamlining knowledge management as the various dimensions of mainstreaming green building.

The analysis then moved to examining how the different drivers interacted in the process of integrating green-building innovations. Organisational publicness was a key characteristic differentiating the roles and concerned interests of network participants in the CMNs. The main finding of the quantitative analysis in the third section was that the extent of organisational drivers' publicness had a direct and significant positive causal correlation with the qualitative outcomes of green-building integration as measured by assessment rating but a relatively minimal effect on the choice of assessment tools and the quantitative outputs of green-building development as measured by assessed gross floor area.

This finding elicited the question: Why does government involvement – or the extent of publicness in drivers – matter in green-building assessment? The answer lay in the incentive structures institutionalised in different sectoral settings, given the recontextualised qualitative analysis in the fourth section. In the private sector, economic-incentive mechanisms (GFA concessions and cash awards) motivated private developers' participation in green-building construction. However, intrinsic and extrinsic barriers to economic incentives hindered the maximisation of qualitative outcomes and fragmented the interactions across societal sectors. Conversely, in the public sector, political-incentive mechanisms encouraged greater involvement from senior government officials, policy entrepreneurs and the political elite in green-building integration. These political incentives serve different purposes in different forms. They can be incorporated in the development of new peri-urban districts to strengthen the capacity and legitimate the leadership of certain government departments. They are also embedded in urban-branding projects to improve the external and internal images of the city and the ruling elite. Gradually overcoming the perceived barriers, these economic and political incentives are driving institutional changes in the urban environmental state by increasing networking and cross-sector collaboration for mutual benefits, including both input and output legitimacy.

Notes

1 Some examples of passive building designs are cross-ventilated layout, clerestory for daylighting, optimised window-to-wall ratio, external shading and climate-responsive built form and orientation.
2 The innovative cooling and dehumidification systems currently comprise underfloor displacement cooling, earth cooling tubes, chilled beams, desiccant dehumidification, ultra-low overall thermal transfer value and adsorption chiller.
3 Some of the 'energy-plus' experimental techniques include regenerative lift systems and biodiesel tri-generation.
4 The legal status of CBGC is stated in the Management Measures of Affiliate Organizations in the Chinese Society for Urban Studies issued on 23 November 2011 [Online]. Available: www.chinasus.org/chinasus/branches/gl/20111123/82202.shtml [Accessed 7 March 2014]
5 In 2010, the year when this mandatory green-building requirement took effect, Shenzhen planned to construct a total of 2.8 million m2 of indemnificatory housing (19.52 percent of the city's total planned housing construction). In the 12th Five-Year Plan period (2011–2015), the city's construction target for indemnificatory housing was raised to a total of 15.36 million m2, or 3.072 million m2 per annum. (Appendix II: The Editorial Committee for Shenzhen Real Estate Yearbook 2010, 2011)
6 These departments include the Buildings Department, Housing Authority and Housing Department, Electrical and Mechanical Services Department, Civil Engineering and Development Department and Environmental Protection Department.
7 Tier-1 actors include Guangdong Building Energy Efficiency Association (GD-BEEA); Guangzhou Management Office of Building Energy Efficiency (GZ-BEEMO), which is closely connected with the *intra*governmental mechanism (Chapter 4); Guangzhou Association of Energy Efficiency and Building Technology (GZ-BEEA); Shenzhen Green Building Association (SGBA); Shenzhen Institute of Building Research (SIBR); BEAM Society; Professional Green Building Council (PGBC); Hong Kong Green Building Council (HKGBC); and CGBC(HK) operating based in Hong Kong.
8 Such as the Surveying and Design Association in Guangzhou (GZ-SDA) and Shenzhen (SZ-SDA) and the Construction Inno-Tech Promotion Station (CITPS) in Guangzhou.
9 Real Estate Association in Shenzhen (REA), the Construction Industry Council (CIC) and the Real Estate Developers Association (REDA) in Hong Kong.
10 For instance, the Shenzhen Planning and Land Development Research Centre (PLDRC) and the Council for Sustainable Development (SDC) in Hong Kong.
11 Tier-3 actors include Guangzhou Energy Saving Association, Guangdong Low Carbon Association based in Guangzhou, Shenzhen Expertise Federation of Energy Conservation, Shenzhen Energy Saving Association, Business Environmental Council (BEC), World Wildlife Fund (WWF), Greenpeace, Friends of the Earth (FOE), Oxfam, and others based in Hong Kong.
12 The Green Building Award is a biannual award to recognise outstanding sustainable building projects inaugurated since 2006, presented at the Green Building Conference organised by HKGBC and PGBC.
13 The assessment tools largely applicable in the three cities are: BEAM and BEAM-Plus developed in Hong Kong; Chinese Green Building Labeling (GBL), Chinese GBL (HK) and the Evaluation Standard for Green Building in Guangdong, which are of the same authority origin; the Evaluation Specification of Green Buildings developed by and applicable in Shenzhen; and LEED, developed by the US.
14 Information on LEED certified projects was retrieved from http://new.usgbc.org/projects, that on BEAM-Plus certified projects was from www.hkgbc.org.hk/eng/BeamPlusDirectory.aspx and that from projects certified by all of the previous versions of BEAM can be found on www.beamsociety.org.hk/en_beam_assessment_project_4_detail.php [Accessed 27 Feb 2013]

15 Project end-use types: 1) private residential, 2) public residential, 3) hotel, 4) commercial (office and shopping mall), 5) mixed use, 6) GIC, 7) industrial and 8) infrastructure (logistics, transport links, stations, roads and utility facilities).

16 Government = 5 > Government-led Body/Corporate = Tertiary Education Institutes = Nongovernmental Organisations = 4 > Public–Private Partnership = 3 > Publicly Held Private Corporate = 2 > Privately Held Company = 1; for the projects with the Hybrid of Privately Held Companies and Government-led Body/Corporate, the publicness score is calculated as $(1 + 4) / 2 = 2.5$.

17 As mentioned in the first section of this chapter, regulatory policies were issued in Guangzhou in 2012 mandating full compliance with green-building standards in all new buildings, in Shenzhen in 2009 for new indemnificatory housing and in Hong Kong in 2012 in new public housing.

18 The GFA of a building is calculated by multiplying the *net site area* by a *plot ratio* applicable in each different development zone. The maximum domestic plot ratios for the existing development area (most of the area for green-building development) are between 7.5 and 10. Detailed regulations are available at www.pland.gov.hk/pland_en/ tech_doc/hkpsg/full/ch2/pdf/ch2.pdf [Accessed 26 November 2013]

References

6, P., Goodwin, N., Peck, E. & Freeman, T. 2006. *Managing Networks of Twenty-First Century Organisations,* Basingstoke, UK, Palgrave Macmillan.

Allen, A. 2003. Environmental Planning and Management of the Peri-Urban Interface: Perspectives on an Emerging Field. *Environment and Urbanization,* 15, 135–148

Bäckstrand, K. 2006. Multi-Stakeholder Partnerships for Sustainable Development: Rethinking Legitimacy, Accountability and Effectiveness. *European Environment,* 16, 290–306

BCA. 2012. *BCA Green Mark Assessment Criteria and Application Forms.* Singapore: BCA [Online]. Available: http://bca.gov.sg/GreenMark/green_mark_criteria.html [Accessed 28 September 2012]

Bozeman, B. & Bretschneider, S. 1994. The "Publicness Puzzle" in Organization Theory: A Test of Alternative Explanations of Differences between Public and Private Organizations. *Journal of Public Administration Research and Theory,* 4, 197–224

Buildings Department. 2013. *Summary of Gross Floor Area (GFA) Concessions in Private Developments* [Online]. Available: www.bd.gov.hk/english/GFA.html [Accessed 27 November 2013]

Bulkeley, H. & Schroeder, H. 2008. *Governing Climate Change Post-2012: The Role of Global Cities – London.* Working paper (123). Norwich, UK, Tyndall Center for Climate Change Research.

Caniëls, M. C. J. & Romijn, H. A. 2008. Actor Networks in Strategic Niche Management: Insights from Social Network Theory. *Futures,* 40, 613–629

EPA, U.S. 2010. *Green Building – Basic Information* [Online]. Available: www.epa.gov/ greenbuilding/pubs/about.htm [Accessed 28 September 2012]

Firey, W. 1946. Ecological Considerations in Planning for Rurban Fringes. *American Sociological Review,* 11, 411–423

Gilchrist, A. 2009. *The Well-Connected Community: A Networking Approach to Community Development,* Bristol, UK, Policy Press.

Héritier, A. 1999. Elements of Democratic Legitimation in Europe: An Alternative Perspective. *Journal of European Public Policy,* 6, 269–282

Hodson, M. & Marvin, S. 2010. Can Cities Shape Socio-Technical Transitions and How Would We Know If They Were? *Research Policy,* 39, 477–485

Kingdon, J. W. 1995. The Policy Window, and Joining the Streams, in *Agendas, Alternatives, and Public Policies* (2nd ed.), New York, NY, HarperCollins College Publishers.

Land, A. E. 1967. Unraveling the Rurban Fringe: A Proposal for the Implementation of Proposition Three. *Hastings LJ,* 19, 421

Li, L. 2011. *Exemplary Showcase: Guangzhou Tower.* Guangzhou Wall Material Innovation & Building Energy Efficiency (vol: 83, 26–35). Guangzhou, Management Office of Wall Material Innovation & Building Energy Efficiency.

Li, W. 2006. Environmental Governance: Issues and Challenges. *Environmental Law Reporter News and Analysis,* 36, 10505–10525

Liu, G., Nolte, I., Potapova, A., Michel, S. & Ruckert, K. 2010. Comparison of Worldwide Certification Systems for Sustainable Buildings. 9th International Conference on Sustainable Energy Technologies. Shanghai, China, Longlife, EU.

Lovell, H. 2007. The Governance of Innovation in Socio-Technical Systems: The Difficulties of Strategic Niche Management in Practice. *Science and Public Policy,* 34, 35–44

McGuire, M. & Silvia, C. 2010. The Effect of Problem Severity, Managerial and Organizational Capacity, and Agency Structure on Intergovernmental Collaboration: Evidence from Local Emergency Management. *Public Administration Review,* 70, 279–288

Miller, S. M. & Moulton, S. 2013. Publicness in Policy Environments: A Multilevel Analysis of Substance Abuse Treatment Services. *Journal of Public Administration Research and Theory,* 24, 553–589

Morris, D. & Gilchrist, A. 2011. *Communities Connected: Inclusion, Participation and Common Purpose* [Online]. Available: www.thersa.org/__data/assets/pdf_file/0011/518924/RSA_Communities-Connected-AW_181011.pdf [Accessed 12 September 2012]

Schot, J. & Geels, F. W. 2008. Strategic Niche Management and Sustainable Innovation Journeys: Theory, Findings, Research Agenda, and Policy. *Technology Analysis & Strategic Management,* 20, 537–554

Skogstad, G. 2003. Legitimacy and/or Policy Effectiveness? Network Governance and GMO Regulation in the European Union. *Journal of European Public Policy,* 10, 321–338

USGBC. 2011. *What LEED Measures* [Online]. Available: www.usgbc.org/DisplayPage.aspx?CMSPageID = 1989 [Accessed 28 September 2012]

Watts, J. 2009. Supertower Offers Glimmer of Hope in Polluted Chinese City: Pearl River Tower in Guangzhou Is Being Billed as a Green Beacon Amid the Pollution of China's Construction Boom. *The Guardian,* 27 May [Online]. Available: www.guardian.co.uk/environment/2009/may/27/china-green-supertower [Accessed 20 March 2013]

Ye, X. & Hua, X. 2009. Central State Commission for Public Sector Reform Approved (Shenzhen Municipal Government Structural Reform Plan). *Shenzhen News,* 1 August [Online]. Available: http://news.sznews.com/content/2009–08/01/content_3948472.htm [Accessed 1 December 2013] (*Zhongyang Bianwei Pizhun <Shenzhen Shi Renmin Zhengfu Jigou Gaige Fang'an>*)

Zeemering, E. & Romero, J. 2012. The Evolution of Sustainable Cities as a Metropolitan Policy Challenge, in Meek, J. W. & Thurmaier, K. (eds.) *Networked Governance: The Future of Intergovernmental Management,* Thousand Oaks, CA, CQ Press.

Zhang, Z. & Lau, S.-Y. 2010. Comparison Study of the Chinese Green Building Label and the USA LEED 2009, in *Conference Proceedings of the 6th International Conference on Green and Energy-Efficient Building & New Technologies and Products Expo,* Beijing, China.

6 Legitimation of electric vehicles – where are the problems?

In this chapter, we shall focus on the sectoral integration of electric vehicles (EVs) for private use in CMNs. While various *incentives* are identified as the major internal driver of innovation integration in the building sector (Chapter 5), *communication and information* are driving the public's acceptance of technological innovations such as EVs for the purpose of innovation legitimation. In this regard, the discussions in Chapters 5 and 6 reinforce the interactive components of CMNs' *collaborative functions* (Chapter 2). Collaborative functions are particularly connected to the process of transferring initiatives for low carbon transitions among CMNs, the communication and information provisions in the networking process and the institutionalisation of incentive structures in the network configurations.

The riddle of electric vehicles

The development of EVs, both hybrid electric-petroleum vehicles and all-electric vehicles, is widely recognised as a significant innovation for low carbon transport, given their emission reduction potential. Encouraging EVs for private use is highly dependent on the level of public acceptance and confidence in this technical innovation, among others. EVs and green buildings are intertwined sectoral innovations. Green building development provides an infrastructural basis for transport innovations. For instance, the trial and general application of EVs rely on easy access to charging stations, which is listed as a criterion in some green building accreditation standards. Both green buildings and EVs are thus part of the sociotechnical transition towards an environmental state with the creation and expansion of technical and market niches.

New-energy automobiles, as part of low carbon transport, have been advocated by the Chinese national government as possible developmental and environmental benefits of this industrial initiative. In early 2009, the State Council recognised the mainstreaming of new-energy automobiles as the primary approach to 'revitalising the automotive industry' (Appendix II: State Council 2009). In the 12th Five Year Plan (released in early 2011), the new energy automotive industry was explicitly chosen as one of the 'emerging strategic industries' that all levels of governments should rapidly develop (Appendix II: State Council 2011). This national strategy was interpreted by both subnational governments and the automobile industry as initiating the wider use of hybrid electric-petroleum and all-electric vehicles for both public and private commuting (Liu 2011). According to the government, these EV models meet most of the criteria for decarbonising road transport. They are more energy efficient than traditional gasoline vehicles[1] and thus reduce the burning of fossil fuels and mitigate the impact of climate change.

Multilevel policies created the technical niches allowing the technological development of EVs, but barriers have emerged in the establishment of market niches, that is, user demand has not stabilised (Caniëls and Romijn 2008; Schot and Geels 2008). To spur technological development and expand the experimental use of EVs, the national and local governments have provided tax relief, established research funds and constructed facilitating infrastructures. Existing EV models have been found to be technically viable for the road conditions of most Chinese cities. Despite these policy and technological developments, there has been limited progress in the promotion of EVs for private use in China. Sales of private electric cars have stagnated, surprising policymakers. Both the government and the private sector are struggling to find ways to expand the use of EVs. They have certain expectations for EVs' environmental and developmental outcomes, but these expectations are misaligned during the networking process, that is, the interactions among public policies, the producer market and citizens' demands. This misalignment has undermined the *output legitimacy* of EVs in the sectoral-integration process.

Therefore, this chapter probes three questions:

- First, how do different network segments *interact* in the sectoral integration process of electric vehicles?
- Second, how are the *drivers* and *barriers* to sectoral integration configured in the interaction process?
- Third, how is the *legitimation* level of electric vehicle (EV) innovation shaped by resource provision and patterns of communication?

Through the literature related to governance networks and based on the proposed theoretical framework of CMNs (Chapter 2: Figure 2.2), the analysis of integrating EVs for private use provides insights into how different collaborative functions come together to enhance the effectiveness of networking in innovation mainstreaming.

Analytical rationale

In this chapter we shall focus on a case study of the expansion of EVs for private use in Hong Kong. With its highly modernised transport system, Hong Kong is perceived as a learning model of sustainable transport for other Chinese cities (Harris 2012). The sectoral integration of EVs in the private sector was initiated in Hong Kong in early 2009, when the government established a Steering Committee on the Promotion of Electric Vehicles and attempted to institutionalise *cross-sector collaboration* (Bryson *et al.* 2006; HKSAR Government 2009).

Hong Kong has taken more EV integration actions than either Guangzhou or Shenzhen. According to government statistics, 310 electric cars had been licensed in Hong Kong by April 2012 (HKSAR Government 2012). In comparison, only two electric cars had been purchased in Guangzhou, although this information was only taken from the media, as no official statistics on private usage are being made available (Lin 2012). Although a major Chinese manufacturer of EVs – BYD Company Limited – is located in Shenzhen, the company focuses on public and commercial EVs rather than private electric cars (Autoreport 2012). Like Guangzhou, Shenzhen has not released statistics on the actual number of private electric cars currently in use in the city. Despite the ambitious policy targets and high subsidies for private purchase, the sale of private electric cars has stagnated in Shenzhen (Liu 2013).

As for the charging infrastructure for EVs, Guangzhou had launched approximately 13 charging points by the end of 2012, while Shenzhen had around 100 charging points for private electric cars by the end of the first quarter of 2013 (Lin 2012; Qiu and Tu 2013). In contrast, 1,005 charging points were in service in Hong Kong by June 2012 (Environment Bureau 2012). Thus, we focus our analysis on Hong Kong to provide more empirical data for the discussion in this chapter.

The transport sector in Hong Kong accounted for 17.8 percent of local greenhouse gas emissions and 32 percent of end-use energy in 2010, the second-largest emission and consumption sector (Appendix II: EMSD 2012; HK-EPD 2013b). In the same year, private cars consumed 32.6 percent (18,097 TJ)[2] of end-use energy in passenger transport, almost equal to the 33.1 percent (18,423 TJ) consumed by public buses – the largest energy consumer among passenger transport modes (Appendix II: EMSD 2012). At similar levels of total energy consumption, it is reasonable to estimate that the per-capita energy consumption of private cars in Hong Kong is much higher than that of public buses, given their vast differences in vehicle capacity. Given the figures on energy consumption and emissions, it is important to integrate private EVs into Hong Kong's low carbon transition.

This chapter does not cover the governance process of technological development but focuses on the *integration of innovation for private electric cars*. It is thus specifically related to the analysis of the 'post-technological development' stage of EVs, the transition from a technical niche to a market niche for EVs. Some scholars have argued that the technological development of EVs in China is dominated by the public sector (Kokko and Liu 2012). Yet for posttechnological development, the innovation integration process involves the governance dynamics for engaging key actors in the nongovernmental space and the market sphere. Whereas the technical niche is a protected space created by the government, the market niche attends to user perceptions and demands (Schot and Geels 2008).

Hillman *et al.* (2011) conceptualise the key stages in the process of technological innovation systems, tailoring them to the study of EVs. They identify seven stages: 1) knowledge development and diffusion, 2) influence on the direction of search, 3) entrepreneurial experimentation, 4) market formation, 5) legitimation, 6) resource mobilisation and 7) development of positive externalities (Hillman *et al.* 2011: 406). Stages 1 through 3 are mainly functions of technological development, the opening of the market niche occurs during Stage 4, and Stages 5 and 6 are concerned with innovation integration. As this chapter is concerned with the networking dynamics that occur during the sociotechnical transition of EVs, the analysis focuses on *legitimation and resource mobilisation*. Legitimation and resource mobilisation serve the essential function of innovation integration and can be regarded as intermediate outcomes resulting from the actions of collaborative functions combined with coordination approaches. This established analytical relationship also fits the theoretical framework proposed in Chapter 2 (Figure 2.2).

As the discussion is centred on issues of *communication, legitimation and resource mobilisation*, this chapter adopts a data collection method based on the textual analysis of newspapers and the public profiles of network organisations, in addition to the semistructured interviews generally applied in this volume (the mixed methods discussed in Chapter 2). Newspapers, as a public, periodic, current and universal source of information, contain articles that reflect cross-sectoral perceptions of the feasibility of EVs. The textual analysis of newspaper articles measures the changing legitimation level of EVs for private use. It can also confirm how specific barriers to innovation integration are configured in governance networks. Further textual analysis of newspaper articles and profiles of network organisations can reveal, first, how the innovation integration process is communicated, and second, to which network segments in particular the innovation is most closely related. Such analysis facilitates further exploration of the connections (and lack of connections) between innovations and governance networks, thus generating explanations for the legitimation level assessed in the textual analysis.

Gearing up for collaboration

Public sector capacity

In the transition to low carbon transport, both the public and private sectors contribute to significant drivers of the integration process of EVs for private use. Carley *et al.* (2013) conducted a survey to assess early consumer interest in plug-in EVs and found that the perceived disadvantages were significant deterrents to consumers' intention to purchase. The authors suggest that these deterrents could be 'addressed by public policy or private investment' (Carley *et al.* 2013: 45). Although it is uncertain to what extent this research (conducted in the United States) is transferrable to the Chinese market, the findings still imply a significant role for public policy instruments and public–private partnerships in China's sectoral integration of EVs. The essential questions remain: How do public–private dynamics shape individual perceptions of EVs to create and sustain their market niche, and how do such interactions transform the collaboration potential in China's environmental state? Answering these questions requires investigation into the intermediate mechanisms – *the communication and legitimation processes* – that drive changes in citizens' opinions.

With the focus on the role of public policy and private investment, the intangible links between citizens and policies advocating the use of EVs seem to be missing.

The *nongovernmental space* could fill this gap and play a more significant role in the integration process, particularly in enhancing the *output legitimacy* of EVs. Chapter 5 demonstrated the expansion of nongovernmental space in the sectoral integration process of green buildings and relevant technological innovations. To persuade the wider public to perform this behavioural change, that is, to switch vehicles, it is important to generate an atmosphere of community, to make individual car owners actually feel the impetus of shifting from burning fossil fuels towards zero emissions. This is the work of the nongovernmental space, allowing communal values to flourish (Gilchrist 2009).

Regarding the current phenomenon – an enthusiastic public sector yet an indifferent general public – observers of the Chinese automobile market have advocated for wider collaboration among multiple segments of society, in addition to government efforts and endorsements (He 2012). The public sector alone, even with solicited investments from the private sector, is not capable of accommodating the governance dynamics of CMNs during the mainstreaming of EVs. Engaging the network segments of nongovernmental space may raise the level of public acceptance of the sectoral innovations required for behavioural changes and thereby speed up the process of legitimation.

In Hong Kong, various public policy instruments have been used to promote EVs for private use. Since 1994, the First Registration Tax has been waived for the purchase of EVs (Appendix II: Gazette 1994). In 2012, road traffic regulations were amended to exempt private EVs from applying for expressway permits, provided the vehicles satisfy certain technical requirements prescribed by leading academics and the Transport Advisory Committee (Appendix II: Transport and Housing Bureau 2012). The government started to procure EVs in 2009, hoping to encourage similar actions by private users (Appendix II: Chief Executive 2009). Charging facilities were launched and managed by various public–private initiatives. The power companies, at the government's request, have been offering a free charging service to EV users since 2009 (CLP 2012; HK Electric 2013). The Housing Authority has also initiated a free parking scheme for charging EVs in selected public-housing car parks (Housing Authority 2012).

However, none of this seems to have had any real effect. Very few EVs are privately purchased in Hong Kong. A 1-week field observation in April 2013 revealed that hardly anyone was using the more than 1,000 EV charging points in Hong Kong. The charging areas in private car parks with small numbers of charging points were blocked from access. Although some government car parks have charging points (around 20–40 in all), the parking spaces with power points are reserved for conventional automobiles due to the low usage rate of recharging points in busy car parks.

Given the existing predicament in creating and sustaining a market niche for EVs, a member of the Steering Committee on the Promotion of Electric Vehicles reflected:

> It [the promotions of EVs] is a tricky area, because basically it's not one in which the government can be particularly proactive. Electric vehicles have an image problem [. . .] and I don't think the government can tackle the image

problem, because when people buy a car, they don't instinctively look for government endorsement – they look at the manufacturer's publicity, they may look at car reviews, etc. [. . .] The committee tries to do things in areas where it has some influence, and tries to ensure that things are moving along in the right direction. But the big breakthrough hasn't come yet.

(Appendix I: 37/HK/07–01–2014)

The supposed 'big breakthrough' would be the creation of a genuine market niche for EVs in Hong Kong. The 'image problem' is basically insufficient legitimation masking the market niche. There has been 'unified support' for the EV switch within the Steering Committee, representing the government, property developers, power companies and academia (Appendix I: 37/HK/07–01–2014). However, the legitimation of EVs does not rely solely on the public sector and big corporate groups. The actual public demands are not met by the expected environmental outcomes for EVs, creating loopholes in the networking for sectoral integration.

Hearing from and giving feedback to the community voice might be a positive and encouraging way of enabling the wider use of EVs by the general public. Societal-wide *communication* plays a significant role here. On 4 April 2009, in the opinion pages section of the *South China Morning Post* (*SCMP*), a reader recommended that the government turn the Discovery Bay residential area into a testing ground for EVs. Just a few days later in the same newspaper, an article explained that the Transport Department was indeed going to test the locally designed EV model in Discovery Bay, which happened to address the reader's suggestion. This public-sector endorsement provided a formal channel for private investment and experimentation by local EV manufacturers. *SCMP* readers and Discovery Bay residents alike gave a warm response to the government's prompt action. This shows that strengthened collaboration with the community is a feasible approach to resource mobilisation and legitimation.

Collaboration and resource mobilisation

There are sufficient resources for technological development, but limited resources for innovation integration. Some policy tools are designed to mobilise the resources invested from the private sector and ultimately encourage individual purchases. The resource mobilisation policies are often designed with underlying incentive mechanisms. Two policy initiatives, incorporating economic and information incentives, have embarked on mobilising resources among private developers and within the transport industry: a *preferential building requirement* to set up charging facilities in new private estate developments and an *earmarked fund* for the transport industry. However, these instruments have not generated a sufficient change in perception to integrate the sectoral innovation.

To provide a wider web of charging infrastructures, full concessions on gross-floor-area calculation have been granted to underground 'electric-vehicle (EV) charging-enabling'[3] car parks in new private buildings. The policy was promulgated by the Buildings Department and took effect from 1 April 2011.[4] Surprisingly, local environmental groups and individual car owners reacted negatively to this policy initiative; they questioned whether the concessions would exacerbate Hong Kong's already inflated property prices (*HKET* 2011). Because it drew negative attention to EV–related facilities, the *resource-mobilisation* instrument jeopardised EVs' *legitimation.* The expected environmental outcomes of this sectoral innovation were still unmet – whether in the government, the private sector or the nongovernmental space – leaving the output legitimacy tangled in the confusing public opinions.

The detrimental public reaction also reduced the perceived information incentives for corporate branding (i.e., building a socially responsible image), which in turn decreased private developers' motivation to construct car parks with charging facilities. To enhance the information incentives, the public sector needs to collaborate with local green-building assessment authorities (i.e., the BEAM Society and Hong Kong Green Building Council). The most recent version of the local green-building assessment method (BEAM-Plus 1.2, launched in November 2012) still does not contain provisions on infrastructural support for EVs in car parks. The current requirements regarding car parks address only ventilation, lighting systems, indoor air quality and public transport connections. Additional credits in the green building assessment scheme should be granted for 'EV charging-enabled' car parks, which would be compatible with government policy initiatives. This would remedy the reduced information incentives, as it is valuable for developers who want to improve their corporate image to gain credits to achieve green building entitlements. Modifying green building assessment methods requires the government to persuade and cooperate with the independent professional community. This will involve *cross-sector collaborations* between the transport and building sectors and across the public, private and professional sectors.

The government introduced an earmarked grant in its 2010 Policy Address – the 'Pilot Green Transportation Fund', an initiative to reduce road transport carbon emissions. However, this pilot fund only subsidises commercial EVs, such as taxis, franchised buses and goods vehicles, which does not directly motivate individuals to purchase EVs for private use. The pilot fund is not just a direct subsidy with an economic incentive. Its implicit objective is rather to prompt foreign vehicle manufacturers and research institutes to import and transfer internationally developed technologies to Hong Kong, which may indirectly facilitate the local integration of newly developed EV technologies (Ming Pao 2010). However, despite a capital investment of HK\$300 million, only a few applications for the grant were submitted in the initial selection rounds (Ng 2011). With no active response from the commercial transport industry, the supposed goals of resource mobilisation have not been achieved yet.

The government, therefore, should be more proactive in persuading the transport industry to set up EV trial schemes. These tactics could involve providing full information on local EV trial operation data and introducing available EV models to industrial operators. Industrial associations and alliances are feasible venues for such persuasion, for example, the Taxi and Mini-bus Concern Group, Hong

Kong Automobile Association and Hong Kong Institute of the Motor Industry. Again, *cross-sector collaboration* efforts are necessary to bridge the public and private sectors within the legitimation environment provided by the nongovernmental space.

Collaboration and legitimation

Resource mobilisation mainly involves hard measures to enable behavioural change by developing environmentally sustainable transport, while legitimation requires soft measures to enable attitude change through persuasion and education (OECD 2004). Legitimation is directly related to the information and communication components identified in the collaborative functions of the CMNs (Figure 2.2). Methodologically, the persuasiveness of EV policy should be reflected in the public media, and hence the level of legitimation can be measured by textual analysis of the major print medium – newspapers.

Hillman *et al.* (2011) provide possible methods for the specific measurement of legitimation in an innovation system. One method is to count the number of articles on EVs in local newspapers and calculate the ratio between positive and negative coverage (Hillman *et al.* 2011). As the media in Hong Kong are more free and independent than in other Chinese cities, it is feasible to conduct research on the legitimation of EVs through Hong Kong newspaper data.

Textual analysis of newspapers serves the purpose not only of measuring the legitimation level but also of discerning the discourse in which EVs are integrated. The analysis identifies the most relevant issues of EV integration. It is conducted by coding keywords extracted from the reviewed newspaper articles, categorising these keywords according to a thematic framework and observing the strengthening/weakening of the conceptual links.

The selection of newspapers was based first on the timeframe of EV development in China. Textual data were retrieved from newspapers issued after 20 March 2009, when the central state outlined the *Automotive Industry Readjustment and Revitalisation Plan*,[5] in which developing 'new-energy automobiles' was designated as a national strategy for the automotive industry. 'New-energy automobiles' refers both to all-electric and to hybrid electric-petroleum vehicles. This policy document is generally regarded as the start of the EV generation in China (Han 2011; Kokko and Liu 2012; Zhao 2010). The end date for the selection of newspapers was 20 March 2013, providing four years of textual data.

The second basis for selection was the accessibility and credibility of the newspaper. The bestselling and most widely distributed newspapers were selected due to their strong influence on public opinion, and a local university's credibility report provided another reference.[6] The division of readership between the general public and commercial investors was also considered, because the purchase of EVs for private use is also heavily affected by reviews of car market conditions. Both Chinese and English newspapers were selected because Hong Kong has a substantial English-speaking population and both languages are widely used, officially and informally.

Taking into account all of these factors, four local newspapers were selected:

1) *Headline Daily* was chosen because it had the highest credibility among the free newspapers (others are *AM730* and *Metropolis Daily*) that are distributed every morning in all Mass Transit Railway stations and some busy pedestrian areas.
2) *Ming Pao* was chosen from among the conventional Chinese print media for its high credibility score and its no-fee accessibility on Yahoo! Hong Kong, a popular local homepage. Although newspapers such as the *Oriental Daily News*, *Apple Daily* and *Sun Daily* were the bestselling ones in Hong Kong, low credibility was recorded among the general public, and thus they were excluded from the data source selection.
3) In the business sector, the *Hong Kong Economic Times* (*HKET*) was the bestselling local financial newspaper and had a high credibility rating.
4) Among the English media, the *South China Morning Post* (*SCMP*) enjoyed the highest credibility and had stable access to the professional community and to business elites.

Only the more conspicuous sections of the newspapers were reviewed and analysed for relevant content, and Table 6.1 summarises the selected sections and counts of relevant articles. The rest of the chapter presents the findings of this newspaper textual analysis on the legitimation level of EVs and on the contexts and network segments in which the voices for or against the EVs were embedded.

Level of legitimation

It is interesting that the number of articles on EVs in these newspapers declined annually from 2009 to 2013 (Table 6.2). A reasonable explanation is that the *notability* of EVs' applicability and usage is *declining* in Hong Kong.

At the same time, the number of EVs in private use was not in any sense rising significantly, despite the exponential increase in the number of local charging points (Table 6.3). It is not a positive sign that before public acceptance of this sectoral innovation has been widely established, the issue has already faded from public attention.

With this seemingly weak public acceptance of EV usage, how has the ratio of positive to negative coverage of EVs in these four newspapers changed over the survey period? The retrieved newspaper articles' nuanced descriptions of EVs were rated on a five-point scale: extremely negative, slightly negative, neutral, slightly positive and extremely positive. A value of 2 was given for articles that were 'extremely positive' or 'extremely negative', 1 for 'slightly positive' or 'slightly negative' and zero for 'neutral'. Table 6.4 illustrates the positive–negative ratios for each of the newspapers, in addition to the *SCMP*'s 'Opinion/ Insight' pages, which contain material from readers rather than from reporters or editors and thus were calculated separately.

Table 6.1 Selection of newspapers for the coverage of electric vehicles (20 March 2009–20 March 2013)

Keyword	Language	Newspaper name	Selected sections	Number of relevant articles
'電動車'	Traditional Chinese	*Headline Daily* (頭條新聞)	News (新聞), *Hong Kong News* (港聞)	69
		Ming Pao (明報)	*Hong Kong News* (港聞)	118
		Hong Kong Economic Times (經濟日報)	*News Snapshot* (新聞特寫), *Hong Kong News* (港聞) and 'Management' Editorial	94
'Electric Vehicles or 'Electric Vehicle'	English	*South China Morning Post*	Headline, Editorial and City	56

Source: Wisenews, accessed through the City University of Hong Kong Library Electronic Database

Table 6.2 Annual changes in newspaper articles on electric vehicles

	HKET		Ming Pao		Headline daily		SCMP		Total (years)	
3/20/2009 – 3/20/2010	36	38%	52	44%	25	36%	31	55%	**144**	**43%**
3/20/2010 – 3/20/2011	22	23%	29	25%	14	20%	13	23%	**78**	**23%**
3/20/2011 – 3/20/2012	17	18%	28	24%	19	28%	9	16%	**73**	**22%**
3/20/2012 – 3/20/2013	19	20%	9	8%	11	16%	3	5%	**42**	**12%**
Total (newspapers)	94		118		69		56		**337**	

Table 6.3 Electric cars and charging points in Hong Kong (2009–2012)

	Private EVs	Newly registered private cars	Government procurement of EVs	Charging points
2009	20	23,783	10	9
2010	54	31,828	10	48
2011	26	34,767	13	260
2012 (by June)	210	13,260	42	1,000

Source: Data retrieved from 2010/11 and 2011/12 Budget Speech, Hong Kong Monthly Digest of Statistics (Census and Statistics Department, June 2012) and Government Press Release (Environment Bureau, 24 May 2012)

It is interesting that the newspapers' evaluations of EVs were quite *diverse*. Nevertheless, the ratio calculated for the *Headline Daily* (*HD*) is not as valid as those for other newspapers because most of the articles were neutral. The *HD*'s message on EVs may have been less influential than others because many of the articles were short, consisting of only two or three sentences. There were more negative than positive articles in the *SCMP* Opinion/Insight pages and *Ming Pao*. The level of legitimation probably varies depending on the recipient groups of the diversified messages conveyed in different newspapers. Frequent readers of *Ming Pao* might be more critical of EVs than readers of the *Hong Kong Economic Times*. Therefore, the level of legitimation is possibly *uneven* among the general public.

Nonscaled measures of the positive/negative ratios in these newspapers over time are reported in Table 6.5. The *HKET* and the *Headline Daily* reduced their positive coverage of EVs, while the negative news in *Ming Pao* remained stable. Both the *HKET* and *Headline Daily* had a generally positive stance on EVs according to the data in Table 6.4, but Table 6.5 demonstrates that their *positive*

Table 6.4 Scaled ratios (+/−) of articles' perceptions of electric vehicles

HKET	2.12	Positive dominated
Ming Pao	0.6	Negative dominated
Headline Daily	3.38	Positive dominated
SCMP	1.56	Positive dominated
SCMP talkback	0.85	Negative dominated

Table 6.5 Annual changes in nonscaled ratios of articles on electric vehicles

	HKET			*Ming Pao*		
	Positive (+)	*Negative (−)*	*Ratio (+/−)*	*Positive (+)*	*Negative (−)*	*Ratio (+/−)*
3/20/2009–3/20/2010	23	5	4.60	9	21	0.43
3/20/2010–3/20/2011	7	5	1.40	9	15	0.60
3/20/2011–3/20/2012	6	3	2.00	12	11	1.09
3/20/2012–3/20/2013	12	7	1.71	3	3	1.00
	Headline Daily			*SCMP*		
	Positive (+)	*Negative (−)*	*Ratio (+/−)*	*Positive (+)*	*Negative (−)*	*Ratio (+/−)*
3/20/2009–3/20/2010	11.75	2.25	5.22	14.4	10.1	1.43
3/20/2010–3/20/2011	2	1	2.00	6.8	1.7	4.00
3/20/2011–3/20/2012	8	2	4.00	3.5	1.5	2.33
3/20/2012–3/20/2013	5.5	2.5	2.20	0	1	0

ratios have *shrunk* in more recent years. This is not a favourable trend for increasing the legitimation level of innovation integration.

Communicating the need for integration: issue context and network segments

To further the analysis of the positive–negative article ratios, this section examines the content of each article in greater depth to identify the issue contexts and network segments in which EVs are most often mentioned. It also identifies which aspects of innovation integration are neglected in the communication process, presumably for the sake of legitimation.

Studies show that the mainstream print media shape people's attitudes on whether they should or should not purchase/use EVs (El Banhawy *et al.* 2012; Lane 2000; Probe and Lavack 2009). The print media, which are 'able to deliver a high level of information' to the target audience group, are part of the 'integrated marketing communication' strategy of EVs, which 'is intended to move the consumer from attention, to information/understanding, to desire, intention and finally action, and is most effective when all tools work together to convey a *single unified message*' (Probe and Lavack 2009: 50). The media and social networks are identified as one of the 'key factors affecting EVs' market penetration', that is, the creation of a market niche, and 'media influence is considered to be one of the key influences on how powerfully and effectively it swiftly conducts and circulates pervasive views and feedback within certain vicinities and metropolitan areas, the potential leverage points of EVs (El Banhawy *et al.* 2012: 4).

In this regard, to conduct the discourse analysis, we mapped the public perceptions to the issue contexts, different network segments in the low carbon transition CMNs and individual preferences with rational choice calculations. We developed a coding scheme accordingly (Chapter appendix: Table 6.1A). We applied this coding scheme to the discourse analysis of textual data retrieved from the *HKET* and *Ming Pao* (*MP*), given their more substantial textual contents and mutually contrasting stances on EVs.

Figure 6.1 displays the context in which EVs and EV integration were mentioned. It appears that *cross-sectoral thinking*, given its significance in resource mobilisation (discussed in the previous section), has *not* been closely associated with EV promotion. Conversely, the various forms of *transnational policy diffusion* were prominent in shaping the issue context of EV integration, which may be related to Hong Kong's positioning as a *global city*.

Conversely, Figure 6.2 illustrates the varying degrees of EV integration in different segments of the local CMNs for low carbon transitions. While communication about EVs was found to be highly integrated with the public sector and the market sphere, loose connections were found with the *nongovernmental space*. The nongovernmental space for the integration of EV innovation is composed of diverse groups – transport and business associations, neighbourhood committees, local green groups, transnational municipal networks, the professional community and academia. These entities generally have weak connections in communicating innovations for broader integration.

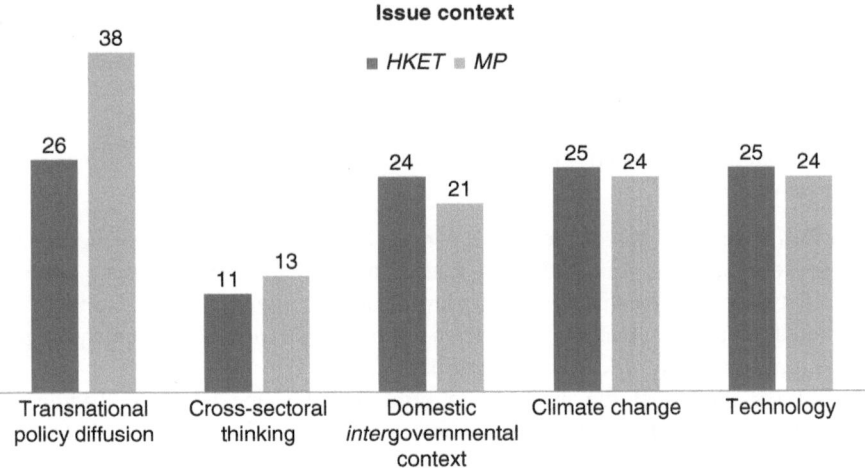

Issue context

■ HKET ■ MP

38

26

24
21

25 24

25 24

11 13

Transnational policy diffusion | Cross-sectoral thinking | Domestic *inter*governmental context | Climate change | Technology

Figure 6.1 Issue context of newspaper coverage on electric vehicles
Source: Compiled by authors

Nongovernmental space
- Industrial associations
- Local ENGOs
- Transmunicipal networks
- Neighborhood
- Academia

Public sector
- Steering committee
- Policy instruments
- Legislature
- Implementation agencies
- Senior administration

Market sphere
- User demands
- Corporate particpants
- Business elites
- Manufactuers

0   10   20   30   40   50   60   70

■ MP ■ HKET

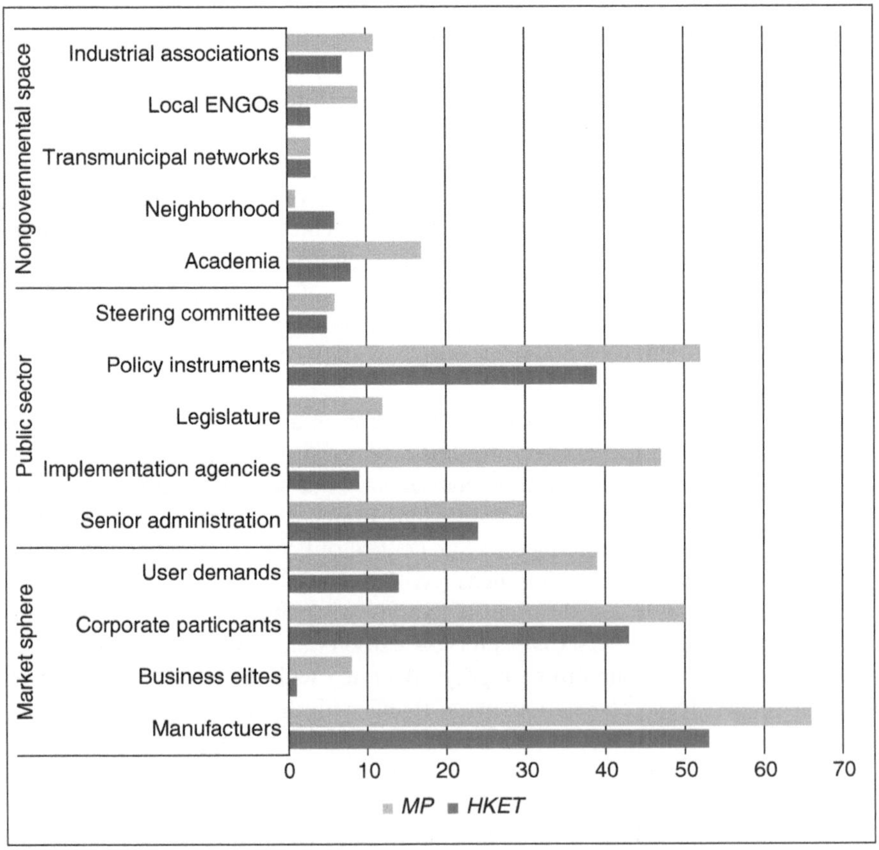

Figure 6.2 Network segments in newspaper articles on electric vehicles
Source: Compiled by authors

Lost in translation: the disconnected network

To integrate a sectoral innovation such as EVs for private use, effective communication with full information about the innovation's accessibility is essential to raising the legitimation level and public acceptance of the innovation. The private purchase and usage of EVs will require the wider public to make behavioural changes that need to be based on a solid societal foundation of legitimation. CMNs for low carbon transitions, as a network governance approach, serve the collaborative function of communication and information provision. This communication function connects and accommodates an array of *network participants* in the public sector, the market sphere and the nongovernmental space centred on the technical and market niches of EVs, which enhances the sectoral innovation's *input legitimacy.*

Given the findings in Figure 6.2, the nongovernmental space is clearly *detached* from the communication networks agglomerated for EV integration. In the four years of newspaper coverage, local environmental nongovernmental organisations (ENGO) appeared fewer than ten times, as summarised in Table 6.6. Most of the mentions were related to broader issues – such as climate change, energy security, (roadside) air quality and comprehensive transport planning – and EV integration appeared only as an indirect solution. In one instance, EV promotion was a secondary issue, in an article describing how the gross-floor-area concessions conferred on private car parks in new buildings with charging facilities might further inflate property prices. Only one ENGO appearance (15 Apr 2011 in the *SCMP*) was directly related to EV promotion.

Table 6.6 Local environmental nongovernmental organisations in newspaper coverage on electric vehicles

Issue	ENGO	Media	Date
Energy security of nuclear power	Civic Exchange, Green Sense	*HKET*	18 Dec 2010
Gross floor area concession in green buildings	Green Sense	*HKET*	16 Dec 2011
Comprehensive transport planning	Friends of the Earth	*HKET*	23 Jun 2012
Air quality: timetable for proposed measures	Not specified	*MP*	24 Jul 2009
Climate change: carbon emission reduction target	Greenpeace	*MP*	17 Dec 2009
Climate change: action plan	Friends of the Earth, Oxfam	*MP*	11 Sept 2010
Roadside air quality: pollutant reduction	Friends of the Earth	*MP*	14 Jun 2012
Area exclusively for electric vehicles	World Wide Fund (Hong Kong) – Climate Programme	*SCMP*	15 Apr 2011
Air quality standards and measures	Civic Exchange, Green Power, Friends of the Earth	*SCMP*	18 Jan 2012

Even the concern about the potential negative environmental impact related to the wider use of EVs was raised not by local ENGOs but by practitioners in the private sector. A general manager of a car dealership company handling government procurement transactions with Mitsubishi in Japan told the journalist that he was made aware of the potential environmental problem of battery waste if EVs became more widely used (Cheung 2009).

It is clear from this print media survey that the nongovernmental space – especially the ENGOs and industrial associations – is not closely involved with the communication and persuasion process of EV integration. It has been discussed in this thesis that the organisational nodes of ENGOs and industrial associations occupying the nongovernmental space are the integral links in the governance networks, particularly in facilitating the process of cross-sector collaboration. Such collaboration is indispensable for innovation integration in low carbon transport. With the missing link of the nongovernmental space in publicising the sectoral innovation, broader integration remains unrealised.

Indeed, the ENGOs are not actively concerned with the sectoral integration of EVs. A discourse analysis of all local green groups' profile statements showed that only a few local ENGOs have expressed explicit opinions about EVs. The innovation integration is disconnected from the established and entrenched networks of local green groups. The ENGOs do occupy a nonofficial seat on the Steering Committee on Promotion of Electric Vehicles, which is currently held by the chairman of the city's World Wide Fund for Nature (WWF). Nevertheless, this is a mere formality, a tiny minority compared to the 7 (out of 11) nonofficial seats occupied by Hong Kong's two power companies and major real estate developers.

It is pertinent to compare the ENGOs' profile statements on EVs with their interest in green buildings (Chapter 5). Table 6.7 presents the finding of this comparative discourse analysis, which covers all local green groups listed in the directory[7] of the Environmental Protection Department, with the addition of a newly established green group – World Green Organisation (WGO) – and the long-standing Business Environmental Council (BEC).

Significantly less attention was paid to EVs than to green buildings. The mean keyword frequency (shown in Table 6.7) related to green buildings in ENGOs' website domains was 12.54, compared with only 5.96 related to EVs. The result of a one-tailed t-test on the numeric data of the two keyword frequencies ($n = 26$) revealed a *statistically significant* difference between the levels of focus on the two sectoral innovations among local green groups, $t(25) = 1.576, p = .063$, with green buildings receiving much more attention from the ENGOs.

Only three ENGOs (Civic Exchange, Clear the Air and the Green Technology Consortium) described EVs as a stand-alone issue covered by their organisations. Among these, only Civic Exchange had an active and prominent position; the other two organisations were not actively engaged in EV-related issues. In comparison, six ENGOs regarded green building as a significant independent issue, and at least half of them were quite active judging by their high keyword frequencies in Table 6.7.

Another round of textual analysis of the organisational documents suggests that many grassroots green groups consider the purchase of EVs to be an unnecessary

Table 6.7 Position of local environmental organisations on green buildings and electric vehicles

Organisation	Green buildings			Electric vehicles		
	Stand-alone issue@	Subordinate to broader issues#	Keyword frequency%	Stand-alone issue@	Subordinate to broader issues#	Keyword frequency%
Business Environment Council (BEC)	Yes	N.A.	108	No	Yes (low carbon economy, not a core focus)	13
Centre of Architectural Research for Education (CARE)	Yes	N.A.	12	No	No	1
Civic Exchange	Yes	N.A.	97	Yes	N.A.	60
Clean Air Network (CAN)	No	Yes (air quality)	11	No	Yes (air quality)	45
Clear the Air	No	Yes (energy & power plants)	1	Yes	N.A.	5
Designing Hong Kong	No	Yes (urban planning & environment)	5	No	No	1
Earth Care	No	No	0	No	No	0
Friends of the Earth – HK (FOE)	No	Yes (energy & climate)	12	No	No	1
Green Council	No	Yes (green label)	34	No	No	16
Green Lantau Association	No	No	0	No	No	0
Greenpeace	No	Yes (climate & energy)	7	No	Yes (reduce air pollution)	3
Green Peng Chau Association	No	No	0	No	No	0
Green Power	No	No	5	No	No	1
Green Sense	Yes	N.A.	5	No	No	0
Green Technology Consortium	Yes (LED Lighting)	N.A.	2	Yes	N.A.	1
Greeners Action	No	No	0	No	No	2

(Continued)

Table 6.7 (Continued)

Organisation	Green buildings			Electric vehicles		
	Stand-alone issue@	Subordinate to broader issues#	Keyword frequency%	Stand-alone issue@	Subordinate to broader issues#	Keyword frequency%
Hong Kong Dolphinwatch Ltd.	No	No	0	No	No	0
Kadoorie Farm & Botanic Garden	No	Yes (sustainable living)	2	No	Yes (sustainable procurement)	1
Ocean Park Conservation Foundation	No	No	0	No	No	0
Produce Green Foundation	No	No	0	No	No	0
Sustainable Ecological Ethical Development Foundation	No	No	0	No	No	1
Tai Po Environmental Association	No	No	1	No	No	0
The Conservancy Association	No	No	2	No	No	1
The Hong Kong Bird Watching Society	No	No	0	No	No	2
World Wide Fund For Nature Hong Kong (WWF)	No	Yes (Climate, Footprint)	12	No	Yes	1
World Green Organisation (WGO)	Yes	N.A.	10	No	No	0

Note:

@ The organisations consider the integration of electric vehicles and green buildings as one stand-alone independent issue on the organisational agenda.

The broader issues may include low carbon development, climate change or sustainable development. If the issue is a stand-alone issue@, this item is regarded as not applicable (N.A.).

% Keyword frequency is counted based on the results of an advanced search in a Web search engine (google.com.hk) confined by the organisational website domain. Keywords for green buildings include 'green building', 'sustainable building/ built environment', 'environmental building' and 'building energy efficiency'; keywords for electric vehicles include 'electric vehicle', 'electric vehicles', 'electric cars', 'environmental vehicles' and 'new energy automobiles/cars/vehicles'. (All data accessed 10 May 2013.)

and environmentally costly 'luxury', which is a misconception that both the public and private sectors need to clarify (Lau 2007). In contrast, local green groups are quite supportive of developing Hong Kong into a green building environment, hence their positive support for green building technologies and policies.

A substantial portion of the local ENGO network seems suspicious about the real environmental effect of integrating EVs. Therefore, to form a stronger connection with the nongovernmental space on EV promotion, actors advocating for EV integration in both the public and private sectors first need to address the hesitations of local ENGOs by providing more *information* to facilitate consensus building. The process of engaging the local ENGO network may be categorised by the *network centrality* of different organisations:

i. **Outlying nodes** – Some green groups are probably irrelevant to these two sectoral innovations, such as the ad hoc issue network organisations formed during crises in urban planning (e.g. the Green Lantau Association and Green Peng Chau Association) and those concerned purely with biodiversity conservation and other environmental issues (e.g. Earth Care, Hong Kong Dolphinwatch Ltd., the Ocean Park Conservation Foundation and Produce Green Foundation).

ii. **Peripheral nodes** – These green groups currently focus on other relevant environmental issues, but it is highly possible to engage them in the dynamic process of sectoral integration: the Conservancy Association, Tai Po Environmental Association, Sustainable Ecological Ethical Development Foundation, Kadoorie Farm & Botanic Garden, Green Power, Green Sense, Green Technology Consortium, Greeners Action and the Hong Kong Bird Watching Society. Other ENGOs currently focusing only on green buildings are also candidates for EV integration, including CARE, Designing Hong Kong and the WGO.

    Green Power: This organisation positively perceives EVs as 'eco-vehicles' that 'should be encouraged by the government . . . to increase competitiveness in the market' (Cheng and Woo 2012: n.p.). It also supports the government's view that 'charging stations for electric vehicles should be set up in different districts' to 'encourage car owners and the commercial sector to opt for electric vehicles' (Cheng and Woo 2012: n.p.). This positive position is quite rare in the local ENGO network. However, neither EVs nor green buildings are major issues for this organisation. Moreover, these sectoral innovations are not directly subordinate to any established organisational main working area. Therefore, deeper engagement and boundary-spanning practices with this organisation are feasible and favourable for the sectoral integration of EVs.

    Greener Action and the Hong Kong Bird Watching Society: Greener Action focuses on food waste, where EVs emerge as a possible means for collecting this waste. In the meantime, the environmental activists of the Bird Watching Society are concerned about pollution from cars running inside Mai Po, the bird-watching nature reserve. They suggest that only EVs should be allowed inside the reserve.[8] Although such organisations do not

focus *per se* on sectoral innovations such as EVs (or any green building technologies), they are supportive of incorporating these sectoral innovations
into their own organisational activities. According to these organisations,
adopting sectoral innovations could enhance the *environmental effectiveness*
of their activities, which would improve the *output legitimacy* of their own
endeavours.

WGO, CARE and Designing Hong Kong: These three organisations have
established positions related to green buildings. It is feasible to further engage
with them in the sectoral integration of EVs, to incorporate the innovation in
their stances on urban planning. The WGO is a newly established and locally
based ENGO, and although it has no established agenda or concern on EVs,
it is important to engage with this new organisation to achieve broader awareness in society.

iii. **Core nodes** – There are several well-established and resourceful ENGOs that
are vocal and influential on various green issues, including the BEC, FOE,
Green Council, Greenpeace and WWF. In addition, two organisations – the
CAN and Clear the Air – are highly concerned with local air quality, which
could lead to a stronger stance on the wider use of EVs. It is critical to get
these core organisational nodes engaged in the consensus building and communication process for sectoral integration.

FOE: As a well-established ENGO, FOE is rather *conservative* in its
position on replacing conventional vehicles with EVs. It has advocated for
actions to be taken to 'phase out dirty vehicles'[9] in its organisational agenda
on air issues. However, the promotion of EVs has not emerged as either a
stand-alone or a subordinate issue for this organisation.

Clear the Air: Despite the fact that EVs have recently been incorporated
into the organisational agenda as a stand-alone issue, the organisation does
not *per se* positively support the integration of sectoral innovation. A video
interview conducted by the *SCMP* in 2009 revealed the organisation's lack
of confidence in EVs, with its vice chair commenting that promoting EVs in
Hong Kong would be 'a distraction' from the problem of roadside air quality and thus should '*not*' be 'the priority'.[10] Such negative perceptions of
nongovernmental green groups and their leaders regarding the environmental
benefits of EVs should have been reversed by years of EV promotion that
have been steered mainly by the public sector in the city.

Green Council: The Green Council website mainly includes reposted news
clips on EVs, without any real organisational efforts to support or oppose
the public policy agenda on EVs. Nevertheless, a discourse analysis of these
reposted news clips demonstrated that the organisation's basic position on
EVs is likely to be positive, which may serve as a foundation for consensus
building and broader engagement.

WWF: The WWF in Hong Kong has opted to legitimate electric bikes
instead of EVs (WWF 2009). Although the organisation commented that EVs
would be 'further away from our day-to-day lives', it has not adopted any
specific position on EVs (WWF 2009). As its chairman is a member of the

administration's Steering Committee on the Promotion of Electric Vehicles, the organisation offers a potential platform for consensus building in the non-governmental space for the public and private sectors to strengthen cross-sector collaboration.

Conclusion

In this chapter, we have argued, first, that cross-sector collaboration is neces-sary for both *resource mobilisation* and *legitimation* – two essential stages of EV integration. In general, the incentive mechanism embedded in the limited policy options (preferential building requirements and earmarked funds) has not been effective in the mobilisation of resources. To avert this, cross-sector collaboration is a key approach, combining transport and building sectoral innovations with a sociotechnical transition strategy to be initiated by the public sector, accommo-dating the private sector, professionals and the community.

We then assessed the *legitimation level and communication process* of EV inte-gration across societal sectors. Legitimation and communication are the integral elements of sociotechnical transitions and are incorporated in the analysis of the governance dynamics in CMNs. The analysis of four years of textual data from four local newspapers with diverse readership indicated that even though EVs have still not gained wide public acceptance, public attention to the subject has reduced. The legitimation level of sectoral innovation conveyed by the main-stream print media was assessed by calculating the ratio between positive and negative articles. The results revealed that EV acceptance levels were uneven across different newspapers, which may have resulted in the varied public per-ceptions of EVs. In addition, despite its aforementioned importance in sectoral integration, cross-sector collaboration has not been incorporated in either policy making or communication of EVs, which was evident from the textual analysis of the retrieved newspaper articles.

The textual analysis of EV integration also suggested that network participants in the nongovernmental space – the political space for academia, ENGOs and transport industry associations – are weakly engaged in the communication pro-cess of sectoral integration. The situation is particularly evident in the limited participation of ENGOs. Therefore, a further analysis of the public profiles of all local ENGOs was conducted, which found that the ENGOs appeared to be mar-ginalised in the sectoral integration process compared with the dominant involve-ment of the public and private sectors. In fact, prominently active local ENGOs were not concerned with EV promotion *per se*, which may explain why the inno-vation integration is interwoven in the governance networks and has attained nei-ther broader legitimation nor wider usage.

Limited engagement is confining the input legitimacy of sectoral innovation in the governance networks. First, the technocrats are engaged in government con-sultation (legislation and policy making) but not in legitimation. Second, there is a lack of representation in the external links of *intra*governmental coordination – only one nongovernmental organisational actor is a member of the Steering

Committee on the Promotion of Electric Vehicles, compared with the extensive representation of property developers and power companies.

Third, the nongovernmental space is largely detached from sectoral integration, as shown by the textual analysis. In this regard, differentiating ENGOs according to their network centrality to EV integration makes it possible to analyse how diverse network segments may interact under different circumstances to communicate and legitimate the sectoral innovation. Managing the market niche of EVs necessarily involves networking, with the selection and identification of actors according to their resources, competencies, innovative ideas and vested interests. The public sector must apply collaborative strategies of cooperation and consensus building in the core organisational nodes, given their rich resources and powerful influence in the nongovernmental space and across societal sectors. The peripheral nodes are also key players to be engaged in building broader societal support for the innovation.

Theoretically, sectoral integration is a process of enhancing input legitimacy that departs from *intra*governmental and hierarchical coordination (Chapter 4) and arrives at cross-sector collaboration in CMNs. Through cross-sector collaboration, *input legitimacy* is transformed into *output legitimacy* by effectively aligning the expected environmental and developmental outcomes across policy and societal sectors.

In this process, interactions are active not only among networking actors, but also between collaborative functions, that is, between the innovation and communication on sectoral integration and legitimation and between the incentives for resource mobilisation and communication for legitimation. These dynamic interactions either mutually improve or collectively impair the collaborative outcomes.

The level of legitimation determines the output legitimacy of sectoral innovations, whereas communication and participation in the governance networks reflects the input legitimacy. The networking for market niches involves the alignment of expectations across the public sector, the market sphere (supplies and demands) and the nongovernmental space, which connects citizens' perceptions with the market and public policies. The output legitimacy refers precisely to the *expected* environmental and development outcomes of the sectoral innovation. These multiple sources of expectations are currently not aligned for the sectoral innovation of EVs, as suggested by the uneven legitimation across societal sectors.

Considering the limited engagement with and poor representation of the sectoral integration process of EVs, it is reasonable to conclude that the input legitimacy of this sectoral innovation has been weak. The nongovernmental space has limited engagement in this process, which, as discussed in Chapter 5, provides a legitimation environment for sectoral innovations such as green building technologies. If the nongovernmental space supporting sectoral innovation is composed of a substantial aggregate publicness of organisations, it also raises the significance of the integration process through the CMNs. Improving the input legitimacy will require enhanced engagement and strategic boundary-spanning activities not only with the core networks but also with those on the periphery of the nongovernmental space for sectoral innovation.

Chapter appendix

Table 6.1A Conceptual mapping and coding scheme for discourse analysis of newspaper articles on electric vehicles

	Conceptual mapping	Codes
Issue context	Domestic *intergovernmental* context	National political agenda [e.g. Developing new-energy automobiles, 12th Five-Year Plan]; Regional development plan; Private sector: mainland
	Climate change threat	Power generation: emission problem; Related issue: climate change; Related issue: low carbon living/ development; Related issue: carbon reduction; Related issue: energy
	Technological development	Technological development
	Transnational policy diffusion	Private sector: global; Mega events; Policy diffusion [e.g. ministerial visits, policy learning]
	Cross-sectoral thinking	Related issue: green building; Public–private initiatives
Market sphere	Manufactures of EVs	Manufacturer: local; Manufacturer: mainland; Manufacturer: overseas
	Business elites	Business elites

(Continued)

Table 6.1A (Continued)

Network segment	Conceptual mapping	Codes
	Corporate participants	Private sector: local [e.g. power companies, property management companies, real estate developers]
	User Demands	EV Applications: Bus EV Applications: Taxi Buyers: individual [e.g. citizen orders] Buyers: institutional [e.g. corporate procurement]
Public sector	Senior administration	Government: senior administration; Political elites
	Implementation agencies	Statutory body; Government: commission/department [e.g. Transport Department, Environmental Protection Department, Innovation and Technology Commission]
	Legislature	Government: legislature [e.g. Legislative Council Members, discussion]
	Policy instruments	Policy supports; Economic incentives
	Intragovernmental steering group	Steering Committee [Steering Committee on the Promotion of Electric Vehicles]
Nongovernmental space	Academia	Scholar; Academic Institute
	Neighbourhood	Owner's corporations
	Transnational municipal networks	TMNs [e.g. C40, Clinton Foundation]
	Local environmental nongovernmental organisations	ENGO [e.g. Greenpeace, Friends of the Earth, etc.]
	Industrial associations of transport sector	Industrial/business associations [e.g. Transport Advisory Committee, Taxi and Mini-bus Concern Group, Hong Kong Automobile Association, Hong Kong Institute of the Motor Industry, etc.]
Rational choice	Environmental factors	Infrastructure [e.g. conventional/express charging stations, payment methods, parking]; Applicability [e.g. local climate, road conditions, etc.] supplies
	Cost calculation	Capital costs; Concurrent costs [e.g. maintenance, battery replacement, time cost for charging]
	Behavioural change	Behavioural change [e.g. commuting habit, consumer behaviour]

Notes

1 It is estimated that electric vehicles convert about 59 to 62 percent of the energy they consume into actual power in the wheels, whereas only about 17 to 21 percent of the energy stored in gasoline is converted to the wheels in conventional-fuel vehicles (EPA, 2011; Miller *et al.*, 2011).
2 Terajoules is a measure of energy: 1 TJ is equivalent to 1 trillion joules.
3 A term used in the Administration's Legislative Council Brief, available at www.legco. gov.hk/yr10-11/english/panels/dev/papers/dev0329cb1-1668-4-e.pdf [Accessed 6 May 2013]. The term specifically refers to within-building parking spaces that are installed with electrical wiring and with provision of sufficient power supply for future installation of standard EV charging facilities.
4 Refer to the government press release [Online]. Available: www.info.gov.hk/gia/general /201101/31/P201101310250.htm [Accessed 6 May 2013]
5 State Council document [Online]. Available: www.gov.cn/zwgk/2009-03/20/content_ 1264324.htm [Accessed 8 May 2013]
6 This tracking study on media credibility is available at www.com.cuhk.edu.hk/ccpos/ en/research/Credibility_Survey%20Results_2010_ENG.pdf, Chinese University of Hong Kong [Accessed 8 May 2013]
7 The checklist of local ENGOs is available at www.epd.gov.hk/epd/english/links/local/ link_greengroups.html [Accessed 9 May 2013]
8 Further discussion is available at www.hkbws.org.hk/cgi-bin/YaBB.pl?board=Discuss ion;action=display;num=1146318153 [Accessed 10 May 2013]
9 See http://www.foe.org.hk/e/content/cat_page.asp?cat_id=41#.UY8yxKL7Apk [Accessed 10 May 2013]
10 The video interview is available at www.cleartheair.org.hk/tc/electric-vehicles.php [Accessed 10 May 2013]

References

Autoreport. 2012. Li Yunfei: Private Electric Cars Are Not the Promotion Focus of Byd [Online]. Autoreport.cn. [Accessed 3 May 2013] (*Li Yunfei: Sijia Diandong Che Bushi Biyadi Tuiguang Zhongdian*)

Bryson, J. M., Crosby, B. & Stone, M. M. 2006. The Design and Implementation of Cross-Sector Collaborations: Propositions from the Literature. *Public Administration Review,* 66, 44–55

Caniëls, M. C. J. & Romijn, H. A. 2008. Actor Networks in Strategic Niche Management: Insights from Social Network Theory. *Futures,* 40, 613–629

Carley, S., Krause, R. M., Lane, B. W. & Graham, J. D. 2013. Intent to Purchase a Plug-in Electric Vehicle: A Survey of Early Impressions in Large US Cites. *Transportation Research Part D: Transport and Environment,* 18, 39–45

Cheng, L. K. & Woo, K. 2012. Car-Free Hong Kong [Online]. Available: www.greenpower. org.hk/html/eng/2012_10.shtml [Accessed 12 May 2013]

Cheung, C. F. 2009. Electric Cars Spur Debate on Battery Waste. *South China Morning Post,* 9 October 2009.

CLP. 2012. *CLP Extends Free Electric Vehicle Charging through 2013.* Hong Kong: CLP Power Hong Kong Limited [Online]. Available: www.clpgroup.com/ourcompany/news/ currentrelease/Documents/20121227_en.pdf [Accessed 11 December 2013]

El Banhawy, E. Y., Dalton, R., Thompson, E. M. & Kotter, R. 2012. *A Heuristic Approach for Investigating the Integration of Electric Mobility Charging Infrastructure in Metropolitan Areas: An Agent-Based Modeling Simulation,* in 2nd International Symposium on Environment Friendly Energies and Applications 25–27 June Newcastle upon Tyne, UK, 74–87

Environment Bureau, H. K. 2012. *Locations of Charging Stations for Public Access.* Hong Kong [Online]. Available: www.enb.gov.hk/en/resources_publications/guidelines/files/charging_points.xls [Accessed 3 May 2013]

EPA, U.S. 2011. *All-Electric Vehicles* [Online]. Available: www.fueleconomy.gov/feg/evtech.shtml [Accessed 3 May 2013]

Gilchrist, A. 2009. *The Well-Connected Community: A Networking Approach to Community Development,* Bristol, UK, Policy Press.

Han, L. 2011. The Policy Improve Domestic Electric Special Vehicle Forward. *Automobile & Parts,* 4, 50. (*Xin Nengyuan Qiche Zhengce: Tuidong Woguo Diandong Zhuanyong Qiche de Kuaisu Fazhan*)

Harris, P. G. 2012. *Transport, Environmental Policy and Sustainable Development in China: Hong Kong in Global Context,* Bristol, UK, Policy Press.

He, L. 2012. Electric Vehicle Market Is Waiting for a Finishing Touch from the Government. *China Business Update,* December, 80–81. (*Diandong Che Shichang Dengzhe Zhengfu Linmen Yijiao*)

Hillman, K., Nilsson, M., Rickne, A. & Magnusson, T. 2011. Fostering Sustainable Technologies: A Framework for Analysing the Governance of Innovation Systems. *Science and Public Policy,* 38, 403–415

HK Electric. 2013. *EV Charging Stations.* Hong Kong: Hongkong Electric Company Limited [Online]. Available: www.hkelectric.com/web/ElectricLiving/EV/EVChargingStations/Index_en [Accessed 11 December 2013]

HKET. 2011. Electric Vehicle Parking Space Exemptions, Whether It Encourage 'Inflation'? *Hong Kong Economic Times,* 16 December 2011. [Accessed 1 June 2013] (*Diandong Chewei Huo Kuanmian, Shifou Guli 'Fashui'?*)

HKSAR Government. 2009. *Appointments to the Steering Committee on the Promotion of Electric Vehicles.* Hong Kong: Government Press Release [Online]. Available: www.info.gov.hk/gia/general/200903/30/P200903300108.htm [Accessed 3 May 2013]

HKSAR Government. 2012. *1,000 Charging Stations for Electric Vehicles to Be in Place by End of June.* Hong Kong: Government Press Release [Online]. Available: www.info.gov.hk/gia/general/201205/24/P201205240382.htm [Accessed 3 May 2013]

Housing Authority. 2012. *Parking Privilege Scheme for Charging of Electric Vehicle.* Hong Kong [Online]. Available: www.housingauthority.gov.hk/en/common/pdf/commercial-properties/leasing-information/parking-spaces/ev_charging_facilities.pdf [Accessed 11 December 2013]

Kokko, A. & Liu, Y. 2012. Governance of New Energy Vehicle Technology in China: The Case of Hybrid-Electric Vehicles, in Nilsson, M., Hillman, K., Rickne, A. & Magnusson, T. (eds.) *Paving the Road to Sustainable Transport: Governance and Innovation in Low-Carbon Vehicles,* London, Routledge.

Lane, B. 2000. Public Understanding of the Environmental Impact of Road Transport. *Public Understanding of Science,* 9, 165–174

Lau, C. F. E. 2007. Just How Eco-Friendly Are Hybrid Vehicles? [Online]. Available: www.foe.org.hk/e/content/cont_page.asp?content_id = 826#.UY8zkKL7Apl [Accessed 12 May 2013]

Lin, H. 2012. Assisted with Policies, Private Purchase of Electric Vehicles Enter the Practical Operation Stage. *Southern Metropolis Daily,* published by Nanfang Media Group. 8 November 2012. [Accessed 15 May 2013] (*Jie Zhengce Dongfeng, Siren Goumai Diandong Che Jinru Shicao Jieduan*)

Liu, W. 2013. New Energy Subsidy Policy Terminated, Private Electric Cars Stagnated Sale. *National Business Daily,* published by National Business Daily. 21 February 2013 [Online]. Available: www.nbd.com.cn/articles/2013–02–21/715587.html [Accessed 3 May 2013] (*Guojia Xinnengyuan Qiche Butie Zhengce Xian Zhenkong, Siren Diandong Che Xiaoshou Tingzhi*)

Liu, X. 2011. Hybrid Electric-Petroleum Vehicles and All-Electric Vehicles Will Be the Mainstream of New Energy Automobiles. *Business Watch Magazine,* 1, 43–44. (*Youdian Hunhe he Chun Diandong Che Jiangcheng Xin Nengyuan Qiche Zhuliu*)

Miller, M. A., Holmes, A. G., Conlon, B. M. & Savagian, P. J. 2011. The Gm 'Voltec' 4et50 Multi-Mode Electric Transaxle. *SAE International Journal of Engines,* 4, 1102–1114

Ming Pao. 2010. Pilot Green Transport Fund Attracts Manufacturers into Hong Kong. *Ming Pao,* published by Ming Pao Group. 7 March 2010. [Accessed 1 June 2013] (*Lvse Yunshu Jijin, Xi Shengchan Shang Laigang*)

Ng, K. H. 2011. Green Transportation Fund Approved All Electric Bus Applications. *Ming Pao,* 15 October 2011. [Accessed 1 June 2013] (*Lvse Yunshu Zizhu, Diandong Bashi Quanpi*)

OECD. 2004. *Communicating Environmentally Sustainable Transport: The Role of Soft Measures,* Paris, Organisation for Economic Co-operation and Development.

Probe, P. & Lavack, A. 2009. *Purchasing Fuel Efficient Vehicles in Canada.* Toronto and Ottawa: Centre of Excellence for Public Sector Marketing and Pollution Probe [Online]. Available: http://pollutionprobe.org/pdfs/PurchasingFuelEfficientCanada.pdf [Accessed 7 November 2014]

Qiu, X. & Tu, C. 2013. Uneven Distribution of Charging Points Confuses 'Blue Taxis', Shenzhen to Build Another 220 Thousand Charging Points. *Shenzhen News,* published by Shenzhen News. 8 April 2013 [Online]. Available: www.sznews.com/news/content/2013–04/08/content_7906854_2.htm [Accessed 3 May 2013] (*Chongdian Zhan Buju Bujun Kunrao 'Landi', Shenzhen Jiang Zengjian 22 Wanyu Chongdian Zhuang*)

Schot, J. & Geels, F. W. 2008. Strategic Niche Management and Sustainable Innovation Journeys: Theory, Findings, Research Agenda, and Policy. *Technology Analysis & Strategic Management,* 20, 537–554

WWF. 2009. Electric Bike [Online]. Available: www.wwf.org.hk/en/news/_m.cfm?5767/Electric-Bike [Accessed 10 May 2013]

Zhao, Y. 2010. The Development Trends and Problems of New Energy Automobiles in China. *China Technology Investment,* 5, 24–26. (*Woguo Xinnengyuan Qiche de Fazhan Qushi ji Wenti*)

7 Institutionalising climate experiments in collaborative municipal networks – what are they leading to?

In Chapter 4, we depicted how the *intra*governmental coordination with external network actors is structured in the three cities. In Chapters 5 and 6, we discussed the expansion of nongovernmental space in the sectoral integration of innovations such as green buildings and electric vehicles, which create the potential for collaboration. In this chapter, we take a disaggregate approach to governance theory to analyse how and why *coordination* and *collaboration*, as two intertwined processes, influence the governing of climate change in China. We also investigate how the emerging interaction between coordination and collaboration affects the established institutional structure of the climate change regime.

The analysis focuses on cases that explain the development and management of pilot city projects with demonstration objectives or experimental functions, which are being integrated into the CMNs on climate change mitigation in the three Chinese cities. Pilot climate initiatives act as political incentives for local government officials. Economic incentives are also generated through various funding schemes appended to pilot initiatives that foster participation in the CMNs for low carbon transition. The nexus of coordination and collaboration can be seen in the *preparation and implementation* of pilot climate initiatives. In this process, coordination is concerned with both vertical *inter*governmental and *intra*governmental relations, while collaborative functions serve to legitimate the imported initiatives and gain consent from formal and informal actors at different levels of governance. The key question of this chapter is:

> Within the dynamic interaction of coordination and collaboration, how and why do the core and peripheral actors in collaborative municipal networks (CMNs) interact to institutionalise climate pilot initiatives in the Chinese environmental state?

Governing climate pilot initiatives in cities

Various pilot projects for climate change mitigation have been launched in Guangzhou, Shenzhen and Hong Kong. These projects are experimental and exemplary public (government) programmes that aim to establish best practices at various levels of governance. Four types of local pilot projects, classified by scale of influence, are found in the low carbon transitions of the building and transport sectors.

The first type refers to *citywide pilot action plans* for low carbon development that require emissions reduction coordination and cross-sectoral collaboration. For instance, upon being designated a National Low Carbon Pilot City and a National Low Carbon Ecological Demonstration City, the Shenzhen Municipal Government initiated two city-wide exemplary projects that were endorsed, respectively, by the National Development and Reform Commission (NDRC) and the Ministry of Housing and Urban–Rural Development (MOHURD).

Second, there are *sectoral pilot schemes.* Some of these initiatives are required by national and subnational governments to be implemented at selected localities. For example, the Ministry of Transportation (MOT) launched the National Low Carbon Transportation System Pilot City, in which Shenzhen and Guangzhou were selected in the first and second batches, respectively, of regional experimentation and national demonstration. The MOHURD also developed various pilot projects in the building sector, including National Demonstration City on Energy Efficiency Monitoring System in Major Public Buildings (2007), National Demonstration City on the Application of Renewable Energy in Buildings (2009) and National Focal City of Public Building Energy Efficiency Retrofitting (2011), all of which are being piloted by the local government of Shenzhen given the municipality's proactive uptake. Sectoral pilot schemes are also being initiated by the private sector with the support of local governments. For example, the Green Freight initiative seeks to try out emission-reduction innovations in the supply chain operation and management of the logistics industry. Guangzhou was the initial experimental site for the Green Freight Pilot Program, which was later expanded to the whole of Guangdong province with the financial and technical assistance of international organisations.

Third, city governments are developing *new town construction projects* at the district level, incorporating low carbon infrastructural features such as green building design and low emission transport. Guangzhou has developed the Nansha New District and Sino-Singapore Knowledge Town, while Shenzhen has established the Guangming and Pingshan New Districts. These districts are located in suburban areas newly incorporated into the development jurisdictions of the two cities. In addition to contributing to environmental protection, these projects are perceived by the local government as models of city branding and regional impact.

The final category of pilot initiatives involves *cross-boundary cooperative schemes* to test the applicability of technological and institutional innovations. For instance, exemplary green building projects have been incorporated into the Shenzhen/Hong Kong codevelopment plan for the Qianhai area. Within the 2010 Framework Agreement on Cooperation between Hong Kong and Guangdong province based on the NDRC's outline development plan for the Pearl River Delta region, Hong Kong's then chief executive, Donald Tsang Yam Kuen, made a commitment to transform the PRDR into a Green Quality Living Area. Cooperation on promoting the use of electric vehicles was also incorporated into the Green

Quality Living Area proposal, leading to a further cooperative scheme between the two cities (Shenzhen and Hong Kong) to experiment with the technological and market compatibility of electric cars.

Pilot initiatives in a semi-authoritarian regime

These low carbon initiatives are embedded in a semi-authoritarian regime, a system context that shapes the dynamics of horizontal *inter*governmental interactions and vertical power structures. These localised low carbon initiatives have also become governing tools in both vertical and horizontal *inter*governmental interactions. With the *decentralisation* of state power in the marketisation process of China, local industries have developed stronger bargaining power with local governments, and thus ministerial-level environmental agencies have become less capable of directly enforcing environmental regulations and standards at the local level (Economy 2006; Rock 2002). City-level governments and their implementation agencies have therefore become key actors in putting forward the national vision of mitigating climate change. With the national government still performing the role of high-level *coordination*, empowered local actors – both state and nonstate – involved in the governing of low carbon pilot initiatives are essential network participants for innovation integration. They bring the potential benefits of *collaboration*, taking into account the regional differences across China and the existing bureaucratic inertia in certain frontline and district-level implementation units (Mah and Hills 2012). However, local governments' capacity for collaboration is constrained by the coordination structure of the vertical *inter*governmental hierarchy. The governing of low carbon initiatives in the semi-authoritarian context is concerned with 'how the central government and the localities can coordinate to better capture the potential additional capacity and additional value that can be generated by locally based collaboration' (Mah and Hills 2012: 93).

Pilot initiatives in economically developed cities in China serve the purpose of localising experiments, which remain centrally administered. Because the development risks and administrative costs are too high to carry out nationwide implementation of environmental innovations, the central administration (government) often looks for local test beds. Conversely, in a context of decentralisation, the central-local government power structure redefines the role of local governments that are implementing pilot initiatives assigned in a top-down manner (Bruton *et al.* 2005). Moreover, developed megacities are inherently experimental. Guangdong has been acting as a pilot region that operates under 'a contingent experimental learning condition of uncertainty' and is perceived by the national government as a 'test bed for experiment' (Bruton *et al.* 2005: 238). This makes the megacities of Guangdong the perfect sites for experimenting with innovations in the transition of the Chinese environmental state (Mol and Carter 2006).

Local leaders perceive the central-local power structures as bringing great *uncertainty*, and hence they are proactive in *intercity competition* for experimental projects. The uncertainties are revealed in the administrative privileges and jurisdictional expansion that shape the allocation and rezoning of land, which leads to municipal authorities' adoption of 'a pragmatic learning approach to planning by experimenting with initiatives that they think could give them edge over their competitors' (Bruton *et al.* 2005: 239). Guangzhou and Shenzhen are proactive in the application of pilot initiatives, while what is 'implicit in their experimental development is the adoption of a competitive stance with each authority attempting to 'win' relative to its own values by attracting more economic development to its area than to other administrative areas' (Bruton *et al.* 2005: 239). The approach of experimenting with initiatives adds to the continual *intra*regional rivalry fuelled by enduring regional competition between urban centres (Zacharias and Tang 2010). The uncertainties in vertical *inter*governmental power structure thus intensify contestation in horizontal *inter*governmental interactions in the CMNs.

However, the perceived *intra*regional rivalry has also given rise to intra*regional alliances* among coastal cities, including Hong Kong and Shenzhen, to strengthen competitive capacity against a more distant urban centre – Guangzhou. As noted by Zacharias and Tang (2010: 215), 'Guangzhou is vying to regain regional pre-eminence as a centre of business and culture', while 'the regional development in support of Guangzhou as an effective administrative and command centre for the region is not in the interest of Shenzhen, which would be rendered a coastal town with much less economic clout in the region'. It is therefore in the interests of Shenzhen and even Hong Kong to build a coastal alliance of cities. For Hong Kong, the possible cost of being cut off from the emerging network of cities in the Pearl River Delta is too high, given that its geographic location in the region is not as central as Guangzhou's (Zacharias and Tang 2010). The emerging need to build *intra*regional alliances has given rise to more innovative and cooperative projects launched by Shenzhen, an insider in the semi-authoritarian regime, to create functional ties with Hong Kong, a special administrative region and thus somewhat an outlier. Through building this *intra*regional alliance, both Hong Kong and Shenzhen are cautiously experimenting with initiatives involving both negotiation and cooperation over their own development interests, even though the administrative bodies of both Shenzhen and Hong Kong accept the ultimate decisions of the central government (Bruton *et al.* 2005).

Analysing the coordination–collaboration nexus

Innovation transfer is one of the major collaborative functions in the collaborative municipal networks (CMNs) for low carbon transition. World cities such as Hong Kong, Shenzhen and Guangzhou play an exemplary role in such a sociotechnical transition, acting as important arenas for developing innovative responses to complex climate change policy issues (Hodson and Marvin 2009). Through governing the various climate pilot initiatives, municipal authorities acquire practical insights into experimental practices that can be transferred to other cities to facilitate broader institutionalisation in the nation-state context.

The governing of these various initiatives requires both effective *coordination* among governmental actors and *collaboration* across policy areas and societal sectors. Moreover, the institutionalisation of innovative initiatives in the environmental state is concerned with the process of *legitimation*. Cross-sector collaboration transforms *input* legitimacy into *output* legitimacy by effectively aligning the expected environmental and developmental outcomes (see Chapter 6). This chapter further analyses how output legitimacy – with regards to the aligned expectations of environmental outcomes – is generated and stabilised to transform the environmental state through coordination and collaboration among the CMNs.

The nexus of coordination and collaboration is embedded in the governing process of climate pilot initiatives. The theoretical framework of CMNs has identified this nexus as the internal driver of the networking process (Chapter 2: Figure 2.2). To further study the coordination–collaboration nexus, in this chapter, we develop an analytical framework, displayed in Figure 7.1, to investigate the dynamics of the governing process in these pilot city projects. The analytical framework is structured to locate the drivers and barriers of coordination and collaboration, which facilitate or impede the institutionalisation of these pilot initiatives in the environmental state.

Figure 7.1 identifies the stages involved in the institutionalisation (preparation, designation, implementation and regeneration) of the municipal governing of pilot initiatives for low carbon transitions. Each stage is specified by its tasks in relation to different resources and contexts. The *tasks* refer to how the implementers perceive and understand the pilot programmes and how these relate to their regular public programmes (Khademian 2002; Li 2011). Qualitative data collected from semi-structured interviews are therefore used to substantiate the implementers' understanding of the low carbon pilot projects.

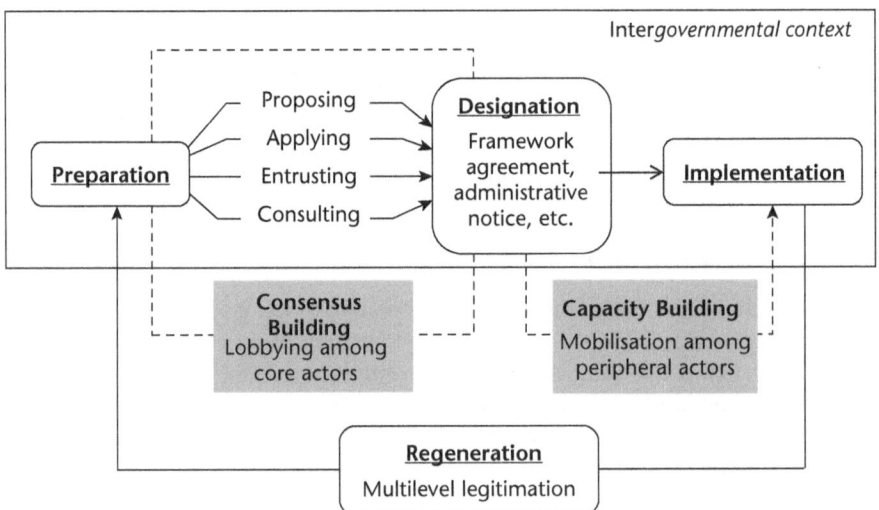

Figure 7.1 Analysing the institutionalisation of climate pilot initiatives
Source: Compiled by authors

Different stages of initiative development are aligned with various *resources*, which cover the network organisations' administrative capacity, financial and human capital, available technical support and the authority and responsibility to carry out the pilot projects (Khademian 2002; Li 2011). Presumably, effective coordination secures a sufficient level of resources before the pilot initiatives are formally designated, that is, at the preparation and designation stages. During the actual implementation and regeneration of so-called exemplary projects, collaboration becomes important because it can harness additional resources not only from the core networking actors through *intra*governmental coordination but also from the peripheral networking actors in the *inter*governmental context and the nongovernmental space.

The analysis of climate pilot initiatives, as illustrated in Figure 7.1, distinguishes the *inter*governmental context from the broader *system context* of a semi-authoritarian state, with the latter discussed in the previous section and identified in the theoretical framework of CMNs (Chapter 2, Figure 2.2). Each stage towards institutionalisation is embedded in a changing networking context of *inter*governmental interactions, environmental and developmental values and regime rules and regulations. The embedded institutionalisation of pilot initiatives is mediated through the nongovernmental space that accommodates epistemic communities. Epistemic communities are important knots that tie together loosely connected social segments to drive programme impacts in governance transitions. In the governance of environmental issues, epistemic communities disseminate scientific knowledge, international standards and regulatory instruments (Braithwaite and Drahos 2000). Organisational members with similar values participating in the epistemic community construct ongoing dialogues and interactions to enable the impact mechanism of an experimental public programme (Haas 1989). In a semi-authoritarian regime such as China's, these interactions and communications are embedded in a society controlled by the state (Diamond 2002; Ho 2007). The proposed analytical framework (Figure 7.1) aims to tackle the question of how collaboration within the epistemic community and the nongovernmental space is structured between the *inter*governmental context and the broader system context, which supports the institutionalisation of pilot initiatives – in particular through consensus and capacity building with lobbying and mobilisation tactics – with stronger input legitimacy.

Local pilot initiatives establish high *output legitimacy* among political elites because they are associated with potential resources and positive performance ratings (Economy 2006; Schreurs 2010). Local administrators regard the pilot initiatives that they have been assigned by upper levels of administration as both a *practical strategy* to foster their cities as 'test beds for new ideas for urban transformation toward low carbon societies' (Schreurs 2010: 97) and as a *normative title* of inherent prestige that is 'highly desirable for local leaders' (Economy 2006: 178). More importantly, the cities' pilot initiatives affect their performance in the institutionalised urban environmental quality evaluation system, which directly affects the priorities and decision making of mayors and governors in cities and provinces, especially for 'the larger, richer, and coastal ones' such as Guangzhou and Shenzhen (Rock 2002).

However, similar levels of *output legitimacy* regarding these various pilot initiatives are not necessarily established in other network segments, for example, among the implementers and collaborators of the initiatives in the nongovernmental space. What matters more in the institutionalisation process is the aligned output legitimacy across societal sectors. The analysis now turns to how the dynamics of the contradictory perceptions and expectations of the various pilot initiatives – through the nexus of coordination and collaboration in network governance – fit into the existing institutional setting and transform the Chinese environmental state.

Towards institutionalisation

As mentioned, the process of governing pilot climate initiatives can be deconstructed into the cyclical stages of preparation, designation, implementation and regeneration, as indicated in the analytical framework (Figure 7.1). Different resources are generated and regenerated in and between these stages. Networking actions (lobbying and mobilisation) are used to shape the implementers' perceptions of the assigned tasks and, ideally, to gain participants' consent for technological and institutional integration.

Preparation and designation

Four governing modes are identified in the preparation and designation stages: i) entrusting national or provincial initiatives in a top-down manner; ii) proposing bottom-up initiatives; iii) applying for top-down pilot schemes with bottom-up administrative effort; and iv) consulting and negotiating in horizontal *inter*governmental settings. The relationships among these four governing modes, the *inter*governmental context and the directions in which initiatives are transferred, are graphically illustrated in Figure 7.2.

Entrusting

An entrusting mode of governing was adopted in the preparation and designation of the National Low Carbon Transportation System Pilot City, an exemplary initiative with sectoral focus in which the demonstration cities were selected by the Ministry of Transport (MOT).

The intercity competition was part and parcel of the governing of this entrusted initiative. The selection was made in two different batches. Shenzhen, in the first batch, was designated in February 2011 (Appendix II: MOT, 2011). Guangzhou, in the second batch, was selected in February 2012 (Appendix II: MOT, 2012). Shenzhen's earlier designation was perceived by the city leadership as evidence of better performance in its low carbon transport planning and closer relationship with the national ministry, compared with Guangzhou.

However, given the rigid targets and overly general guidelines in the entrusted initiatives, there was not much room for manoeuvring local support or collaboration.

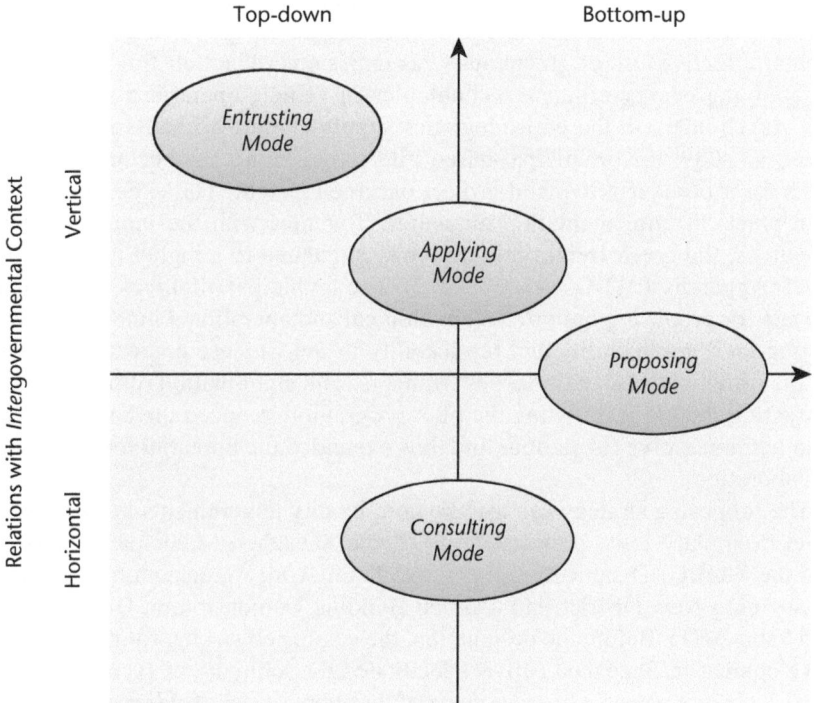

Figure 7.2 Governing modes of pilot-project initiation
Source: Compiled by authors

An official in the Shenzhen Transport Commission (SZ-TC) who undertook the implementation of this pilot scheme reflected that the entrusted tasks only tightened the top-down requirements of the regular environmental-protection activities already initiated by the local government (Appendix I: 21/SZ/24–05–2012). The low carbon component was perceived merely as an extended function of ongoing activities. The entrusted initiative may have strengthened vertical coordination, but it failed to overcome bureaucratic inertia in giving a significant push to either horizontal collaboration or institutional reforms. With limited horizontal engagement, the input legitimacy of the initiative had not yet been established at the designation stage.

Proposing

Another sectoral low carbon initiative – the Green Freight Pilot Project in Guangzhou – combines bottom-up proposing with top-down entrusting. This initiative is part of the broader Green Freight Demonstration Project in Guangdong province, supported by funding from the World Bank and by technologies transferred from the Smartway Project of the Environmental Protection Agency in the United States. Its basic mandate is to demonstrate the global and local environmental effectiveness of greenhouse gas emissions reduction from road freight by applying energy efficient technologies in vehicle operations and engaging key stakeholders in the entire logistics supply chain (World Bank 2013). Initiated in 2009, this municipal-based pilot programme has become a working basis for a broader provincial project on green freight. The experiment generated practical implementation outcomes. Together with the input of external resources, the green freight initiative was expanded to a higher administrative level (Appendix I: 11/GZ/17–02–2012). The municipal pilot was designated as the site for proposing imported technological and operational innovations while testing their applicability and replicability in the Chinese context as entrusted by the provincial and national governments. The combination of proposing and entrusting modes in initiating the pilot programme reduced the barrier of stringent administrative regulations and thus expanded the potential for cross-sector collaboration.

The proposing strategy can also be used by city governments in their district-level exemplary town projects. In 2007, the Shenzhen Municipal Government and the MOHURD signed a Cooperative Framework Agreement on Developing Guangming New District into a Green Building Demonstration District (Huang and Yang 2011). Before the designation, the actors networking for green building development in Shenzhen actively facilitated the lobbying of (vertical) central-local coordination and a broader range of (horizontal) stakeholders across sectors.

An official in the Shenzhen Housing and Construction Department (SZ-HCD) recalled the process of lobbying the national administration:

> It was not our department [SZ-HCD] that signed the agreement, but our department made the whole thing happen. We went on facilitating the tasks, communicating with the [Shenzhen] municipal government, liaising with the construction ministry [i.e. MOHURD] and connecting them together. Consequently, they sat down together in Beijing, where they signed this cooperative framework agreement.
>
> (Appendix I: 20/SZ/23–05–2012)

This resulted in a positive outcome for the local government:

> The Green Building Demonstration District in Guangming has impressed the national government: the construction ministry [MOHURD] is quite familiar with the demonstration district, so on a lot of public occasions when he [the minister of MOHURD] enumerated the National Green Building Demonstration Districts, he mentioned Guangming New District. Thus, in a way, the Guangming district has been lifted to the attention of the ministerial level, for they do know about this.
>
> (Appendix I: 20/SZ/23–05–2012)

District management authorities and industrial entrepreneurs in the green building sector have also proposed exemplary town projects through bottom-up initiatives. Projects in designated districts have encouraged similar initiatives in other districts, eventually leading to city-wide responses: 'Currently, individual districts have gradually started experimenting [with green building demonstrations]; like in Guangming and Pingshan [districts], the authorities have already launched trial implementations. Then right now, over the other side in Nanshan [district], there is a high probability of continuing the trial' (Appendix I: 17/SZ/11–05–2012). These local actions – not only systematically planned initiatives but also improvised ones – were placed under the coordination of the city's Urban Planning, Land and Resources Commission (SZ-UPLRC). 'In some districts, the initiatives were systematically proposed and carried out collectively by the entire district . . . within other districts, however, several demonstration projects emerged from time to time; it was somewhat piecemeal but through it, the exemplary district is making progress' (Appendix I: 17/SZ/11–05–2012).

The new town development approach has also been adopted in Guangzhou, although it is perceived as taking an implicit competitive stance against Shenzhen (Bruton *et al.* 2005). Various relevant demonstration initiatives are taking place at the district level, proposed by the public and private sectors and endorsed by the municipal government. As stipulated in the municipal policy on the Rapid Development of Green Building promulgated in January 2012, it is compulsory for all new buildings constructed within newly developed districts to meet green-building standards. Guangzhou has also proposed a new pilot scheme to the national government, referencing the successful designation of Shenzhen's Guangming district (Liao *et al.* 2012).

A coordinating official on the Guangzhou Development and Reform Commission stressed that the commission supports governing through consensus building and policy coordination among the core actors involved in the new town initiatives, that is, that the tasks related to developing low carbon demonstration districts (in particular Nansha New District and Sino-Singapore Knowledge Town), are 'not only concerned with any single aspect of either transport or building, but an integrated planning process for low carbon development [transition], considering the realities [capabilities] of various parties' (Appendix I: 9/GZ/08–02–2012). However, the development ideas actually originated in the private sector. Singbridge, the developer of and investor in the Sino-Singapore Knowledge Town, is a foreign corporation wholly owned by Temasek Holdings of Singapore. One of these 'realities' is that the initiatives proposed by actors external to the local administrative system are often constrained by procedural barriers before formal designation. To bypass these barriers, private sector actors seek collaboration with liaison officers internal to the local government. Mobilising governmental participation is a prerequisite to the formal launch of such initiatives involving land use allocation. Officials with decision-making power in the city government can be persuaded by the expected incentives of generating an exemplary impact and improving the municipal image in the eyes of higher-level officials.

Interestingly, the input legitimacy of the new town initiatives adopting a pro-posing strategy appears to have been generated based on a 'first come, first served' principle. Governing through coordination and collaboration are simultaneously used by ambitious local governments and resourceful actors in the private sector, who are eager to exploit the emerging market of innovation, land and develop-ment opportunities. Together, they are driving entrepreneurship in low carbon transitions. Network actors have negotiated for aligned output legitimacy with lobbying and mobilisation tactics.

Applying

In developing low carbon districts, the Guangzhou government also intended to apply for nationally endorsed pilot city projects. As explicitly stated in a 2010 municipal policy – *Guidelines on Rapid Development of a Low Carbon Economy* – 'the further planning of Sino-Singapore Guangzhou Knowledge Town, Nansha Low Carbon Town . . . [is] an effort of building demonstration areas on low carbon economic development and green quality living zone, *to proactively compete for the application* of National and Provincial pilot areas of the low carbon economy' (Appendix II: Guangzhou Municipal Government, 2010: 8).

The designation that the Guangzhou government eyed so keenly was an initia-tive recently launched by Shenzhen. An official of Shenzhen's Development and Reform Commission (SZ-DRC) explained its successful application to the city's past environmental performance and existing industrial structure: '[Shenzhen] is already very good: our energy consumption always remains around half of the national average; our water consumption is about one-tenth of the national aver-age; then our industrial structure is, you know, comparatively better – thus in every aspect, we are standing on a better foundation [than Guangzhou] to carry out low carbon undertakings' (Appendix I: 16/SZ/07–05–2012).

This official, who himself participated in Shenzhen's application process for the national pilot, considered that the designation decision was based on local environmental conditions, supported by the national administration:

> It was around 2009 when we were actively fighting for the project. However, because our work is closely connected with the national government and we need to report our regular tasks to the national administration annually, the national government was indeed able to judge which place around the coun-try has larger potential to be a low carbon pilot, and which one has a better foundation or larger impact in demonstration.
>
> (Appendix I: 16/SZ/07–05–2012)

In addition to maintaining close interactions in vertical *inter*governmental coor-dination, the local government also made proactive administrative efforts to apply for this nationally endorsed initiative: 'There are national factors involved, as well as factors that we have been active in the application; what we can do in active application is to show them, the national administration, the sound performance we had in the past' (Appendix I: 16/SZ/07–05–2012).

Internal lobbying and consensus building locally among core actors, that is, within the *intra*governmental coordination structure (Chapter 4), were key approaches in building and reinforcing Shenzhen's track record in the minds of national officials:

> We definitely need to mobilise other departments to work together. This low carbon task is not only about us – we are merely the city-wide coordinator. Each local authority, each district and each functional department, they all have their respective achievements and highlights. These are things that we need to gather not only during the competition period, but at regular times of the year we may come together to discuss the municipal development holistically. We need to maintain active communication with other departments to carry out this task.
>
> (Appendix I: 16/SZ/07–05–2012)

The Shenzhen government has also been actively applying for sectorally focused experimental initiatives. Since 2007, the SZ-HCD has successfully applied for and managed five low carbon building pilot projects with fiscal support from the central budget.[1] The application process was explained by the SZ-HCD implementer:

> Actually it was the construction ministry [MOHURD] who set up this task, the demonstration city, so we went on applying, and thus were chosen in the first batch of national demonstration cities. Subsequently the national government [the Ministry of Finance] gave us some money while the municipal government provided some matching funds. So we started doing this thing, with four more initiatives that followed.
>
> (Appendix I: 20/SZ/23–05–2012)

Initiatives proposed through the application strategy require a combination of vertical *inter*governmental and *intra*governmental coordination initiated by the municipal government while collaboration is incorporated during the implementation stage. Before the successful application, the primary departments (i.e. SZ-DRC and SZ-HCD) established internal legitimacy within the core of the CMNs through their functional jurisdiction and authority. Other relevant municipal departments were also engaged to ensure a sufficient level of input legitimacy in achieving the goal of being designated for nationally endorsed initiatives. The expected developmental and environmental outcomes of these initiatives are often predefined by the national administration in the application process for the reference of interested parties at the municipal level, which makes it easier to reach a coherent level of output legitimacy among the core network participants.

Consulting

The consulting strategy was used before the designation of the Qianhai code-velopment area, a pilot cross-boundary cooperative initiative that incorporates green building and low carbon transport demonstration components. Formally stipulated in the 2010 Qianhai Overall Development Plan, the Qianhai area was envisioned as an 'energy-efficient environmental town symbolic of low energy consumption, low pollution and low emissions' (Appendix II: Shenzhen Municipal Government, 2010b). The green building demonstration aspect of this project was to be carried out by the SZ-HCD and the newly established Qianhai Management Authority; the local government wants to turn the strategic area of Qianhai into a green building exemplary zone with national impact (Dou 2012).

Before realising this vision, the Shenzhen government went through a process of consulting and negotiating during the preparation stage. Given Qianhai's special geographic and strategic location, gaining the consent and participation of the Hong Kong government was a prerequisite to formal startup. Initially, however, neither the government nor the public in Hong Kong were as supportive as their counterparts in Shenzhen. In this regard, formal consultations with stakeholders in Hong Kong were constantly initiated by the Shenzhen government, which was also actively lobbying for the central government's endorsement of the same initiative (Fang 2012). Nevertheless, with the promising vision of importing the management strategies from Hong Kong to Qianhai, the Shenzhen government was much more proactive in initiating this cooperative experiment. It is evident that effective communication, based on face-to-face meetings and leading to mutual understanding of the expected benefits to be gained by participating parties, is beneficial during the consulting stage as a means of generating internal legitimacy among the core participants of the CMNs.

After reaching a development consensus with Hong Kong, the Shenzhen government compiled the overall plan and submitted it to the national government. To confirm the innovative and exploratory significance of the Qianhai experiment, the State Council issued a formally approved designation document that was officially circulated top down through the NDRC, the Hong Kong government, and the Guangdong Provincial Government (Appendix II: State Council 2010). This formal designation secured the participation of the negotiating parties after the consultation process and provided institutional preparation for implementation.

The consulting strategy was also embodied in the policy development of cooperation between Hong Kong and Shenzhen on experimenting with electric vehicles. In 2009, the Steering Committee on the Promotion of Electric Vehicles was set up within the Hong Kong government (Chapter 6). Upon internal consultation within this *intra*governmental coordination structure (Chapter 4), the committee agreed that the promotion of electric vehicles should be steered towards exploring collaboration possibilities with various vehicle manufacturers, particularly with regards to conducting trials on available models of electric vehicles (Appendix II: Environment Bureau 2009). Meanwhile, the Hong Kong government was planning to try out an electric vehicle manufactured by BYD, a private corporation founded in Shenzhen and facilitated by the Shenzhen Municipal Government. In 2010, this initiative was incorporated into the Guangdong-Hong Kong Cooperative Framework Agreement on the environment. Embedded in a broader interactive *inter*governmental setting, the political significance of this cross-boundary initiative was enhanced. The municipal administrations on both sides were then more obliged to strengthen their existing collaboration on the integration of electric vehicles.

Eventually, in 2011, the experimental initiative was institutionally designated when a Memorandum of Understanding was signed between BYD Company Limited, representing Shenzhen, and the Automotive Parts and Accessory Systems Research and Development Centre (APAS), representing Hong Kong (HKPC and APAS, 2011). The APAS was established and directly funded by the Hong Kong government as a government-organised nongovernmental organisation incorporated to the Productivity Council.[2] After two years of research and consultation, the institutionalisation of collaborative ties marked the formal designation of this cross-boundary cooperative initiative on electric vehicles.

Coordination and collaboration interactions are found in the process of consultation on initiative generation in CMNs. Internal discussions are activated within the *intra*governmental coordination structure to mobilise support from the core actors. Bottom-up submissions of proposed initiatives are agreed upon after reaching a consensus through horizontal *inter*governmental and cross-sectoral collaboration; top-down endorsements then follow in the vertical *inter*governmental coordination.

Inter*governmental context for preparation and designation*

Figure 7.2 illustrates how *inter*governmental relations and the diffusion of the pilot initiatives shape the governing strategies towards formal endorsement. When the initiative is developed in a top-down manner in the vertical *inter*governmental structure, an entrusting mode is adopted until the actual launch. Conversely, although the initiative is developed in a top-down manner, a consulting mode (rather than an entrusting mode) is applied if the initiative is embedded in a horizontal *inter*governmental setting.

The *inter*governmental structure also stimulates intermunicipal contestation. In the project development of Nansha New Town, Guangzhou highlighted the geographical proximity of and transport links between the Nansha area and the Special Administration Regions of Hong Kong and Macao (Du and Luo, 2013). Shenzhen adopted a similar tactic when it was lobbying for the central government's endorsement of the codevelopment (with Hong Kong) of Qianhai. Adopting a similar strategy, Guangzhou has tried to compete with Shenzhen for the resources provided by the central administration and for a more significant regional impact. Given the special status and administrative privileges conferred on Shenzhen by the national government in recent decades, Guangzhou has been at a disadvantage in this intermunicipal competition. In the entrusting process of the National Low Carbon Pilot City, Guangzhou again lost out to Shenzhen.

Shenzhen's rise has clearly diminished Guangzhou's regional impact. Thus, Guangzhou has become proactive in fighting for political endorsement by proposing and developing more bottom-up initiatives with demonstration potential in the overall low carbon transition.

Implementation

Upon successful designation, various pilot climate initiatives are mapped out in an *inter*governmental setting, moving forward to implementation. To carry out these initiatives, increased workloads, changing policies and institutional design are inevitably imposed on the implementers and implementing authorities. Meanwhile, administrative barriers and institutional challenges emerge. To overcome these obstacles, the implementers use horizontal mobilisation – not only among the core actors internal to the CMNs but also among peripheral network actors – to beef up their implementation capacity.

Purporting change

The Shenzhen government's regular governing activities were changed by the policies it implemented as a National Low Carbon Pilot City. Substantial carbon dioxide emission reductions were targeted: 39 percent per unit of GDP by 2015, with a minimum of 45 percent by 2020, based on 2005 levels (Appendix II: SZ-DRC, 2012). The local government was therefore in need of unconventional and vanguard means to adjust to the new target and steer the municipal-wide carbon transition:

> Because we have now become the pilot, we need to walk one step ahead of the others . . . Since our city already had low energy consumption, what we are now working on is to make larger [reduction] efforts within more limited space.
> (Appendix I: 16/SZ/07–05–2012)

Given the limited reduction potential of the city, the local government tried out solutions for its low carbon transition, such as optimising the industrial structure, strengthening the supervision of energy-saving and emission-reduction activities, engaging in district level demonstration projects and increasing low carbon awareness by promoting the concept of low carbon living in the city's annual Energy Saving Promotion Week.

Concrete changes in local policies were also initiated to streamline the implementation. The SZ-DRC devised a *Medium and Long Term Plan of Low Carbon Development* (2011–2020) for Shenzhen, with clearly specified sectoral mitigation actions and targets. In the National Low Carbon Transportation System Pilot Cities (both Shenzhen and Guangzhou), implementation plans were prescribed by the Ministry of Transport (MOT), the entrusting agency that provided a specific policy template for each pilot city (Appendix II: MOT 2011). Upon the formal submission of an implementation plan, the local authorities were held responsible for the sectoral reduction targets they had developed and committed to. In this way, the tasks originally concentrated on the central government were localised through a web of selected local administrations.

The national government specifies and quantifies reduction requirements during the implementation stage of pilot initiatives. As specifically prescribed by the requirements of the National Demonstration City on Energy Efficiency Monitoring Systems in Major Public Buildings, Shenzhen was to construct seven million square metres of new buildings equipped with solar water heating systems and to retrofit four million square metres of existing buildings. However, the local implementers at the SZ-HCD perceived these additional requirements as merely imposing an extra workload on top of their regular tasks (Appendix I: 20/SZ/23–05–2012). A local implementation plan was developed to cope with this additional work, although it was not prescribed by national authorities:

> After the assigned target of putting up solar water heaters, an implementation plan was released by the [Shenzhen] Municipal Government, stating that the original policies were no longer applicable [for the current targets]. This mandatory policy had been expanded to cover all buildings. So long as you are using heated water and your building fulfils the installation conditions, you have to do it [putting up solar water heaters].
>
> (Appendix I, 20/SZ/23–05–2012)

The local implementers understood that this policy change was intended, first, to fulfil the building-related emissions-reduction targets required by the national government (which had set up the pilot scheme) and, second, to explore and experiment with innovative means of increasing the energy efficiency of buildings for the city's own development needs.

Nonetheless, the transformed local policies and emissions-reduction practices are constrained by the limited institutional capacity of local implementers. The scope of supervision and management are expanded with new self-administered tasks on the low carbon agenda. The mandatory energy-efficiency levels for buildings and transport require additional manpower, supplies and technical assistance for auditing and inspection. Cross-sectoral coordinative low carbon actions are fundamental to fulfilling the emissions-reduction promises. To beef up the institutional capacity of local implementers, the implementation of low carbon initiatives entails external resource inputs, a process of expanded mobilisation not only with core internal actors but also with peripheral ones.

Expanded mobilisation

The lack of institutional capacity in local environmental agencies in China has been frequently discussed in the literature as the major challenge in enforcing environmental activities (Francesch-Huidobro *et al.* 2012; Li 2006; Li and Li 2012). Resource mobilisation is closely related to political opportunities in the Chinese semi-authoritarian context of environmental governance (Ho 2007).

Expanded mobilisation with organisations external to the implementation stage potentially reduces the institutional deficit arising from administering low carbon initiatives. Mobilisation in this context is not only about 'mustering financial, material, and personal resources' (Ho 2007: 196) but also about engaging with the broader societal influence of civil society and the epistemic community (Haas 1989). The power of these nongovernmental organised forces in resource mobilisation is evident in the climate advocacy coalitions in the Guangzhou area (Francesch-Huidobro and Mai 2012). Combined with the globalising functions of the two cities, local low carbon initiatives are subject to the broader mobilisation of a wider range of resources and societal influence.

Because the local government lacks the authority to engage with its counterparts in other regions, it is often difficult for local implementers to expand mobilisation beyond their own administrative jurisdictions. This weak institutional basis of mobilisation was pointed out by an implementer of the National Low Carbon Transportation Pilot Initiative in Shenzhen:

> Currently, in terms of horizontal coordination it is a relatively difficult task for each city – indeed all our selected cities originally envisaged that if the municipal leaders, people in mayors' offices and their leaders, would act as the coordination agency, the strength [of authority] would be greater. However, at the moment, we probably have not reached that step yet. It is because if we have not lifted the issue to the coordination at the municipal level – this horizontal communication appears to lack 'qualification', and the implementation [of convening a coordination group across cities] is difficult.
>
> (Appendix I: 20/SZ/23–05–2012)

Despite the difficulties in broadening mobilisation within the horizontal *inter*governmental context, local industry has been engaged in capacity building for the Guangzhou Green Freight Pilot. The external financial input for this initiative created partnerships among government agencies, logistics companies, freightstation operators and expert consultants. By designating logistics companies as demonstration units, the local government enforced the energy-efficiency requirements in their various logistics operations, including freight stations, trucks and suppliers' offices. To coordinate the various responsibilities in emissions reduction, the implementers organised meetings to engage stakeholders from local government, industry and external sponsors. These face-to-face meetings were an indispensable communication tactic for strengthening implementation capacity, with the local government being more open to external suggestions and input from practitioners.

The strategy of 'demonstrations within demonstrations' also secured commitment to the pilot initiative. For example, the Green Freight Pilot initiative was a demonstration project in itself, but there were numerous 'demonstration units' within this pilot project. The Green Freight implementer conferred the title 'demonstration unit' on industrial participants. Shenzhen also hosted smaller-scale pilot projects, separate from its National Low Carbon Pilot City plan. Together, these experiments mobilised the participation of the industry and the public by generating positive information incentives.

Inter*governmental context for implementation: constraints and opportunities*

To enforce the implementation plan for the National Low Carbon Transportation Pilot Initiative, *inter*governmental coordination emerged in the vertical transport administrative system. A 'four-tier working mechanism' of pilot initiatives was devised and included in the national policy document, distributed between the MOT, the provincial-level transportation authority, the pilot city transportation authority and the implementation actors of the pilot projects. This institutional framework has reinforced top-down supervision and limited the bargaining power of individual state-endorsed transportation companies, which are the major GHG emitters at the local level. Vertical *inter*governmental management for low carbon development was considerably enhanced, which provided structured opportunities to enforce the pilot initiative. An implementation official commented on the *inter*governmental effect:

> Even without these pilot initiatives, we have already been carrying out these activities [prescribed by the pilot schemes] and the operational guidance provided by the MOT also existed. However, the substance of the pilot city project [i.e. the concept of low carbon transport] did penetrate our regular activities.
>
> (Appendix I: 21/SZ/24–05–2012)

The *inter*governmental context also adds political constraints if an initiative is implemented without significant leadership support. As discussed, there was a weak institutional basis for mobilisation across the entrusted cities of the National Low Carbon Transport Pilot Initiative. There was more horizontal interaction among the local governments designated as National Low Carbon Pilot Cities, with regular face-to-face meetings engaging representatives of all pilot cities (Appendix I: 16/SZ/07–05–2012). In the transport initiative, the implementer had difficulty soliciting the support of mayors, while local commitments to cooperation with mutual benefits were not built among entrusted cities. The national government, including the State Council and the NDRC, frequently emphasised the significance of pilot implementation in official correspondence and official press releases, which made a real difference to political support.

Inter*governmental resources: keeping the wheels turning*

Substantial financial and human resources are embedded in the central-local structure. Unlike the model environmental city scheme of the 1990s, the low carbon city initiatives are no longer solely dependent on externally generated resources such as international financial assistance, foreign direct investment and opportunities to host international events (Economy 2006; Schreurs 2010). Instead, domestic resources are generating strong economic incentives for the preparation and implementation of low carbon initiatives.

Financial resources

In particular, generous earmarked grants set up by the national government are attached to local governments upon their designation of pilot city projects. Although certain bottom-up initiatives are still initiated and sponsored by the funding and technological support of international organisations or multinational corporations, abundant domestic resources embedded in the *inter*governmental context are available for initiatives such as the National Low Carbon Pilot City and various sectoral focused national pilot projects in building and transportation.

After attaining the National Low Carbon Transportation System Pilot City, Shenzhen received additional financial resources from the national government, considering the lack of relevant resources at the municipal level and the inability to apply for provincial resources. The local implementer described the dilemma in its department's current resource flow:

> Above all, we have been fighting for a special fund for our sector. Currently, only our recurrent expenditures are covered cumulatively . . . speaking of some rewards or incentives for projects and some supporting funds for dem- onstration schemes, there is no separate appropriation yet. We are actively seeking a special fund to be set up together with the municipal finance depart- ment. But then it is only within our ministry [MOT] that there is a Trans- portation System Energy Efficiency and Emission Reduction Special Fund being set up. This special fund comes up every year with a certain amount of support to reward those demonstration projects as a means of 'replacing subsidies with awards' (*yijiang daibu*).
>
> (Appendix I: 21/SZ/24–05–2012)

The 'replacing subsidies with awards' approach is a way of offering financial support to individual project owners. It aims to encourage bottom-up initiatives proposed by the epistemic community and local industry.

The MOT fund is a 'policy inclination' of the national government towards the *entrusted* low carbon transport pilot cities (Appendix I: 21/SZ/24–05–2012). A certain portion of the Special Fund is intended exclusively for the pilot cities, as stated in the relevant national policy (Appendix II: MOT, 2011). In the same policy document, the national government encourages pilot cities to actively seek financial support from the same administrative level while prescribing the respon- sibilities for transportation commissions at both the provincial and municipal lev- els. Nevertheless, the national policy requirements have not yet been fulfilled by local administrative tiers. In this case, the integration of top-down initiatives should be compatible with similar levels of local commitment.

When *applying* for the national pilot initiatives in the building sector, the local implementers in the SZ-HCD shared an understanding that additional financial resources would potentially be appended upon designation. Normally, subsidy grants are available if the local demonstration project is in collaboration with the Ministry of Finance and when it is the ministry that issues the application notice and documents (Appendix I: 20/SZ/23–05–2012). During the implementation stage, financial resources are allocated on the principles of 'first come, first served' and 'available while stocks last' (Appendix I: 20/SZ/23–05–2012). Such central-local financial arrangements provide the epistemic community with short-term economic incentives for innovation integration.

International financial and technical assistance is still widely available for *proposing* low carbon initiatives. International agencies such as the World Bank and the UNDP still use financial assistance to directly influence local reforms in Asian countries (including China) in the globalised knowledge-building process (Haque 2013). In the early design and actual implementation of Guangzhou's Green Freight Pilot Project, external financial sponsorship was provided by the World Bank while practical guidelines were imported from the Clean Air Initiative of Asia and the American Smart Way Project. The local government perceived this bottom-up initiative as generating significant political incentives due to its exemplary demonstration and positive regional influence. Government endorsement was therefore matched with external resources. From the perspective of local officials, the freight pilot project was an opportunity to broaden their collaborative networks with the industry and the epistemic community (Appendix I: 11/GZ/17–02–2012). The initiative was smoothly implemented, with the combined support of foreign financial assistance and domestic political elites.

In the cross-boundary collaboration between Hong Kong and Shenzhen on the wider use of electric vehicles, the participating governments provide financial incentives to local car owners and the epistemic community, which indirectly strengthens their collaboration. In Hong Kong, the Vehicle First Registration Tax has been waived for buyers of electric vehicles; an Innovation and Technology Fund provided sponsorship to a local vehicle manufacturer to launch an electric vehicle model; and the distribution of the Pilot Green Transport Fund encourages the integration of electric vehicle technologies (Chapter 6). Shenzhen's new energy industrial development plan set ambitious targets of reaching the capacity to manufacture 200,000 new-energy automobiles by 2015[3] and to supply at least 24,000 new-energy automobiles to the city's public and private transport sectors by 2012, thereby reaching a total of 100,000 vehicles[4] by 2015 (Appendix II: Shenzhen Municipal Government 2009a). To reach these top-down–imposed targets, a Special Fund of New Energy Industrial Development was established in Shenzhen, generating RMB500 million for seven consecutive years to support research into and application of technology in areas including the development of electric vehicles (Appendix II: Shenzhen Municipal Government 2009b). In addition, since July 2010, private purchasers of new hybrid electric vehicles in Shenzhen receive a subsidy of RMB50,000 from central (national) finance and a further RMB30,000 from municipal finance; those purchasing private all-electric vehicles receive RMB60,000 from the Ministry of Finance and another RMB60,000 from the city (Jiang 2010). With the two collaborating governments having reached a consensus on the integration of electric vehicles, the financial resources mobilised by both sides serve to synchronise the initiative's implementation across the horizontal *inter*governmental context.

Human resources

Apart from financial resources, human capital is integral to local implementers' capacity building. Although financial resources are sometimes abundant, the deployment of departmental human resources for implementing pilot initiatives is often limited. During the implementation of the National Low Carbon Transportation System Pilot City, there was a clear lack of regular staff members being deployed as task coordinators. Only one manager was appointed as a coordinator in the city's Transport Commission (SZ-TC), and he was responsible for *all* of the implementation and communication tasks related to the pilot project. This turned out to be a constraint on the implementation of the initiative. The assigned manager had difficulty soliciting collaborative opportunities across the transport sector because he was preoccupied with administrative work for internal coordination, which limited the scope of expanded mobilisation.

A human resource deficit also exists in employment conditions and training. The Qianhai Management Authority was established to facilitate the Qianhai experimental implementation, but it was registered as a nongovernmental 'institutional unit (*shiye danwei*)'. The employment conditions at the institutional unit are generally inferior to those at governmental establishments, yet they are also subject to travel restrictions like regular civil servants. Because employees at the management authority were redeployed from their original governmental positions with civil servant benefits, the inferior employment conditions reduced their morale to manage the pilot initiatives in Qianhai.

To cope with the human resource deficit in the implementation of pilot initiatives, manpower also clusters around newly established and state-funded research facilities commissioned to gather and retain qualified human resources in the local epistemic community. One example is the APAS Research and Development Centre, set up by the Hong Kong government in 2011 to facilitate collaboration with the Shenzhen government on the wider use of electric vehicles. The centre serves as a government funding platform for research projects and as an institutionalised communication point for the Hong Kong–Shenzhen collaboration.

Because those assigned to the job do not necessarily possess technical knowledge about low carbon development, relevant on-the-job training is arranged in the agencies newly designated for low carbon experiments. The national government strengthened training in local governments designated as National Low Carbon Pilot Cities. The national training was organised by the NDRC and convened all of the implementers in the five provinces and eight cities selected as pilot sites. Low carbon development was an unfamiliar concept to new developers, who viewed it as just one more official slogan (Appendix I: 16/SZ/07–05–2012). The national training thereby provided a knowledge-transfer platform for the local pilot implementers to learn about technological innovations while sharing implementation experiences and progress, particularly in unfamiliar areas such as establishing carbon emissions trading and making citywide carbon inventories. Nevertheless, similar learning and collaborative platforms for low carbon initiatives are still institutionally weak, infrequent and with limited knowledge support. It is unclear who is obliged to provide training and how to make it work. Top-down organisation may ensure a high turnout rate and active participation in a command-and-control manner. Mobilisation of local resources in the epistemic community or the private sector can generate educational and societal outcomes in a more effective and longer-term way.

Regeneration cycle: What is next for pilot initiatives?

The ultimate goal when implementing exemplary and experimental climate mitigation initiatives is to broaden the impact and further the integration of climate governance innovations in terms of institutional design, policy setting or technological advancement. Therefore, it is desirable that the implementation outcome is capable of going through a regeneration cycle with regenerated resources and enhanced institutional capacity to facilitate the local integration and multilevel legitimation of the pilot schemes.

Regenerated resources

Resources are clearly essential throughout each stage of a pilot scheme, including the postimplementation cycle. A successful pilot initiative makes it easier to secure resources. The promising results from the implementation of the Green Freight Pilot project in Guangzhou resulted in additional international financial assistance to expand the project to a provincial scale. The funding was regenerated based on the financial mechanism of the Global Environmental Facility, launched and managed by the World Bank and the United Nations.

In Shenzhen, resources were regenerated across various building energy-efficiency pilot projects, based on a departmental financial mechanism set up in an *inter*governmental context. This mechanism was instituted within the city's Wall Reform Foundation, financed by developers through wall reform financial deposits. Its guidelines were promulgated and took effect in 1995, when the national government made a commitment to raise nationwide awareness and apply new technologies to the construction materials for walls. These materials generate less pollution, improve energy efficiency through better insulation and reduce construction costs through higher resource efficiency (Appendix II: Shenzhen Construction Department 1995). It was therefore a top-down effort imposed from the national level (Appendix I: 20/SZ/23–05–2012). According to a local implementer of pilot initiatives in the building sector, this financial mechanism was a resource generator for the operations of all completed pilot projects on building energy efficiency (Appendix I: 20/SZ/23–05–2012). Funds were collected from the deposits made by project developers who pledged to use new and energy efficient wall materials. Real estate developers were able to reclaim their deposits if they fulfilled the requirements written into the new management measures. It turned out that some developers violated the written rules and thus lost their deposits, and this unclaimed money was transferred to the Wall Reform Foundation, adding to the financial support for building energy efficiency innovations, particularly research projects, the setting of municipal standards and demonstration projects implemented by municipalities from the national government. According to an implementer's estimate, the Wall Reform Foundation put a total of RMB50 million into demonstration and research projects (Appendix I: 20/SZ/23–05–2012).

The Wall Reform Foundation eventually closed down in 2012 because the newly established Energy Efficient Development Special Fund performs a similar function with a better funding mechanism and is more compatible with the current low carbon policy direction and objectives. The new fund took over the remaining balance of the former Wall Reform Foundation, in addition to an initial appropriation of RMB30 million from the municipal budget. The new fund adopted a similar strategy of resource generation to ensure continuous support for building energy-efficiency demonstration projects in Shenzhen.

Local integration

Ensuring the sustainability of resources achieved the further goal of locally integrating tested experimental initiatives and promising exemplary projects. The local integration is part of the institutional efforts to localise the global impact of climate change and global practice of mitigation rather than directly transferring or transplanting existing global best practices.

However, this local integration process has encountered several dilemmas in the institutional landscape, including social norms, accepted cultural practices and established rules and regulations. The management authority in the Qianhai codevelopment area was conceived as an institutional innovation embedded in low carbon pilot initiatives, which carried the vision of localising the global concept of the rule of law, with an attempt to overrule the supremacy of administration in the existing semi-authoritarian *inter*governmental structure of the Chinese environmental state (Fang *et al.* 2012).

This institutional design was concerned with establishing a statutory body with legal responsibility for managing the codevelopment plan rather than relying on the intervening power of mayors or higher administrators. A research officer in the Shenzhen municipal government perceived it as strengthening autonomy, legalising operations and increasing the predictability of administrative behaviour by reducing arbitrariness. However, these institutional reforms met strong resistance in this semi-authoritarian context. When registering the Qianhai management authority, the category of 'statutory body' did not exist, so the organisation was listed as a public institution unit (*Shiye Danwei*). Because a public institution unit does not possess a high administrative ranking, the Qianhai Management Authority is not empowered to coordinate senior municipal administrative units such as the city's development and reform commission. It also lacks the resources to engage with and achieve shared consensus with the epistemic community.

Therefore, authoritarian components still prevail in the operations of the Qianhai initiative, deviating from the original vision of institutional reform.

Strengthening the institutional capability of local integration requires continuous trial practices and reexamination in a cyclical manner. To encourage the wider use of electric vehicles, the Hong Kong government has extended the trials of vehicle models from different global manufacturers to seek a suitable model that accommodates local market demands and is compatible with the congested, compact road situations (Appendix II: Environment Bureau 2009).

During the implementation stage of the Guangzhou Green Freight Pilot, constant discussions and negotiations examined the applicability of integrating foreign best practices in the specific locality of Guangzhou. After the municipal pilot, research reports were produced by internal and external consultants, including the Clean Air Initiative for Asian Cities, various local logistics firms and the study group within the Guangdong Transportation Bureau (GD-TB, 2010). These reports suggest the possibility of expanding the municipal green freight pilot to the provincial and/or national scale. Coordination and collaboration, both internal and external to the CMNs, are critical in enabling these regular assessments and discussions.

Multilevel legitimation

Beyond the consensus building with internal lobbying in the preparation stage and expanding mobilisation in the implementation stage, the regeneration cycle involves a process of multilevel legitimation to gain consent on trial practices (either to demonstrate or to experiment) from actors at the subnational, national and transnational levels. Legitimation is defined as 'the process by which institutionalisation is rendered valid by the consent of the stakeholders concerned and the public at large' (Francesch-Huidobro 2012: 796). This process of validating consent among network actors is needed, along with adequate and effective institutional design, to align the expectations of the environmental and developmental outcomes of the chosen pilot initiatives, that is, the output legitimacy (Chapter 2), and to translate this acquired climate knowledge and experience into 'politically savvy' actions (Francesch-Huidobro 2012).

Given their economic scale and population size, Guangzhou, Shenzhen and Hong Kong are regarded as globalising or global cities (Chapter 3). The more a city is globalised, the greater the possibility that the global effect on the city can equal its domestic hierarchical power, despite the political reality that the city is located in a semi-authoritarian regime. Through the mechanism of decentralisation, local governments in China are gaining more and more autonomy (if not greater capacity) in decision making and governing. Therefore, it is essential to gain consent from formal and informal organisations (network actors) at global, national and local levels for the institutional integration of climate pilot initiatives.

The process of integrating climate initiatives in a 'politically savvy' way involves a dynamic nexus of coordination and collaboration. The collaboration approach predominates in the global and domestic local contexts, while coordination is the major approach at the domestic national level. In both global and domestic local contexts, voluntary agreements and actions are institutionalised as the key components in the integration of climate initiatives. At the global level, there are constant negotiations among international networking actors over the appropriate emissions-reduction targets and feasible climate action plans. At the domestic local level, awareness raising and actual experiences have shaped the mindset of citizens towards climate change. At the initial stage of legitimation, the general consent of citizens on the climate action plans proposed by formal and informal organisational actors are built not by mandatory forces but through a voluntary approach, that is, the consent is not legally binding among network actors. What makes the climate pilot initiatives valid is the collaboration among formal and informal actors to achieve mutually shared goals.

Conversely, at the domestic national level, top-down mandatory administrative controls are the most effective approach. The national concerns over climate action have two facets. First, the national government, among all of the legitimate players in the global context, needs to respond to the international pressures posed by global negotiations over emissions reductions. Second, the national government is also concerned with balancing a healthy pace of economic development with its global responsibility for emissions reduction. Given these concerns, specific targets have been set and substantial economic/environmental plans have been made through the coordination of the national government and the vertical *inter*governmental hierarchy. These top-down national actions are seen as legitimate mainly through *inter*governmental coordination, with consent gained from formal organisational actors across different levels of the administrative system in a semi-authoritarian context.

However, the current institutional design of climate governance in China is still insufficient to accommodate this dynamic nexus of coordination and collaboration in the multilevel legitimation process. First, weak communication of the outcomes of pilot initiatives to stakeholders and peripheral network actors has limited the way horizontal engagement can broaden the impact of such initiatives. Second, communication about the horizontal *inter*governmental relations connecting the city governments is often indirect. Given their high competitiveness and hidden agendas, the governments of economically developed cities prefer to appeal directly to the central government before seeking possible ways to build collaborative ties with their horizontal counterparts. Third, the rigid administrative structure has hampered any attempts at institutional innovation in China.

Conclusion

By analysing how climate mitigation pilot initiatives are being institutionalised in Chinese cities, this chapter has discussed the contextual influence of the *inter*governmental structure in the establishment of governance networks, the dynamic interactions among network actors in the CMNs and the nexus of coordination and collaboration in governing strategies to achieve aligned output legitimacy. The different stages in the institutionalisation of climate pilot initiatives within the CMNs are specified in the analytical framework presented in Figure 7.1, including preparation, designation and implementation in an impact regeneration cycle.

Each stage has been discussed with regards to its financial and human resources, the perceptions and interactions of its participants and the *inter*governmental and network context in which each stage is located.

The analysis revealed that before the designation of climate initiatives, there are four governing modes (entrusting, applying, proposing, consulting), which are determined by the nature of climate initiatives and their *inter*governmental relationships (Figure 7.2). Internal lobbying is carried out during the *preparation stage* to achieve consensus building and align output legitimacy regarding the concerned initiatives among the core actors in the CMNs. During the *implementation stage*, institutional changes are carried out to transform the status quo of the governance structure. Stronger external mobilisation is therefore necessary to build consensus among the core and the peripheral actors, representing a further step in establishing output legitimacy across societal sectors. Because these climate pilot initiatives experiment with and demonstrate sectoral innovations from either an institutional or technological perspective, the implementation outcomes are presumably used by local political leaders to generate national and global impact from these improvised experiments by extending the practices and influences of the CMNs. Changes might be imposed on the climate action programme in the pilot regions (for example, other places in China coordinated by the national administration) in an impact regeneration cycle. Resource regeneration, local integration and multi-level legitimation are three important dimensions of this regeneration cycle.

Different contextual constraints and advantages embedded in *inter*governmental structures give rise to the nexus between coordination and collaboration throughout the process of institutional integration for climate initiatives. We find that this dynamic nexus is a defining feature of the collaborative municipal networks (CMNs) for low carbon transitions configured in a semi-authoritarian regime. Before moving on to the analytical and theoretical conclusions of this volume, this chapter has identified that by reinforcing the collaborative functions within a semi-authoritarian regime, the CMNs for climate mitigation are gradually transforming the Chinese environmental state into a more pluralistic and legitimate structure. The emerging CMNs are empowering local actors in the transformation process, despite the uncertainties and complexities in the converging global practices, the effects of climate negotiation and the changing national concerns during the institutionalisation of new functions in the environmental state.

However, the existing institutional design is still inadequate to accommodate such a major transformation. Enhanced channels of communication need to be instituted to facilitate the institutionalisation of sociotechnical innovations for low carbon transitions. Despite the release of low carbon policies and plans, the administrative structure and procedures need to be streamlined to remove the remaining barriers to institutional innovation.

Notes

1 The five pilot projects are: 1) National Demonstration City on Energy Efficiency Monitory System in Major Public Buildings (since 2007), 2) National Demonstration City on the Application of Renewable Energy in Buildings (since 2009), 3) National

Demonstration City on Construction Engineering Standardisation (since 2010), 4) Key City in National Public Building Energy Efficiency Retrofit (since 2011) and 5) Pilot City on Comprehensive Utilisation of Construction Waste (since 2012).

2 The Productivity Council is a statutory body to promote productivity excellence by achieving a more effective utilisation of resources and enhancing the value-added content of products and services.

3 The local manufacturing capacity covers exports in addition to local supplies.

4 Only for local supplies.

References

Braithwaite, J. & Drahos, P. 2000. *Global Business Regulation,* Cambridge, Cambridge University Press.

Bruton, M. J., Bruton, S. G. & Li, Y. 2005. Shenzhen: Coping with Uncertainties in Planning. *Habitat International,* 29, 227–243

Diamond, L. J. 2002. Thinking about Hybrid Regimes. *Journal of Democracy,* 13, 21–35

Dou, Y. 2012. Qianhai: A Green Building Exemplary Zone to Be. *Shenzhen Special Zone Daily,* 3 July 2012 [Online]. Available: http://sztqb.sznews.com/html/2012–07/03/content_2108895.htm [Accessed 25 March 2013] (*Qianhai Jiang Jiancheng Lvse Jianzhu Shifan Qu*)

Du, R. & Luo, A. 2013. Nansha New District: Docking Hong Kong and Macau. *People's Daily,* published by People's Daily. 22 February [Online]. Available: http://leaders.people.com.cn/n/2013/0222/c58278–20563703.html [Accessed 25 March 2013]

Economy, E. 2006. Environmental Governance: The Emerging Economic Dimension. *Environmental Politics,* 15, 171–189

Fang, K. 2012. With the Flag of the SEZ, Play the Cards of Hong Kong: Qianhai HK-SZ Cooperative Experiment. *Nanfang Weekend,* 31 August 2012 [Online]. Available: www.infzm.com/content/80227 [Accessed 13 September 2013] ('*Ju Tequ de Qi, Da Xianggang de Pai' – Qianhai de Shengang Hezuo Shiyan*)

Fang, K., Wang, F. & Shi, X. 2012. How Advanced Is the Qianhai Reform? *Nanfang Weekend,* 2 September 2012 [Online]. Available: www.infzm.com/content/80228 [Accessed 13 September 2013] (*Qianhai Gaige Youduo 'Qian' – Huopi Liang Zhounian, 'Tequ Zhong de Tequ' Dandang Tanlu Jianbing*)

Francesch-Huidobro, M. 2012. Institutional Deficit and Lack of Legitimacy: The Challenges of Climate Change Governance in Hong Kong. *Environmental Politics,* 21, 791–810

Francesch-Huidobro, M. & Mai, Q. 2012. Climate Advocacy Coalitions in Guangdong, China. *Administration & Society,* 44, 43S–64S

Francesch-Huidobro, M., Lo, C. W.-H. & Tang, S.-Y. 2012. The Local Environmental Regulatory Regime in China: Changes in Pro-Environment Orientation, Institutional Capacity, and External Political Support in Guangzhou. *Environment and Planning – Part A,* 44, 2493–2511

GD-TB 2010. *Project Brief of GEF-Funded Guangdong Provincial Green Freight Exemplary Program.* Guangdong Transportation Bureau. Guangzhou, Guangdong [Online]. Available: www.gdlshy.com/Front/Introduce/ShenQingDetail.aspx [Accessed 13 March 2013] (*Quanqiu Huanjing Jijin Zengkuan Guangdong Sheng Lvse Huoyun Shifan Xiangmu Jianbao*)

Haas, P. M. 1989. Do Regimes Matter? Epistemic Communities and Mediterranean Pollution Control. *International Organization,* 43, 377–403

Haque, M. S. 2013. Public Administration in a Globalized Asia: Intellectual Identities, Challenges, and Prospects. *Public Administration and Development,* 33, 262–274

HKPC & APAS. 2011. *Signing of Memorandum of Understanding by BYD Company Limited and Automotive Parts and Accessory Systems R&D Centre Limited* [Online]. Available: www.apas.hk/index.php?option=com_content&catid=&view=article&id=250&lang=tw [Accessed 25 March 2013]

Ho, P. 2007. Embedded Activism and Political Change in a Semiauthoritarian Context. *China Information,* 21, 187–209

Hodson, M. & Marvin, S. 2009. 'Urban Ecological Security': A New Urban Paradigm? *International Journal of Urban and Regional Research,* 33, 193–215

Huang, W. & Yang, X. 2011. Guangming New District: Striving for National Green Building Demonstration. *Nanfang Daily,* 28 December 2011 [Online]. Available: http://epaper.nfdaily.cn/html/2011–12/28/content_7043166.htm [Accessed 14 March 2013] (*Guangming Xinqu Zhengchuang Guojia Lvse Jianzhu Shifan Qu*)

Jiang, S. 2010. Shenzhen Pioneered the Introduction of Local New Energy Subsidies Policy. *News Morning Post (Xinwen Chenbao)* [Online]. Available: http://auto.ifeng.com/roll/20100825/406309.shtml [Accessed 5 February 2014] (*Shenzhen Shuaixian Chutai Difang Xin Nengyuan Butie Zhengce*)

Khademian, A. M. 2002. *Working with Culture: How the Job Gets Done in Public Programs,* Washington, DC: CQ Press.

Li, W. 2006. *Informational Environmental Regulation in Practice.* PhD Dissertation, Virginia Polytechnic Institute and State University.

Li, W. 2011. Self-Motivated Versus Forced Disclosure of Environmental Information in China: A Comparative Case Study of the Pilot Disclosure Programmes. *The China Quarterly,* 206, 331–351

Li, W. & Li, D. 2012. Environmental Information Transparency and Implications for Green Growth in China. *Public Administration and Development,* 32, 324–334

Liao, Y., Cheng, X. & Wang, H. 2012. Guangzhou Took the Lead in the Full Implementation of Green Building in New Urban Districts. *China News, Guangdong,* 9 February 2012 [Online]. Available: http://big5.chinanews.com:89/gate/big5/www.gd.chinanews.com/2012/2012–02–09/2/177935.shtml [Accessed 25 March 2013] (*Guangzhou Shuaixian Zai Xin Chengqu Quanmian Shi'shi Lvse Jianzhu*)

Mah, D. N. Y. & Hills, P. 2012. Collaborative Governance for Sustainable Development: Wind Resource Assessment in Xinjiang and Guangdong Provinces, China. *Sustainable Development,* 20, 85–97

Mol, A. P. J. & Carter, N. T. 2006. China's Environmental Governance in Transition. *Environmental Politics,* 15, 149–170

Rock, M. T. 2002. Integrating Environmental and Economic Policy Making in China and Taiwan. *American Behavioral Scientist,* 45, 1435–1455

Schreurs, M. A. 2010. Multi-Level Governance and Global Climate Change in East Asia. *Asian Economic Policy Review,* 5, 88–105

World Bank. 2013. *GEF Guangdong Green Freight Demonstration Project* [Online]. Available: www.worldbank.org/projects/P119654/gef-guangdong-green-freight-demonstration-project?lang=en [Accessed 17 March 2013]

Zacharias, J. & Tang, Y. 2010. Restructuring and Repositioning Shenzhen, China's New Mega City. *Progress in Planning,* 73, 209–249

Part III

Challenges of climate change governance in China

8 Beyond coordination and collaboration

A carbon reduction implementation strategy[1]

Introduction

We have argued that sectoral *intra*governmental coordination structures display fragmented *intra*governmental capacity in low carbon transitions in Hong Kong. As such, collaborative municipal networks may encounter obstacles at the implementation stage of this transition. In this chapter, we propose a carbon reduction implementation strategy (CRIAS) which is firmly grounded in the political and technical feasibility of taking up specific carbon reduction policy tools in Hong Kong. While climate mitigation governance is complicated by the need for governments to choose, evaluate and deploy tools to implement their political goals (Bressers 1998; Gupta *et al.* 2007; Jordan *et al.* 2011), neither the general literature on policy tools (Bailey 2007; Dahl & Lindblom 1953; Hood 1983, 2007; Stavins 1997) nor the specific literature on climate change mitigation tools (Carley 2011; Goers *et al.* 2010; Gupta *et al.* 2007; Igielska 2008; Jordan *et al.* 2003, 2011; Oikonomou & Jepma 2008; Twomey 2012; Wurzel *et al.* 2012; Zito *et al.* 2003) provides much clarity on how these tools function in practice.

With greenhouse gas (GHG) emissions from Chinese cities becoming an issue of global concern (International Energy Agency [IEA] 2012; Liu *et al.* 2012), this chapter argues that the pros and cons of the selection, evaluation and deployment of carbon reduction policy tools in Chinese cities deserve more detailed critical analysis than they have thus far received. Hong Kong is expected to contribute significantly to reducing the country's climate footprint. Evaluation of the political and technical feasibility of alternative GHG-reduction strategies in Hong Kong is thus of broad significance.

The strategy proposed by the Environment Bureau (2010) to achieve Hong Kong's carbon reduction targets is to revamp the fuel mix for electricity generation by increasing the proportion of nonfossil sources (nuclear energy in particular), whereas our theoretical, methodological and empirical analysis shows strong support from both the professional community and general public for demand-side management (DSM)[2] measures instead. However, the introduction of these measures and corresponding policy tools needed to enable them are likely to encounter a range of sociopolitical and technical challenges (Hoogwijk & Graus 2008).

Accompanying the support for DSM is strong opposition to any increase in nuclear energy use. Such opposition stems from such accidents as the 1979 Three

Mile Island accident in the US and 1986 Chernobyl disaster in the former USSR and, more recently, the 2011 Fukushima Daiichi nuclear disaster in Japan, which triggered widespread public support for reducing GHG emissions without relying on nuclear power (*The Japan Times* 2013; *South China Morning Post* 2013). The IEA estimates that the additional nuclear generating capacity originally planned for completion by 2035 will be reduced by about 50 percent as a result of the Fukushima accident (*The Economist* 2011).

As an alternative to altering the fuel mix, DSM provides a variety of technological, organisational and behavioural solutions for modifying consumer energy demand to reduce electricity consumption and generation and thus GHG emissions (Boshell & Veloza 2008; McPherson 1993; Molderink 2012). As a GHG reduction policy measure, DSM currently is more politically viable than nuclear power and more technically feasible than renewable energy (RE). Evidence also suggests that DSM can improve energy efficiency (Auffhammer *et al.* 2007; Cooper *et al.* 2013; Lee *et al.* 2012; Papagiannis *et al.* 2008; Ramchurn *et al.* 2011). Although proponents argue that 'DSM is not only good practice for reducing energy consumption – it has also co-benefits such as improved wellbeing and comfort of end-users' (Bonneville & Rialhe 2006: 10), its use is not without constraints and challenges (Strbac 2008).

Making choices between and within fuel remix and DSM programs depends on national and local circumstances and varies across sociopolitical and technical lines. What is clear, however, is that no GHG reduction program can be implemented effectively without the deployment of appropriate policy tools (Jaccard *et al.*1997). Moreover, the effective and flexible deployment of policy tools provides governments with a 'policy window' (Kingdon 1995) for adopting a particular target-reduction strategy when another is politically or technically infeasible.

Conceptually, this chapter draws on these theorisations of DSM and typologies of policy tools, climate policy tools in particular. Empirically, it critically applies them to Hong Kong as a test bed of how such tools are deployed in practice. We build competing GHG reduction scenarios, propose an integrated reduction strategy (i.e. the Carbon Reduction Implementation and Assessment Strategy, or CRIAS) and conclude that a DSM strategy can achieve reduction targets without the need for heavily weighted changes to the fuel mix.

In adopting a tools perspective, we concur with Jordan et al. (2011) that tools are better signifiers of political commitment than mere political pronouncements, as their deployment requires money, time and energy. However, we remain mindful of the inherent challenges to tool selection, such as the limitations arising from their innate characteristics (e.g. cost effectiveness), the sociopolitical and institutional contexts in which they are to be deployed (and why they are often not deployed) and the politics involved in deciding who determines tool choice, who wins and who loses.

For our analysis, we select tools currently being used or slated for use in Hong Kong to achieve GHG reduction targets and those that have worked in other policy areas such as urban regeneration and housing (see Scott 2005, 2007). We assess variations in the effort required to deploy these tools and their environmental effectiveness and impact. We assume that such variations can be explained by a variety of factors, including their sociopolitical, policy and institutional context and the entrepreneurial potential of the decision makers and other stakeholders who have to muddle through the problems, experiences, reactions and interactions between the tools and their target populations.

We focus on three sectors of the economy: electricity generation for *buildings* (2009 emissions: 29.1 million metric tons of CO_2 equivalent [CO_2-eq]), fuel use in *transport* (2009 emissions: 7.3 metric tons of CO_2-eq) and other fuel end use for *electricity generation* (2009 emissions: 2.9 million metric tons of CO_2-e) (Carbon Disclosure Project 2012, C1.6), because they are the greatest GHG emitters in Hong Kong (e.g. 67 percent of total emissions in 2008 came from electricity generation, of which 90 percent was used in buildings) and have the greatest potential for reduction. We combine quantitative and qualitative methods of analysis and propose three reduction scenarios that we test on key stakeholders, a theoretical, empirical and methodological first in the study of GHG reduction in a Chinese city. The combined results show that the advantage of selecting and deploying a broad range of commonly used policy tools is that predictable effects can be publicly disclosed for stakeholder review and approval.

The remainder of the chapter proceeds as follows. We first set the conceptual and contextual scene by introducing the sociopolitical, policy and institutional context of GHG reduction efforts in China and Hong Kong, along with a brief critical review of the literature on climate policy tools and tool selection criteria. We then discuss the methodology used in our quantitative and qualitative analyses. Thereafter, we report the findings of our empirical analysis by first identifying the measures proposed in the HKSAR-ERM (2010) and WWF-HK (2010) studies and then matching them to a set of currently and potentially available tools selected according to definitions and criteria proposed in the existing literature and our own explanation of why different tools might be selected and used in different contexts. We then analyse stakeholder views of the reduction measures suggested by the two reduction studies and by our calculations and analysis, particularly their views of the environmental effectiveness and impact of deploying individual tools and a combination thereof. Finally, we conclude that a DSM strategy can achieve GHG reduction targets without the need for more heavily weighted changes to the fuel mix and consider the implications for future research. Our aim in this chapter is not only to further understanding of the dynamics of tool selection and provide a broader assessment of innovative policy mechanisms and their use in practice but most importantly to fill the gap in research on GHG reduction implementation strategies in Chinese cities.

Policy tools and GHG emission reduction

In articulating climate change policy goals, governments not only set long-term strategies and corresponding short-term targets. They also identify the specific measures through which those policies and targets are to be realised and the tools, methods and techniques necessary for their implementation. As the climate strategies and targets and corresponding measures being proposed for Hong Kong, which are discussed in greater depth in the next section, lack the *policy tools* essential for GHG reduction, we focus here on this critical decision-making ingredient.

Sociopolitical, policy and institutional context

The climate policy of the Chinese Central People's Government (CCPG) is a major driver of the HKSAR government's climate agenda. In 2009, the CCPG announced a voluntary national target: a 40 to 45 percent reduction (from 2005 levels) in CO_2 production for each yuan of national income (carbon intensity) by 2020. The target was to be achieved by intensifying efforts to conserve energy and improve energy efficiency, developing renewable and nuclear energy, increasing forest coverage and stepping up efforts to develop a low carbon economy (Environment Bureau 2010, para. 4.8). Hong Kong is duty bound to support the national target, as application of the United Nations Framework Convention for Climate Change and Kyoto Protocol was extended to Hong Kong in 2003 in accordance with Article 153 of the Basic Law, Hong Kong's mini-constitution. The HKSAR government's approach to tackling climate change thus intensified after the national target's announcement. It committed itself not only to supporting the target but also to exceeding it by achieving a more aggressive 50 to 60 percent reduction in GHG emissions by 2020 based on the understanding that an economically developed and politically liberal city such as Hong Kong should make a more significant contribution to the reduction nationwide.

Hong Kong's commitment resulted in the proposal of three reduction scenarios by Environmental Resources Management (ERM 2010: 28–36), all of them based on enhancing energy efficiency, using cleaner fuels, relying less on fossil fuels and promoting a low carbon economy. ERM recommended 'Scenario 3', an aggressive approach focused on altering the fuel mix for electricity generation, as the most suitable for Hong Kong because it would also promote the substantial development of nuclear energy in mainland China. Accordingly, the government published a consultation document, *Hong Kong Climate Change Strategy and Action Agenda 2010* (HKSAR-ERM), which was essentially the full adoption of Scenario 3 (Environment Bureau 2010 part 1, section 5). HKSAR-ERM suggests altering the electricity generation fuel mix by maximising the use of natural gas (NG) and increasing its share to about 40 percent by 2020 and by increasing nuclear power (NP) intake from mainland China to supply 50 percent of Hong Kong's electricity by 2020, compared with 23 percent in 2009 (Environment Bureau 2010: 43, part 1, section 5). It could be argued that HKSAR-ERM is premised upon such assumptions about mainland China's energy policy as the further development of nuclear energy rather than on the need to tackle climate change.

Whatever the government's rationale, environmental nongovernmental organisations (NGOs) including World Wildlife Fund Hong Kong (WWF-HK) have been highly critical of the HKSAR-ERM proposals, not only because of safety concerns over nuclear power but, more importantly, because of their lack of DSM strategies in the form of measures to reduce peak electricity demand within the electricity grid and end-user energy consumption without sacrificing service quality and user comfort (WWF-HK 2010; Appendix I: 15/HK/18–08–2012). As an alternative, WWF-HK (with its consultant, Arup) put forward *Road Map 2020*, arguing that it is possible to reduce carbon emissions without increasing nuclear power generation (WWF-HK 2010).

These two proposals – HKSAR-ERM and WWF-Arup – form the basis of the empirical analysis in this chapter. Their proposed carbon reduction measures are summarised in Table 8.1A (Chapter appendix). We argue that those measures cannot be implemented effectively without the support of appropriate government policy tools (see also Francesch-Huidobro 2012). The following section critically examines the literature on policy tools, climate policy tools in particular, assesses the range of tools for implementing carbon reduction measures and considers several tool selection criteria.

Policy tools

Policy tools are the range of instruments available to governments for implementing their policy objectives (Hood 2007). Researchers define these tools by providing typologies, such as *regulations*, which are used in hierarchical, prescriptive governance modes; *market-based tools*, which are used in governance modes that offer societal choice; *information*, which enables societal actors to choose how they wish to be governed; and *voluntary agreements*, which bring governmental and nongovernmental actors together to address problems in a formalised but nonhierarchical manner (see Gupta *et al.* 2007; Hood 2007; Peters & Nispen 1998). Table 8.1 presents the range of policy tools available for implementation.

Tool choice criteria

The effort required to deploy a policy tool, its effectiveness and its effects on reducing GHG emissions all depend on the presence of appropriate conditions for tool uptake, which highlights the importance of determining the criteria used to choose tools for implementing carbon reduction measures in a particular setting. Following Gupta et al. (2007) and the IEA (2006), we propose four key criteria and their suggested weighting.

Environmental effectiveness is the extent to which a policy tool meets its intended objective and its positive environmental outcome. The amount of CO_2 saved (in percent) as a result of a tool determines whether it has a high, medium or low degree of effectiveness. Such assessments are possible only when baseline data are available, although even then the weighting of the CO_2 amount saved must take into account the temporal, sectoral and spatial scales of emission reduction. Accordingly, our weighting of this criterion is based on expert advice gathered during stakeholder interviews and the Measures for the Rational Use of Energy (MURE) database, which assigns *low* effectiveness to tools with a reduction potential of 0 to 0.1 percent, *medium* effectiveness to those with a 0.2 to 0.5 percent

Table 8.1 Deploying policy tools

Tools	Description	Analysis of literature
1 Regulations and Standards (regulatory)	They specify the actions a firm or individuals must undertake to achieve environmental objectives.	Regulatory tools have rather predictable outcomes, however, Jaffe et al. and Sterner (cited in Gupta *et al.* 2007: 754) note that when they are prescribed, there are few incentives for firms and individuals to be innovative in searching for more efficient methods to reduce GHG, which may lead to a lack of technological change. Empirical evidence also shows that it is essentially the political will behind them that is most significant for their effective use (Buchner et al. 2011; Hoffman 2005).
2 Subsidies (market economy)	Grants, low-interest loans and tax credits are used by governments to stimulate development and diffusion of new technologies. However, their economic cost is generally high, especially when it becomes a long-term government commitment.	Governments frequently use subsidies to stimulate the development and transfer of new technologies and assist in offsetting market externalities (Gupta *et al.* 2007; UNEP 2012), but their economic costs are generally high, especially when they become long-term government commitments.
3 Taxes (market economy)	These allow governments to set limits on how much is spent on environmental policy. They can be 'positive' (e.g. tax reduction/allowance) or 'negative' (e.g. charges/fees). They are generally more cost effective but cannot guarantee a particular level of emissions.	Oikonomou and Jepma (2008) identify three types of taxation or 'negative penalties': *emission charges/taxes* that are direct payments based on actual measurements of the quantity and quality of pollutants discharged; *user charges* that are payments for the cost of a collective service; and *product charges/taxes* which are applied to products that generate pollution when they are manufactured, consumed or disposed of. The use of taxation eliminates the need for regulatory measures, thereby reducing the costs of compliance and enforcement (Carley 2011; Stavins 1997: 10–11); however, it can be difficult to implement when political barriers are encountered.
4 Public Benefit Funds (public economy)	PBFs, contributed from a variety of funding sources, often function in a similar way to that of subsidies.	When administered effectively, PBFs can have a positive effect and are largely well received (C2ES 2011; Nadel and Kushler 2000).

5	**Government Expenditure** (public economy)	It involves the use of public money to develop infrastructure, e.g. energy-efficient government buildings/ public facilities; installation of EVs recharging stations, etc.	It has been used for developing low carbon infrastructure and supporting research and development measures to reduce GHG emissions (IEA 2012; IPCC 2012). When direct government expenditure is used effectively and efficiently, it can create a positive image of the government, which may then help to push through other environmental measures (Bruce *et al.*, 1996: 414).
6	**R&D** (public economy)	It can stimulate technological advances, reduce costs and enable progress toward GHG reduction.	Further R&D to make these technologies market ready requires government policies to nurture investment (OECD 2011, 2012).
7	**Public Information** (regulatory/ voluntary)	Public disclosure requirements and awareness/education campaigns allow consumers to make better-informed choices.	Firms may view public disclosure policies as overly burdensome and argue that voluntarily provided information is sufficient (Sterner 2003).
8	**Voluntary Agreements** (voluntary)	These are agreements between a government authority and one or more private parties to achieve environmental objectives or to improve environmental performance.	These are often conducted in a formalised manner. The distinction between them and a 'pure regulation' is blurred. Local governments gain two main benefits from adopting VAs: the ability to develop and improve their relationship with the public, as voluntarily implementing measures generates positive perceptions and improves their reputation as public leaders, and a lower potential for legal action and consequently reduced legal costs (Gupta *et al.* 2007).

Sources: Jaffe *et al.* and Sterner, cited in Bruce *et al.*, 1996: 414; Buchner *et al.* 2011; C2ES 2011; Carley 2011; Gupta *et al.* 2007: 754; Hoffman 2005; IEA 2012; IPCC 2012; Nadel and Kushler 2000; OECD 2011, 2012; Oikonomou and Jepma 2008; Stavins 1997: 10–11; Sterner 2003.

potential and *high* effectiveness to those with a potential > 0.5 percent (MURE 2012). The MURE database was developed by a team of European scholars led and coordinated by the Institute of Studies for the Integration of Systems in Rome and the Fraunhofer Institute for Systems and Innovation Research in Germany. It is widely used as a critical source of data and analysis in research on energy efficiency and the effectiveness of energy policy tools (Bigano *et al.* 2011; Bossoken 1999; Eichhammer 2008; Filippini *et al.* 2013; McCormick & Neij 2009; Mundaca *et al.* 2010; Petrichenko 2010).

Cost effectiveness is the extent to which a policy tool achieves its GHG reduction objectives at a minimum cost to society and is calculated as dollars per ton of CO_2 saved with reference to a base year. The effectiveness grades in our analysis are based on indicative ranges and expert advice: low cost effectiveness is > 25\$CO$_2$ equivalent (CO$_2$eq), medium is 0 to 25\$CO$_2$eq and high is <0\$CO$_2$eq.

The final two criteria are the policy tool's *distributional impact* and *institutional feasibility.* The former is defined as the tool's societal consequences and includes such dimensions as fairness and equity (Gupta *et al.* 2007). Even when a policy tool is effective in meeting a reduction goal at minimum cost, it may encounter political opposition if it disproportionately benefits or disadvantages certain groups. *Institutional feasibility* is the extent to which a tool is likely to be viewed as legitimate, gain acceptance and be adopted and implemented (see also Francesch-Huidobro 2012). The sociopolitical context within which the tool will be used and the drivers of and barriers to its implementation must also be weighed in tool selection.

We consider all four criteria in our selection, although our primary focus of analysis is *effectiveness* at both the output and impact levels.

Data sources and research methods

Our construction of the proposed CRIAS implementation strategy draws on the MURE database and in depth analysis of the views of a select group of Hong Kong stakeholders on the suitability of given policy tools for the measures proposed in the HKSAR-ERM and WWF-Arup reports and our CRIAS strategy.

Quantitative data: MURE database

MURE (2012) gathers information from 27 countries on energy efficiency trends (the ODYSSEE database) and policy tools (MURE), thus allowing the simulation and comparison of the tools' potential impact. MURE categorises policies by their 'status' (ongoing, completed, proposed and unknown) and demonstrated 'impact', that is, low (L), medium (M), high (H) and unknown (U). To reduce uncertainty, 'unknown' elements were omitted from our data analysis.

MURE provides information only on the semiquantitative impacts of related policies (e.g. L, M and H). To further quantify the data, we assumed a scale of 1 to 99 percent, with L = 1 to 33 percent, M = 34 to 66 percent and H = 67 to 99 percent and assigned an average value to each level: $I_H = 0.83$, $I_M = 0.50$ and $I_L = 0.17$. We then matched the MURE measures in the categories of 'Household', 'Transport' and 'Cross-cutting' with our three focal sectors: 'Buildings', 'Transport' and 'Energy'.

Qualitative data: stakeholder views

The views collected during stakeholder interviews and discussion forums provided our qualitative data. Krueger (1988) notes that such qualitative techniques are low cost, speedy, flexible and socially oriented with high face validity. Interviewees were selected for their extensive knowledge and practical experience of various areas of global and local climate change governance. Although they do not represent a statistically meaningful population, they allowed us to explore the suitability of our reduction implementation strategy. The discussion sessions also gave us an opportunity to crucially observe at first hand the process of people discussing the pros and cons of identifying potential tools for the implementation of GHG reduction measures.

The stakeholder discussions considered the HKSAR-ERM and WWF-Arup action plans and reduction measures, along with our three scenarios and the policy tools proposed for implementation of the various measures. All interview data were recorded and transcribed and provide the basis for the qualitative reflections on our quantitative effectiveness estimates.

Analysing policy tools for GHG reduction

MURE database and library

Our analysis of the MURE database revealed several policy tools similar to those reviewed herein, for example, legislative tools (standards and regulations), financial tools (grants and subsidies), fiscal tools/tariffs (taxes) and information/ education and cooperative measures. It was thus fitting and possible to match the MURE measures and tools with those considered in our study to create an MURE Library, thereby allowing the potential impact of each tool on each measure to be quantified on the basis of the average impact estimated by the sum of the policies and measures in the database.

For example, the first of the measures in CRIAS is the Building Energy Codes (BEC) in the Building sector. The impact of introducing this measure through regulation was equated to policies related to energy performance standards in the MURE database, which reports 37 policies of high (H) impact, 19 of medium (M) and 7 of low (L). The overall impact of the implementation strategy was thus estimated to be:

$$((H_xL_H) + (M_xL_M) + (L_xL_l)) / SUM(H, M, L) = ((37 \times 0.83) + (19 \times 0.50) + (7 \times 0.17)) / (37 + 19 + 7) = 0.66 \text{ or } 66\%.$$

$$((H_xL_h)' (M_xL_m) + (L_xL_l)) \div \sum (H,M,L) = ((37 \times 0.83)) + (19 \times 0.50) + (7 x \times 0.17)) \div (37 + 19 + 7) = 0.66) \text{ or } 66\%).$$

Similarly, the impacts of introducing the BEC through financial tools, fiscal tools and voluntary agreements (VAs) were given values based on the equivalent figures in the MURE database. The same process was used for all measures until the library was fully populated. Calculating the potential impact of each tool by the sum of the policies and measures in the MURE database is arguably simplistic because different measures within each category in the database are often incomparable in terms of scope, impact and funding amount, and counting the number of measures in each group can be imprecise. However, because this estimate reflects the presence and number of reduction measures and the estimated average impact across 27 EU countries, it provides a point of reference for Hong Kong (and possibly other Chinese cities), where no such database exists. Bigano et al. (2011) adopted a similar approach. Using data from MURE, they created dummy variables for subcategories of policy measures to analyse the relationship between energy efficiency and energy supply security. Similarly, in their study of energy policy instruments' effects on energy efficiency, Filippini et al. (2013) also created dummy variables to ensure that each measure in a policy measure subcategory carried equal weighting.

HKSAR-ERM and WWF-Arup measures library

We also created a Measures Library comprising reduction measures from the HKSAR-ERM and WWF-Arup documents and the expected GHG reduction they assigned to each measure. Although some of the measures appear to have zero reduction potential owing to differences in the naming of the respective measures used and the grouping together of some measures in the estimated reduction analysis, they are nevertheless assumed to be important. Pedestrianisation is one such measure.

Live gameboards

Finally, we created two live gameboards, one for the HKSAR-ERM proposal and one for WWF-Arup. Matching the various policy tools to the measures in these documents allows decision makers to use our CRIAS implementation strategy to test the possible impact of each policy tool on each measure. The live gameboards also permit users to test various combinations of tools and measures and develop a bespoke scenario for each measure's expected GHG reduction target. The bespoke scenario presents the measures and their expected impact as detailed in the HKSAR-ERM and WWF-Arup studies, and the live gameboards then incorporate the quantified expected impact that each tool would have, as estimated from the MURE database. By selecting 'Yes', 'No' or 'N/A', a gameboard user can create different scenarios for implementation and test their effectiveness in achieving the overall carbon targets set in the two studies. Live gameboards also allow the user to create a customised strategy for achieving the expected carbon savings from each measure.

We used the gameboards to generate our three proposed scenarios: Baseline, Low and High (Tables 8.2, 8.3 and 8.4).

Baseline scenario

The Baseline Scenario (Table 8.2) reflects the current situation in Hong Kong, proposing the deployment of tools that will enable the implementation of reduction measures that are already in place or that will be implemented in the near future (Environment Bureau 2013).

The following paragraphs analyse the various reduction measures in Table 8.1A supplemented by the views of the interviewees.

Table 8.2 Baseline scenario

Sector	Measures	Policy tools								Expected impact	Estimated reduction in absolute carbon emissions[a]			
		Regulation	Tax incentives	Subsidies	Voluntary agreements	Public information	Public benefit funds	Govt expenditure	R&D		HKSAR/ERM		WWF/ARUP	
											Possible	Achieved	Possible	Achieved
Buildings	Building Energy Codes	Y								66%	0.00%	0.00%	6.62%	4.35%
	District Cooling							Y		43%	0.00%	0.00%	0.00%	0.00%
	Water-cooled A/C					Y			Y	42%	0.00%	0.00%	0.00%	0.00%
	OTTV	Y								73%	0.00%	0.00%	0.00%	0.00%
	EE Systems					Y			Y	85%	0.00%	0.00%	0.00%	0.00%
	EE Appliances	Y			Y	Y				100%	0.00%	0.00%	2.22%	2.22%
	Power Plants ESS									0%	0.00%	0.00%	4.33%	0.00%
	EE Behaviour					Y				42%	0.00%	0.00%	3.40%	1.42%
Transport	Alternative Fuels			Y					Y	100%	0.00%	0.00%	1.00%	1.00%
	Fleet Efficiency		Y	Y						98%	0.00%	0.00%	1.40%	1.37%
	EVs		Y	Y				Y	Y	100%	0.00%	0.00%	1.00%	1.00%
	Pedestrianisation									0%	0.00%	0.00%	0.00%	0.00%
	Biofuels		Y							55%	0.00%	0.00%	0.00%	0.00%
Energy	WtE							Y		43%	0.00%	0.00%	2.44%	1.04%
	RE									51%	4.00%	2.04%	1.15%	0.59%
	Fuel Mix			Y						43%	29.00%	12.36%	13.39%	5.70%
										Total Achieved	33%	14%	37%	19%
										Target for 2020	19–33%		37%	

Note:

Y = 'Yes'

[a]Compared to 2005

Table 8.3 Low scenario

Sector	Measures	Policy tools								Expected impact	Estimated reduction in absolute carbon emissions[a]			
		Regulation	Tax incentives	Subsidies	Voluntary agreements	Public information	Public benefit funds	Govt expenditure	R&D		HKSAR/ERM		WWF/ARUP	
											Possible	Achieved	Possible	Achieved
Buildings	Building Energy Codes	Y								66%	0.00%	0.00%	6.62%	6.62%
	District Cooling									0%	0.00%	0.00%	0.00%	0.00%
	Water-cooled A/C									0%	0.00%	0.00%	0.00%	0.00%
	OTTV	Y								73%	0.00%	0.00%	0.00%	0.00%
	EE Systems					Y				42%	0.00%	0.00%	0.00%	0.00%
	EE Appliances	Y			Y	Y				100%	0.00%	0.00%	2.22%	2.22%
	Power Plants ESS									0%	0.00%	0.00%	4.33%	0.00%
	EE Behaviour					Y				42%	0.00%	0.00%	3.40%	2.63%
Transport	Alternative Fuels				Y					41%	0.00%	0.00%	1.00%	1.00%
	Fleet Efficiency				Y					41%	0.00%	0.00%	1.40%	0.59%
	EVs				Y					41%	0.00%	0.00%	1.00%	0.83%
	Pedestrianisation									0%	0.00%	0.00%	0.00%	0.00%
	Biofuels				Y					41%	0.00%	0.00%	0.00%	0.00%
Energy	WtE									0%	0.00%	0.00%	2.44%	0.98%
	RE			Y					Y	51%	4.00%	2.04%	1.15%	1.15%
	Fuel Mix			Y	Y					83%	29.00%	24.03%	13.39%	11.10%
										Total Achieved	33%	26%	37%	27%
										Target for 2020	19–33%		37%	

Note:

Y = 'Yes'

[a] Compared to 2005

Building Energy Codes (BEC) have been a mandatory requirement for all new buildings and major renovations since September 2012. A focus group (Appendix I: 1FG/HK/28–05–2012) participant considered them 'a good first step' but noted that further improvements are necessary. For example, companies are required to perform energy audits only once every ten years and are not technically required to make any improvements. Accordingly, 'the energy audits are ineffective in regulating building energy usage', the participant concluded.

In January 2011, the government approved a budget to implement a district cooling (DC) system at the Kai Tak residential project and has since investigated the potential for further development. However, the discussion group participants agreed that DC is impractical in Hong Kong owing to its climate. A focus group (Appendix I: 1FG/HK/28–05–2012) participant noted that 'one by-product of DC is waste heat, and due to Hong Kong's position in the sub-tropics and its sun angle, there [is] no way to utilise this waste heat in a meaningful way'. Guidance notes on a code of practice for water-cooled air-conditioning (WCAC) systems were issued in November 2011, but the measure has yet to be implemented, prompting another participant (Appendix I: 1FG/HK/28–05–2012) to question the lack of impetus, particularly as the payback period of WCAC implementation is likely to be just three or four years.

The Buildings Regulation (Energy Efficiency), Cap 123M stipulates controls on the building envelopes of commercial and hospitality buildings through overall thermal transfer value (OTTV). However, the NGO participants suggested that the government should regulate both commercial and residential buildings with reference to OTTV, particularly as 'many residential buildings are now built in a similar fashion to commercial buildings ([i.e.,] they are all high-rise)', an interviewee (Appendix I: 2FG/HK/29–05–2012) said.

From 2009 to 2012, the Building Energy Fund scheme sponsored residential, commercial, industrial and mixed-use buildings to improve their energy efficiency systems (EES), and the government has invested in the Hong Kong Science and Technology Park as a platform for companies with a portfolio of EES. The Mandatory Energy Efficiency Labelling Scheme covers such appliances as room coolers, refrigerators, washing machines and dehumidifiers, whereas other appliances are covered by the Voluntary Energy Efficiency Labelling Scheme. The government has also launched a host of energy-efficiency campaigns to encourage energy-efficient (EE) behaviour. In the view of one focus group participant (Appendix I: 1FG/HK/28–05–2012), a 'passive design' would be more practical for EES than a performance-based system that focuses on regulating energy use, as it would avoid 'the mechanical heating and cooling of a building through the use of natural ventilation, solar heat gain, solar shading or better insulation'.

With regard to alternative fuels and fleet efficiency, to encourage the use of highly fuel-efficient vehicles for Hong Kong's commercial fleet, reductions in the first registration tax are offered to buyers of newly registered environmentally friendly commercial vehicles. In addition, the Pilot Green Transport Fund launched in March 2011 provides subsidies to transport operators to purchase low-emission vehicles. The Hong Kong Automotive Parts and Accessory Systems

R&D Centre (APAS) was established in 2006 to develop hybrid vehicle technologies. The HKSAR-ERM study proposes that 30 percent of private cars and 15 percent of buses and goods vehicles will be hybrids (running on petrol and diesel blended with 10 percent ethanol and biodiesel, respectively) or electric vehicles (EVs) by 2020. However, a focus group (Appendix I: 2FG/HK/29–05–2012) participant highlighted the lack of necessary implementation tools: 'For both measures, we would need tax incentives, subsidies, public information, and government expenditure'.

A profit tax deduction is available for any capital expenditure on environmentally friendly vehicles, including hybrids and EVs, and one of the foci of the APAS is to develop EV–compatible technologies. The Steering Committee on the Promotion of Electric Vehicles was established in April 2009, and the government has announced plans to ask all franchised bus companies to test zero-emission electric buses and to provide funds to assist them in purchasing such vehicles. However, one focus group (Appendix I: 2FG/HK/29–05–2012) participant pointed out that although EV use can reduce air pollution, 'more R&D is required to see their actual impact on GHG reduction'.

The final three measures in the Baseline Scenario are biofuels, waste-to-energy (WtE) and renewable energy (RE). A duty-free arrangement for the use of biodiesel as a motor vehicle fuel was introduced in 2007, and a planned organic waste treatment facility will produce 14 million kWh of electricity annually from 200 tonnes of organic waste. In addition, a sludge-management facility, which was completed in 2013, is powered by the heat produced during sludge incineration. Finally, under the current Scheme of Control (SOC) agreement with Hong Kong's two power companies (China Light and Power and Hong Kong Electric Co.), an 11 percent increase in the permitted return is offered for investment in RE facilities, and a 0.01 to 0.05 percent bonus in the permitted return is awarded depending on the extent of RE used in electricity generation.

Low scenario

The Low Scenario (Table 8.3) would result in implementation strategies that could be put in place with minimal effort and cost but would have a fairly insignificant impact. The tools proposed in the HKSAR-ERM document were chosen as the minimum set of tools necessary to achieve the 19 to 33 percent reduction in absolute carbon emissions by 2020 (from 2005 levels), with a particular focus on those that would require little government expenditure. The main policy tool in this scenario is voluntary agreements (VAs), which have minimal cost implications. In addition, several existing tax incentives for promoting sustainable transport are maintained, and subsidies are introduced to promote changes in the fuel mix. Together, these tools can achieve a 26 percent reduction in absolute carbon emissions.

To move the resulting carbon emission savings closer to the high WWF-Arup target of 37 percent the implementation strategy must include a wider selection of tools allowing, more demand-side measures. Again, in developing the Low Scenario, easily implementable tools with few economic implications were chosen.

Although this strategy still fails to achieve the full reduction target, it moves closer to the low (27 percent) target in the WWF-Arup study. VAs are also key in this scenario, with some provision for public information and R&D activities.

High scenario

The High Scenario (Table 8.4) includes the deployment of tools that would require greater effort and a higher cost but lead to a more significant reduction. As the majority of savings in the HKSAR-ERM study come from the fuel mix, introducing regulation and subsidies in combination with some support for RE options would theoretically be adequate to support the full implementation of this scenario. In reality, however, this would be an inappropriate strategy given the importance of controlling the demand side and offering support to other sectors. The NGO interviewees in focus group discussion (Appendix I: 2FG/HK/29–05–2012) made a link between the proposed EE measures and DSM, suggesting that the current Scheme of Control (SOC) governing the two power companies, China Light & Power and Hong Kong Electric, is insufficient to reduce GHG emissions in Hong Kong. The SOC, which came into force in January 2008, has three aims: reducing tariffs, reducing emissions and opening the market to competition (Environment Bureau 2008). One NGO participant suggested that DSM is the way forward, as it would 'reduce the amount of energy [used] without sacrificing the quality or the enjoyment of the energy that we currently have', but the first step is a further review of the SOC. An interviewee from the WWF explained that one aim of the WWF-Arup proposal was to list various measures for revising the SOC and to argue for more effort to be devoted to DSM.

Another focus group (Appendix I: 2FG/HK/29–05–2012) participant cited three overseas examples: the cases of the US state of California, the UK and Hong Kong. A scheme in the former, the participant said, grants subsidies to individual households to purchase EE appliances, whereas the UK requires 40 percent of energy-efficiency measures to be conducted in low-income groups, and electricity companies in Taiwan offer discounts to customers who can prove they have reduced their electricity consumption. Such schemes would be possible in Hong Kong if the government were to promote DSM, the participant added. It was apparent to the majority of the interviewees that the greatest proportion of the GHG reduction proposed in the HKSAR-ERM was expected to come from a change in the fuel mix (10 percent coal, 40 percent natural gas, 3 to 4 percent RE and about 50 percent nuclear). A power company representative in focus group discussion (Appendix I: 1FG/HK/28–05–2012) explained that most of the reduction targets in that consultation chapter centred on the energy sector, 'as it is easy to regulate energy companies in Hong Kong'. The representative further remarked that 'we would actually prefer . . . more regulations', which 'would drive the energy sector to look for more innovative solutions and build a better image for the industry'.

However, as one of the NGO representatives pointed out, the government 'is really reluctant to provide incentives for the public to save energy', although it provides incentives for utility companies to do so. Resolving the situation will require political will. Another NGO interviewee made an interesting point

Table 8.4 High scenario

Sector	Measures	Policy tools								Expected impact	Estimated reduction in absolute carbon emissions[a]	
		Regulation	Tax incentives	Subsidies	Voluntary agreements	Public information	Public benefit funds	Govt expenditure	R&D		Possible	Achieved
						HKSAR/ERM						
Buildings	Building Energy Codes	Y		Y						100%	0.00%	0.00%
	District Cooling				Y	Y		Y	Y	100%	0.00%	0.00%
	Water-cooled A/C				Y	Y		Y	Y	100%	0.00%	0.00%
	OTTV	Y		Y						100%	0.00%	0.00%
	EE Systems	Y	Y			Y				100%	0.00%	0.00%
	EE Appliances	Y	Y			Y				100%	0.00%	0.00%
	Power Plants ESS	Y		Y	Y					100%	0.00%	0.00%
	EE Behaviour		Y		Y		Y			100%	0.00%	0.00%
Transport	Alternative Fuels		Y	Y		Y	Y			100%	0.00%	0.00%
	Fleet Efficiency	Y	Y	Y		Y	Y			100%	0.00%	0.00%
	EVs		Y			Y		Y		100%	0.00%	0.00%
	Pedestrianisation				Y	Y		Y		100%	0.00%	0.00%
	Biofuels		Y			Y		Y	Y	100%	0.00%	0.00%
Energy	WtE									0%	0.00%	0.00%
	RE	Y		Y		Y		Y	Y	100%	4.00%	4.00%
	Fuel Mix	Y		Y						100%	29.00%	29.00%
										Total Achieved	33%	33%
										Target for 2020		19–33%

WWF/ARUP

Category	Measure										100%		
Buildings	Building Energy Codes	Y				Y					100%	6.62%	6.62%
	District Cooling				Y	Y	Y	Y			100%	0.00%	0.00%
	Water-cooled A/C				Y	Y	Y	Y			100%	0.00%	0.00%
	OTTV	Y				Y					100%	0.00%	0.00%
	EE Systems	Y	Y			Y					100%	0.00%	0.00%
	EE Appliances	Y	Y			Y					100%	2.22%	2.22%
	Power Plants ESS	Y		Y							100%	4.33%	4.33%
	EE Behaviour		Y		Y	Y					100%	3.40%	3.40%
Transport	Alternative Fuels		Y		Y	Y					100%	1.00%	1.00%
	Fleet Efficiency	Y	Y			Y					100%	1.40%	1.40%
	EVs		Y			Y					100%	1.00%	1.00%
	Pedestrianisation				Y	Y			Y		100%	0.00%	0.00%
	Biofuels			Y		Y			Y		100%	0.00%	0.00%
Energy	WtE				Y	Y		Y			100%	2.44%	2.44%
	RE	Y			Y	Y					100%	1.15%	1.15%
	Fuel Mix				Y	Y		Y			100%	13.39%	13.39%
	Total Achieved											**37%**	**37%**
	Target for 2020										**37%**	**37%**	

Note:

Y = 'Yes'

[a]Compared to 2005

concerning RE use, noting that RE is cheaper than nuclear technology, although there are no suitable places in Hong Kong to construct either type of facility. The representative continued: 'We are always urging our government to find RE opportunities in mainland China, [which would] be much cheaper than [having such facilities built] by China Light & Power or Hong Kong Electric'. This interviewee concluded, however, that the most promising way forward might be RE certificates.

Interviewees generally agreed that regulatory tools are useful but considered that they could also be applied to measures other than the fuel mix. One NGO focus group (Appendix I: 2FG/HK/29–05–2012) participant suggested that the government adopt a similar approach to the EU's issuance of command-and-control directives, whereby 'certain reduction measures are made mandatory, and each country, company, or organisation is given a [degree] of flexibility in how to achieve the target'. Also discussed was the use of performance-based standards as a basis for the BEC. A participant in another focus group (Appendix I: 1FG/HK/28–05–2012) suggested that such standards 'may be key in regulating energy usage in Hong Kong', noting that the Business Environment Council is currently working 'on a set of recommendations, one of which is to implement performance-based standards that are tightened over time and laid-out in advance so that investors [and] facilitators . . . have adequate time to plan and prepare'. Another interviewee (Appendix I: 1FG/HK/28–05–2012) said that the Hong Kong Green Building Council is also working on a performance-based benchmarking system focused on regulating the amount of energy used per square meter. Another added that ongoing changes in technologies point toward the efficacy of a performance-based approach to the BEC. Finally, a business-sector interviewee cited regulation as the best means of implementing such measures as the BEC, WAC, EES and EE appliances, the expansion of alternative fuels including biofuels and the improvement of vehicular fleet efficiency and pedestrianisation.

Taking a more holistic approach than the other two scenarios, the High Scenario also places greater emphasis on providing financial incentives such as tax incentives and subsidies for policy implementation, as these tools can have a significant influence on the success of a given measure. Of the various market-based tools available, the interviewees generally welcomed tax relief for power company investment in RE infrastructure and operations. A focus group participant (Appendix I: 1FG/HK/28–05–2012) suggested that such tax relief could also be extended to owners/developers who incorporate EES in building design. Another concurred, pointing out that 'if tax relief is given to a developer for incorporating EE technologies, not only [will] the value of his/her property goes up, but so will the value of the whole area'. Interestingly, whereas such tax incentives were seen by the business-sector representatives as effective implementation tools for the promotion of EVs, the NGO participants saw them as effective promoters of such DSM measures as EES for buildings and power plants and EE appliances and behaviour.

To clarify interviewees' views on the role the SOC plays in reducing GHG, we reminded them of the SOC stipulations. Since a new SOC was agreed between the

HKSAR government and the two power companies in 2008, penalty arrangements have been added to encourage them to cap emissions: their permitted monetary rate of return on net fixed assets (a maximum of 9.99 percent) will be reduced by 0.2 to 0.4 percentage points depending on their actual emission levels, with a maximum penalty of HK$200 to $300 million. If the emissions of all pollutants fall below the specified caps, the companies will be entitled to an award of 0.1 to 0.5 percentage points in their permitted return. Another incentive is an 11-percentage-point return to power companies that invest in RE facilities and a further bonus depending on the extent of RE used in generating electricity. Energy conservation is encouraged by rewarding the power companies for the number of energy audits they conduct for their clients and the resultant actual energy saved. The maximum award is a 0.02-percentage-point increase in the permitted return. Finally, both power companies have established funds to promote energy efficiency and provide public education on energy conservation (Environment Bureau 2010).

This review of the SOC stipulations prompted one interviewee to suggest that a better approach to reducing electricity consumption would be consumer- rather than power company–oriented incentives, such as a lower tariff for those who consume less. However, one concern is whether such an incentive would apply to smaller households/businesses only or be extended to larger entities such as mass transport operators or very large buildings that necessarily consume a great deal of energy. Another concern is whether it would be a sufficiently robust incentive for large companies that can afford higher bills. The focus group (Appendix I: 1FG/HK/28–05–2012) interviewees suggested that subsidies would be effective implementation tools for the BEC, WAC, EES and OTTV and for the promotion of alternative fuels and EVs.

The interview and discussion session findings indicated that more attention should be paid to increasing public information and researching new and improved technologies through public expenditure and R&D activities. A focus group (Appendix I: 1FG/HK/28–05–2012) interviewee noted that public information is an essential tool for pushing forward GHG reduction measures whether they are mandatory or voluntary. The presence of a communication gap was also noted. Those working in the business sector felt that the government had pushed through the HKSAR-ERM without giving adequate details of how the GHG reduction targets were calculated. It was also felt that DSM measures would benefit most from public information and measures to curb emissions through the promotion of pedestrianisation, with a focus group (Appendix I: 2FG/HK/29–05–2012) participant noting that although utility companies and large organisations are major GHG emissions contributors, 'it is also important to educate the public about GHG emissions and bring about change in how ordinary people can [change] their everyday behaviours to limit their environmental impact'.

In relation to public expenditure, the government-established Environmental Conservation Fund and Sustainable Development Fund support projects that contribute to environmental education, research and technology, with support granted on a competitive basis. It is thought that such funds can reinforce the use of two other tools: public information through education and knowledge transfer through

R&D. Many of the discussion group participants suggested that all of the proposed DSM measures would benefit from these funds. Government expenditure was also considered essential for implementation of new technologies such as WAC and DC and the promotion of alternative fuels, pedestrianisation and EVs.

One participant suggested that R&D funding for new technologies such as energy storage and smart grids had to come from government and that more research was needed to identify the most effective way to promote the BEC and other such building-related measures as OTTV, WAC and DC. R&D would also further the beneficial use of biofuels and EVs and help to negate the drawbacks of using the aforementioned sources.

One stakeholder in a focus group discussion (Appendix I: 1FG/HK/28–05–2012) also posited that there is room for VAs in certain cases, noting that 'if [the reduction measures or technologies] are not yet commercially viable, then VAs or subsidies that will stop after a certain period can help bridge the gap between where we are and where we want to be in energy efficiency'. The use of VAs was also suggested as a means of incentivising the power utilities to become more efficient. Drawing upon a personal study of the overseas experience, particularly that in locales in which the electricity grid is monopolised by certain companies, a focus group (Appendix I: 2FG/HK/29–05–2012) participant noted the efficacy of assigning carbon reduction targets to utility companies. As the companies are monopolies, they face no competitive disadvantage from such targets and are forced to communicate them to their clients. This interviewee believes that adoption of such a tool in the SOC would be 'an effective way to utilise DSM'.

The implementation strategy for the WWF-Arup approach in the High Scenario achieves the 37 percent reduction target. However, as the proportion of the reduction attributed to the fuel mix is lower than that in the HKSAR-ERM case, this strategy avoids the introduction of additional regulations in the energy sector and focuses more on providing financial incentives to help control the demand side.

In summary, our proposed implementation strategy, CRIAS, comprises several key elements (see Tables 8.2–8.4), which are discussed critically in the following final section of the chapter.

Conclusion

City governments need innovative and appropriate policy-implementation tools to mitigate GHG emissions. We have assessed how such tools work in practice in Hong Kong, the variations in the effort required for their deployment and their environmental effectiveness and impact. We have contextualised these variations sociopolitically and institutionally, taking into account the entrepreneurship capacity of the decision makers and other stakeholders involved. We have taken a tools approach to our analysis of competing carbon reduction scenarios owing to our view that the task of tool uptake and deployment is not the technocratic, parsimonious, and nonpolitical exercise the literature suggests. Rather, it is a highly contextualised and contested task revealing governance challenges that may be exacerbated when confronted with 'wicked problems' such as climate change

(Rittel *et al.* 1973). Theoretically, our analysis supports the argument that descriptive accounts of particular policy tools whose aim is to provide ideal guidance on the choice, evaluation and deployment of those tools are of limited usefulness (Jordan *et al.* 2011) unless the politics of tool selection are also factored in. Empirically, our findings challenge the commonly held assumption, which is also held by the Hong Kong government, that changes to the fuel mix hold the answer to lowering carbon emissions. We propose instead that a DSM strategy would achieve the same, if not better, results with less risk to the city's energy security.

In our quantitative analysis of the MURE database, we proposed three scenarios: Baseline, Low and High. In the Baseline Scenario, we showed that in the current situation in Hong Kong, a government strategy focused on the fuel mix would achieve a 14 percent reduction in absolute carbon emissions, whereas an alternative demand-side strategy would achieve approximately the same or a slightly greater reduction. This finding supports the notion that DSM can be as effective as fuel mix control in reducing carbon emissions. By adopting such low-impact tools as VAs and public information, the Low Scenario would achieve greater reductions but fall short of the proposed targets. This result was not unexpected because voluntary efforts are generally of limited use and cannot be relied upon given that society tends to be driven by self-interest. The more interesting result was observed in the High Scenario, in which a mix of regulation, tax incentives and government expenditure resulted in a much greater emission reduction when individuals are given clear instructions on what to do (or suffer penalties). Again, it was seen that a DSM strategy could achieve the desired target without the need for more heavily weighted measures to the fuel mix. This is a crucial point in Hong Kong's case, as the city's fuel mix will depend heavily upon the availability of nonfossil fuels should the government's policy direction be followed. Currently, the only viable source of such fuels is nuclear energy from mainland China. A DSM strategy offers an alternative option that also mitigates concerns over energy security. Obtaining support for implementation of such a strategy will be challenging, however, as the stakeholders involved have different interests and motivations.

Our qualitative analysis of stakeholder views on our three proposed scenarios suggests that most support the heavier-handed approach of the High Scenario, as it makes clear which tools are needed and which should be eliminated. In this scenario, business players would abide by regulations and standards and find innovative ways to gain competitive advantage while reducing their carbon emissions. Users would then exercise their purchasing power to buy low carbon products and services. Moreover, setting standards would sift out the weak performers in the market, an aspect likely to prove effective in Hong Kong's market-driven economy. Driving energy-inefficient equipment and appliances out of the market would also have the added benefit of improving the quality of the environment by, for example, reducing air pollution.

In conclusion, the CRIAS implementation strategy proposed herein is an innovative decision-making mechanism that can be used to guide carbon reduction strategies. CRIAS is supported not only by a critical analysis of the literature on policy tools uptake but also by a strong the rationalisation for studying Hong Kong. China is the world's leading GHGs emitter and Hong Kong is expected to make important contributions to reducing China's climate footprint; thus evaluation of the technical feasibility of alternate strategies for reducing GHG. If we are to push for stronger, more effective climate policies, then focusing on policy tools as they are used in practice is the way forward. However, we remain aware that our approach does not constitute unequivocal guidance for public policy makers and theorists, nor is it intended to. Although CRIAS uses Hong Kong as the application case, it can be generalised for use in other cities whose governments are currently formulating and/or evaluating a climate change plan and action agenda. Future research should focus on the institutional feasibility of comparable carbon reduction strategies and questions of distributional justice.

Chapter appendix

Table 8.1A Measures library

| Sector | Measure | Measure ref | Descriptions | | Estimated reduction in absolute carbon emissions (compared to 2005) | |
			HKSAR-ERM	WWF-ARUP	HK-SAR / ERM	WWF / Arup
Building	Building energy codes	BEC	Expanding the scope and tightening the requirements of the Building Energy Codes (BEC), such that by 2020, major electrical equipment in all new commercial buildings will be up to 50% more energy efficient as compared with 2005 building stock (ERM 2010).	40% penetration of BEC in commercial buildings (approximately 3.6M sq. meters); 45% reduction by complying the BEC; 50% improved efficiency of new buildings (WWF-HK 2010).	0.00%	6.62%
	District cooling	District cooling	Expanding the use of district cooling or water-cooled air conditioning (WAC), such that by 2020, up to 20% of all commercial buildings will be up to 50% better in refrigeration performance compared with buildings using regular air conditioners (ERM 2010).		0.00%	
	Water-cooled aircon	Water-cooled A/C	Expanding the use of district cooling or water-cooled air conditioning (WAC), such that by 2020, up to 20% of all commercial buildings will be up to 50% better in refrigeration performance compared with buildings using regular air conditioners (ERM 2010).		0.00%	

(Continued)

Table 8.1A (Continued)

	Descriptions	Estimated reduction in absolute carbon emissions (compared to 2005)
Overall thermal transfer value (OTTV) / Green roofing (GR) — OTTV	Reducing energy demand in new buildings by various means such as tightening the overall thermal transfer value (OTTV) standards and promoting wider adoption of green roofing (GR), such that by 2020, all new commercial buildings will reduce their energy demand by up to 50% as compared with new buildings in 2005 (ERM 2010).	0.00%
Energy efficient (EE) systems — EE Systems	Improving energy efficiency in commercial buildings through good housekeeping, information technology products and intelligent building environmental management systems, such that by 2020, 25% of existing commercial buildings can be 15% more energy efficient compared with 2005 (ERM 2010).	0.00%
EE Appliances — EE Appliances	Expanding the scope and tightening the energy-efficient electrical appliance standards for domestic use, such that by 2020, all appliances sold in the market will be 25% more energy efficient compared with 2005 (ERM 2010). 75% penetration of energy efficient appliances	0.00%
	35% improvement in efficiency of appliances (WWF-HK 2010).	2.22%

Sector	Category	Measure description		
	Power plants energy saving scheme (ESS)	Power plants ESS	Require Hong Kong power companies to set an aggressive DSM energy saving target and launch relevant energy-saving scheme (ESS) for customers; revise the tariff system in a way that heavy consumers, who consume more electricity, should pay a higher price. The measure can compensate the profit loss of power companies and reduce the burden of increased tariff for more energy-efficient end users; the measures contribute to 15% of total energy consumption reduction (WWF-HK 2010).	0.00%	4.33%
	EE Behaviour	EE Behaviour	Switch off office equipment and appliances after working hours, set up individual lighting and air-conditioning zones. Apply energy-saving films to windows if at all possible. Turn off the PC monitor during group meetings. Install smart meter to understand the source of electricity consumption and the status of usage. Every household in Hong Kong saves 500 kWh per year, including say No to standby power, switch off unnecessary appliances, implement more green tips and take part in Earth Hour (WWF-HK 2010).	0.00%	3.40%
Transport	Alternative fuels	Alternative fuels	Fuel efficiency of private cars and buses improve 25%; high penetration of fuel-efficient cars (100%); 100% of buses GV etc. upgraded; efficiency of buses uplift by 25% (WWF-HK 2010). Wider use of motor vehicles running on alternative fuel such that 30% of private cars, 15% of buses and goods vehicles are hybrid and EVs or other vehicles with similar performance by 2020 (ERM 2010)	0.00%	1.00%
	Fleet efficiency	Fleet efficiency	Implementation of importers' average fleet efficiency standards such that new vehicles will be 20% more energy efficient than the 2005 market average (ERM 2010).	0.00%	1.40%

(Continued)

Table 8.1A (Continued)

		Descriptions	Estimated reduction in absolute carbon emissions (compared to 2005)	
Electric vehicles (EVs)	EVs	The adoption of greener transportation requires communitywide support. The government has been encouraging major EV manufacturers and agents around the world to introduce a greater variety of EVs to Hong Kong and collaborating closely with other organisations to expand the charging network for EVs (ERM 2010).	0.00%	1.00%
Pedestrianisation	Pedestrianisation	Greening road transport cannot solely be a government initiative. We have to look to the general public to join in this green effort by leading a low carbon lifestyle. We encourage people to walk as far as practicable, and if not, to take public transport. If a private car is needed, we suggest that an environment-friendly, zero- or low carbon-emitting one be considered (ERM 2010).	0.00%	
Biofuels	Biofuels	Following the international trend and technological improvement, the consultants consider that our reliance on fossil fuels for motor vehicle use may be further reduced by 2020 by requiring petrol and diesel to be blended with 10% of ethanol and biodiesel respectively. In particular, we will look into the possibilities of better utilising waste cooking oils in producing biodiesel locally. In the international arena, the European Union (EU) has already mandated 10% renewable energy (mainly through use of biofuels) in its transport fuels by 2020. This will boost the global production of biofuels and enable Hong Kong to have access to adequate supply of biofuels by 2020 (ERM 2010).	0.00%	

Electricity					
Waste to energy	WtE				2.44%
Renewable energy	RE	To substantially increase the share of nonfossil low carbon fuels, such that renewable energy would make up about 3–4% of the fuel mix (ERM 2010).	Renewable energy: 300MW Wind farm; Renewable energy: 0.05% of HK area covered with PV (photovoltaic for solar electricity; WWF-HK 2010).	4.00%	1.15%
Coal, gas, nuclear	Fuel mix	In view of the highly polluting and high carbon nature of coal, to suppress the percentage of coal-fired power in our fuel mix and keep coal-fired power plants at a very low utilisation rate or as reserve, such that coal would account for no more than 10% of the fuel mix (ERM 2010). Taking into account the supply of natural gas secured under the MOU between Hong Kong and the Mainland, to maximise the use of natural gas and increase its share in the fuel mix to around 40% (ERM 2010). The balance of about 50% would be met by imported nuclear power (ERM 2010).	50% LNG (WWF-HK 2010).	29.00%	13.39%
			Total Achieved	33%	37%
			Target for 2020	19–33%	37%

Notes

1 We acknowledge the contributions of Thomas S.K. Tang and Electra Stratigaki to this chapter. An earlier version was presented at the International Conference on Regional Energy, Kuala Lumpur, Malaysia, 5–6 December 2012. The research undertaken was funded by AECOM and GARC (City University of Hong Kong project n. 9220060).
2 Supply-side measures are concerned with modifying the generation (for example, from fossil fuels to renewables), transmission and distribution of energy in order to cut emissions. On the other hand, demand-side measures aim at modifying consumer demand for energy, allowing the consumer to control usage (for example, through smart grids), save money and also cut emissions.

References

Auffhammer, M., Blumstein, C. & Fowlie, M. 2007. *Demand-side management and energy efficiency revisited* [Online]. Available: http://escholarship.org/uc/item/1hj0983z [Accessed 14 December 2013]

Bailey, I. 2007. Market Environmentalism, New Environmental Policy Instruments, and Climate Policy in the United Kingdom and Germany. *Annals of the Association of American Geographers,* 97(3), 530–550

Bigano, A., Arigoni, O.R., Markandya, A., Menichetti, E. & Pierfederici, R. 2011. The Linkages Between Energy Efficiency and Security of Energy Supply in Europe, in Galarraga, I., González-Eguino, M. & Markandya A. (eds.) *Handbook of Sustainable Energy,* Northampton, UK, Edward Elgar Publishing Ltd.

Bonneville, E. & Rialhe, A. 2006. *Demand Side Management for Residential and Commercial End-Users* [Online]. Available: www.leonardo-energy.org/Files/DSM-commerce.pdf [Accessed 14 December 2013]

Boshell, F. & Veloza, O.P. 2008. *Review of Developed Demand Side Management Programs Including Different Concepts and Their Results* [Online]. Available: http://ieeex plore.ieee.org/xpl/login.jsp?tp=&arnumber=4641792&url=http%3A%2F%2Fieeexpl ore.ieee.org%2Fxpls%2Fabs_all.jsp%3Farnumber%3D4641792 [Accessed 14 December 2013]

Bossoken, E. 1999. *Case Study: A Comparison between France and the Netherlands of the Voluntary Agreements Policy.* INESTENE [Online]. Available: www.mure2.com/doc/Case3.pdf,‎ [Accessed 9 December 2013]

Bressers, H. 1998. The Choice of Policy Instruments in Policy Networks, in Peters, B.G. & van Nispen, F.K.M. (eds.), *Public Policy Instruments: Evaluating the Tools of Public Administration,* Cheltenham, UK, Edward Elgar.

Bruce, J.P., Lee, H. & Haites, E. F. (eds.) 1996. *Climate Change 1995: Economic and Social Dimensions of Climate Change,* Cambridge: Cambridge University Press (published for the IPCC).

Buchner, B., Falconer, A., Herve-Mignucci, M., Trabacchi, C. & Brinkman, M. 2011. *The Landscape of Climate Finance,* Venice: Climate Policy Initiative.

Carbon Disclosure Project. 2012. *Measurement for Management: CDP Cities 2012 Global Report,* London, Carbon Disclosure Project.

Carley, S. 2011. The Era of State Energy Policy Innovations: A Review of Policy Instruments. *Review of Policy Research,* 28(3), 265–294

Center for Climate and Energy Solutions (C2ES). 2011. *Public Benefit Funds* [Online]. Available: www.c2es.org/what_s_being_done/in_the_states/public_benefit_funds.cfm [Accessed 16 April 2012]

Cooper, S. J. G., Dowsett, J., Hammond, G. P., McManus, M. C. & Rogers, J. G. 2013. Potential of Demand Side Management to Reduce Carbon Dioxide Emissions Associated with the Operation of Heat Pumps. *Journal of Sustainable Development of Energy, Water and Environment Systems,* 1(2), 94–108

Dahl, R. & Lindblom, C. 1953. *Politics, Economics and Welfare,* New York, NY, Harper and Brothers.

Economist, The. 2011, April 28. *Gauging the Pressure* [Online]. Available: www.economist.com/node/18621367 [Accessed 3 December 2013]

Eichhammer, W. 2008. *Distinction of Energy Efficiency Improvement Measures by Type of Appropriate Evaluation Method.* Fraunhofer Institute for Systems and Innovative Research [Online]. Available: www.evaluate-energy-savings.eu/emeees/en/events/final_conference.php [Accessed 8 December 2013]

Environment Bureau. 2010. *Hong Kong's Climate Change Strategy and Action Plan: Consultation Document* [Online]. Available: www.epd.gov.hk/epd/english/climate_change/files/Climate_Change_Booklet_E.pdf [Accessed 16 April 2012]

Environment Bureau. 2013. *Sectoral Measures* [Online]. Available: www.epd.gov.hk [Accessed 7 June 2013]

Environmental Resources Management. 2010. *A Study of Climate Change in Hong Kong – Feasibility Study* [Online]. Available: www.epd.gov.hk/epd/tc_chi/climate_change/files/1_CC_Final_Report_Eng.pdf [Accessed 16 April 2012]

Filippini, M., Hunt, L. C. & Zoric, J. 2013. *Impact of Energy Policy Instruments on the Estimated Level of Underlying Energy Efficiency in the EU Residential Sector.* Surrey Energy Economics Discussion Chapter Series, School of Economics, University of Surrey [Online]. Available: www.seec.surrey.ac.uk/research/seeds.shtml [Accessed 8 December 2013]

Francesch-Huidobro, M. 2012. Institutional Deficit and Lack of Legitimacy: The Challenges of Climate Change Governance in Hong Kong. *Environmental Politics,* 21(5), 791–810

Goers, S. R., Wagner, A. F. & Wegmayr, J. 2010. New and Old Market Based Instruments for Climate Change Policy. *Environmental Economics and Policy Studies,* 12: 1–30

Gupta, S., Tirpak, D. A., Burger, N., Gupta, J., Höhne, N., Boncheva, A. I., Kanoan, G. M., Kolstad, C., Kruger, J. A., Michaelowa, A., Murase, S., Pershing, J., Saijo, T. & Sari, A. 2007. Policies, Instruments and Co-Operative Arrangements, in Metz, B., Davidson, O. R., Bosch, P. R., Dave, R. & Meyer, L. A. (eds.), *Climate Change Mitigation. Contribution of Working Group III to the Fourth Assessment Report of the Intergovernmental Panel on Climate Change,* Cambridge/New York, NY, Cambridge University Press.

HKSAR-ERM 2010. *A Study of Climate Change in Hong Kong - Feasibility Study,* Hong Kong: Environmental Resources Management.

Hoffman, A., 2005. Climate Change Strategy: The Business Logic Behind Voluntary Greenhouse Gas Reductions. *California Management Review,* 47(3), 21–46

Hood, C. 1983. *The Tools of Government,* London, Macmillan.

Hood, C. 2007. Intellectual Obsolescence and Intellectual Makeovers: Reflections on the Tools of Government after Two Decades. *Governance,* 20(1), 127–144

Hoogwijk, M. & Graus, W. 2008. *Global Potential of Renewable Energy Sources: A Literature Assessment* [Online]. Available: www.ecofys.com/en/publication/global-potential-of-renewable-energy-sources [Accessed 3 December 2013]

Igielska, B. 2008. Climate Change Mitigation: Overview of the Environmental Policy Instruments. *International Journal of Green Economics,* 2(2), 210–225

Intergovernmental Panel on Climate Change (IPCC). 2012. *Section 6.1.2. Types of Policies, Measures and Instruments* [Online]. Available: www.ipcc.ch/ipccreports/tar/wg3/index.php?idp=226 [Accessed 16 April 2012]

International Energy Agency (IEA). 2006. *Energy Statistics of Non-OECD Countries,* Paris, IEA.

International Energy Agency (IEA). 2012. *Addressing Climate Change: Policies and Measures Databases* [Online]. Available: www.iea.org/textbase/pm/?mode=cc [Accessed 16 April 2012]

International Energy Agency (IEA). 2012. *Key World Energy Statistics 2012* [Online]. Available: www.iea.org/publications/freepublications/publication/kwes.pdf [Accessed 3 December 2013]

Jaccard, M., Failing, L. & Berry, T. 1997. From Equipment to Infrastructure: Community Energy Management and Greenhouse Gas Emission Reduction. *Energy Policy,* 25(13), 1065–1074

Japan Times, The. 2013, November 19. *Cut Emissions without Nuclear Power* [Online]. Available: www.japantimes.co.jp/opinion/2013/11/19/editorials/cut-emissions-without-nuclear-power/ [Accessed 3 December 2013]

Jordan, A., Benson, D., Wurzel, R. & Zito, A. 2011. Policy Instruments in Practice, in Dryzek, J. S., Norgaard, R. B. & Schlosberg, D. (eds.), *The Oxford Handbook of Climate Change and Society,* Oxford, Oxford University Press.

Jordan, A., Wurzel, R. K. W. & Zito, A. R. (eds.) 2003. *New Instruments of Environmental Governance,* London, Frank Cass.

Kingdon, J. W. 1995. The Policy Window, and Joining the Streams, in *Agendas, Alternatives, and Public Policies,* 2nd ed., New York: HarperCollins College Publishers.

Krueger, R. 1988. *Focus Groups,* Newbury Park, CA, Sage.

Lee, J., Kim, H., Park, G. L. & Kang, M. 2012. Energy Consumption Scheduler for Demand Response Systems in the Smart Grid. *Journal of Information Science and Engineering* 28: 955–969

Liu, Z., Geng, Y., Lindner, S., Guan, D. 2012. Uncovering China's Greenhouse Gas Emissions from Regional and Sectoral Perspectives, *Energy Policy* 45: 1059–1068

McCormick, K. & Neij, L. 2009. *Experience of Policy Instruments for Energy Efficiency in Buildings in the Nordic Countries,* Lund, Sweden: International Institute for Industrial Environmental Economics (IIIEE), Lund University.

McPherson, E. G. 1993. Evaluating the Cost Effectiveness of Shade Trees for Demand Side Management. *The Electricity Journal,* 6(9), 57–65

Molderink, A. 2012. Demand-Side Energy Management, in Iniewski K. (ed.) *Smart Grid Infrastructure & Networking,* New York: McGraw-Hill Professional.

Mundaca, L., Neij, L., Worrell, E. & McNeil, M. 2010. Evaluating Energy Efficiency Policies with Energy-Economy Models. *Annual Review of Environment and Resources,* 35, 305–344

MURE Database. 2012 [Online]. Available: www.muredatabase.org [Accessed 20 April 2012]

Nadel, S. & Kushler, M. 2000. Public Benefit Funds: A Key Strategy for Advancing Energy Efficiency. *Electricity Journal,* 13(8), 74–84

OECD. 2011. *OECD Environmental Outlook to 2050* (Chapter 3: Climate change) [Online]. Available: www.oecd.org/dataoecd/32/53/49082173.pdf [Accessed 16 April 2012]

OECD. 2012. *The OECD Environmental Outlook to 2050: Key Findings on Climate Change* [Online]. Available: www.oecd.org/dataoecd/21/30/49089652.pdf [Accessed 16 April 2012]

Oikonomou, V. & Jepma, C. J. 2008. A Framework on Interactions of Climate and Energy Policy Instruments. *Mitigation and Adaptation Strategies for Global Change* 13: 131–156

Papagiannis, G., Dagoumas, A., Lettas, N. & Dokopoulos, P. 2008. Economic and Environmental Impacts from the Implementation of an Intelligent Demand Side Management System at the European Level. *Energy Policy,* 36(1), 163–180

Peters, B.G. & Nispen, F. (eds.) 1998. *Public Policy Instruments,* Cheltenham, Edward Elgar.

Petrichenko, K. 2010. Towards Higher Energy Efficiency: Greece Case. *Analytical,* 6: 6380

Ramchurn, S., Vytelingum, P., Rogers, A. & Jennings, N. 2011. Agent-Based Control for Decentralised Demand Side Management in the Smart Grid. *Tenth International Conference on Autonomous Agents and Multiagent Systems (AAMAS 2011),* Taipei, Taiwan, May 2–6, 2011.

Rittel, H.W.J. & Webber, M.M. 1973. Dilemma in a General Theory of Planning. *Policy Sciences,* 4(155), 169–169

Scott, I. 2005. *Public Administration in Hong Kong: Regime Change and Its Impact on the Public Sector,* Singapore, Marshall Cavendish.

Scott, I. 2007. Legitimacy, Governance and Public Policy in Post-Handover Hong Kong. *Asia Pacific Journal of Public Administration,* 29(1), 29–49

Stavins, R. N. 1997. *Policy Instruments for Climate Change: How Can National Governments Address a Global Problem?* Discussion Chapter 97–11, University of Chicago Legal Forum, Washington, DC, University of Chicago Law School.

Sterner, T., 2003. *Policy Instruments for Environmental and Natural Resource Management,* Washington, DC, Resources for the Future Press.

Strbac, G. 2008. Demand Side Management: Benefits and Challenges. *Energy Policy,* 36(12), 4419–4426

South China Morning Post. 2013, December 9. *Power Reliability: A Priority for Hong Kong.*

Twomey, P. 2012. Rationales for Additional Climate Policy Instruments under a Carbon Price. *Economic and Labour Relations Review,* 23(1), 7–32

Wurzel, R.K., Zito, A.R. & Jordan, A. 2012. *Environmental Governance in Europe: A Comparative Analysis of 'New' Policy Instruments,* Cheltenham, UK, Edward Elgar.

WWF-HK. 2010. *Appendix: Carbon Emission Reduction Roadmap–WWF/ARUP Road Map 2020* [Online]. Available: http://assets.wwfhk.panda.org/downloads/appendix_eng.pdf [Accessed 11 May 2012]

Zito, A.R. 2003. Introduction to the Symposium on 'New' Policy Instruments in the European Union. *Public Administration,* 81(3), 509–512

9 Conclusions

Governing climate change mitigation through dynamic interactions within collaborative municipal networks (CMNs) is an approach to governance that, as we have argued throughout this volume, is emerging in Chinese cities. This approach to governance relies on interactive resource exchange, communication and consensus building among the organisational actors networked within and across cities. To steer through the governing challenges imposed by the complex problem of climate change, this form of governance is embedded in two dimensions of transition. The first is a sociotechnical shift that involves innovation transfers and integration and which is internally driving the second dimension, the transformation of governance practices in response to a complex and cross-sectoral policy problem that requires both coordination and collaboration inside and outside the government arena.

We focused our study on the mitigation responses in the building and transportation sectors in three cities of southern China because cities are essential sites of response to climate change mitigation, and their building and transportation sectors have the largest potential for carbon emissions reduction. Exposure to globalisation in the implementation of climate mitigation action agendas has encouraged low carbon innovation transfer activities in these world cities. The unique Chinese political system has enabled drivers but also imposed barriers to the urban governing of low carbon transitions. Identifying and analysing these factors enables us not only to address the dearth of knowledge with regards to urban climate governance in China but also to confront the absence of explanatory theoretical models other than those that have been applied to collaborative and network governance across the political and system contexts of the Global North.

The study used both *deductive* analysis, because it anticipated outcomes for the CMNs, and *inductive* analysis, because it extrapolated the implications for the Chinese environmental state. The anticipated outcomes of interactions in collaborative network arrangements were identified in the review of existing studies of subnational climate governance and theories about collaborative governance, interactive governance and governance networks. The propositions made on anticipated outcomes were critically elucidated from the literature and incorporated in the theoretical framework constructed in Chapter 2. Key aspects of research were then identified, in particular *intra*governmental and *inter*governmental

coordination, horizontal and cross-sectoral collaboration, the process of legitimation and the arenas of networking such as the nongovernmental space. The
analytical frameworks applied in Chapters 4 to 8 (Figures 4.2, 4.3, 4.4, 5.1, 5.4,
7.1 and 7.2 and Table 4.3) explored and examined the empirical realities of CMNs
and helped to refine the overall theoretical framework proposed in Chapter 2. The
integrated process of CMNs was deductively studied within the particular case
context of Chinese cities that have embarked on climate mitigation responses.
These empirical studies then enabled inductive extrapolations about the restructuring of the Chinese environmental state.

Collaborative municipal networks in practice

Our study took into account the dynamic interplay of factors in the governing of
climate mitigation responses in China, including state authority, political space,
committed leadership and sectoral engagement. We found that CMNs are adjusting their municipal governance practices to accommodate these sociotechnical
changes and/or mitigation responses.

Sociotechnical perspective: fragmented integration

At the heart of the notion of low carbon development is the integration of socio-
technical transitions in the restructuring environmental state. The term 'sociotechnical transitions' denotes a gradual shift across systems of technology, supporting
policies and regulations, user practices and preferences, markets and demands,
facilitating infrastructure and maintenance and supply chains (Geels 2004). To
govern such system change, our study found that interactions across societal and
policy sectors have emerged along the CMNs. Transitions in the building and
transport sectors use advanced technology and are supported by recently promulgated policies and regulations, but the expected outcomes of these policies have
not yet been fulfilled in terms of user perceptions, awareness and market preferences. These unfulfilled outcomes are beyond the scope of control and regulation
of local governments, which then resort to the nongovernmental space and the
private sector through networking and collaboration. The empirical discussions
in Chapters 5 and 6 identified a public–private dichotomy in the development of
green buildings, due to the particular institutional arrangement of incentives and a
stagnated integration of private electric vehicles, and because of the limited cross-
sectoral collaboration that undermines legitimation and engagement. We explain
this further.

Although the idea of *green buildings* as a sectoral solution to climate change
mitigation has rapidly developed and gained acceptance in Guangzhou, Shenzhen and Hong Kong within the past five years, there exists a dichotomy between
the public and private sectors in qualitative green-building performance once
development projects have been finalised and evaluated. Projects with higher
publicness and more government involvement are more likely to exhibit better
qualitative performance. Political incentives for the public sector (e.g. promotion

opportunities for municipal authorities and position-related performance), in addition to the economic incentives for the private sector, have motivated actors affiliated with municipal government agencies to put more effort – including resources, expertise and monitoring – into ascertaining better green building performance. We can therefore conclude that strategically positioning and combining diverse sources and forms of incentives in CMNs is an integral part of engaging broader participation in innovation integration. This ultimately strengthens the institutional capacity of China's environmental state.

The integration of private *electric vehicles* has been hampered in Chinese cities. The consumer market has shown a lack of responsiveness to this sectoral innovation, despite government provision of strong infrastructure and capital investment in its initial application and subsequent research and development. Resource mobilisation and the legitimation of sectoral innovation depend on collaboration among the private sector, professional/industry associations and the community. Yet the nongovernmental space – comprising local environmental nongovernmental organisations, academic research institutes and transport associations – has been weakly engaged in the communication process and largely detached from government-initiated action in the sectoral integration of electric vehicles. This is evident in the case of Hong Kong (Chapter 6). This finding thus suggests that cross-sectoral engagement through effective communication strategies is still lacking in the governance of mitigation strategies. This fragments the emerging CMNs and diminishes the networks' influence on the restructuring of China's environmental state.

A transition to sociotechnical approaches to climate mitigation has not yet occurred. Sectoral integration is hampered by the boundaries of policy portfolios and governance spheres. Nevertheless, the ongoing integration of sectoral innovations in China's environmental state through CMNs has triggered institutional changes in the governing structure and in cross-sectoral relations.

Municipal perspective: stifled empowerment

Guangzhou, Shenzhen and Hong Kong are globally connected with the world economy and instrumental to domestic development and thus are ideal sites for experimenting with low carbon initiatives for both China and the world. We suggest that a discussion on the governance of global cities should not be concerned merely with the simplistic ranking of world cities based on quantitative indicators but also with more substantive benchmarking such as the dynamics underpinning their connectivity – the *inter*governmental and cross-sectoral interactions that take place as they compete for leadership in a global context and on a global issue (climate change).

Global cities' governance strategies are shaped by multilevel demands. Despite claiming a more prominent presence in international negotiations, we found that the globally connected Chinese cities in our study were still largely confined within national boundaries in terms of sustaining the bargaining power and authority to improve vertical *inter*governmental relations and expanding networking boundaries in the building of horizontal *inter*governmental collaboration. However, in most policy areas, Hong Kong has the inherent advantage of autonomy from Beijing due to its 'special administrative region' status, which requires vertical *inter*governmental relations only when it is seeking to build collaborative ties with Chinese cities or provinces.

The governing of the sociotechnical responses to low carbon development in Chinese cities is intertwined with and influenced by their administrative systems. We suggest that the development of horizontal *inter*governmental collaboration is not as prominent in reality as it is advocated to be, because horizontal *inter*governmental interaction emphasises collaboration between network actors of equal administrative rank. For instance, the Environment Bureau in Hong Kong should be speaking to provincial-level governmental agencies such as the Guangdong Development and Reform Commission in the governing of collaborative initiatives on low carbon development, but this is not the case. This implicit social norm in the administrative culture of China tends to put pressure on innovation and policy entrepreneurship. Potential network contributors of resources and expertise that do not have a high administrative rank are consistently barred from entering the municipal networks because they are considered outright outsiders.

With this limited room for innovation, city-level network actors are seeking opportunities to carve out their own space. We found that what has truly motivated the burgeoning of low carbon initiatives in Chinese cities is the competitive spirit of city administrators in striving for land and financial resources, preferential policies and local empowerment. In this regard, ambitious local governments have preferred to resort directly to the central government before consulting their horizontal counterparts. It is the enormous influence of vertical *inter*governmental relations that has overridden the weakened ties of horizontal interactions connecting the cities through low carbon transitions.

Global and globalising cities are competing for prominence in state development and regional influence, which triggers *intra*regional rivalries in places such as the Pearl River Delta, where these cities are located. Such rivalries also give rise to *intra*regional alliances, such as that between Shenzhen and Hong Kong. More innovative and cooperative projects have been launched by Shenzhen, an insider in the semi-authoritarian regime, to tie up functionally with Hong Kong, an outlier in China's administrative system and, to a certain extent, political regime. The need for low carbon transitions brings optimal opportunities for strengthening such an alliance. The Shenzhen government regularly consults and negotiates with its Hong Kong counterpart while at the same time lobbying the national government to support experimental projects such as the Qianhai Co-development Plan and the promotion of electric vehicles. Thus, we suggest that the Hong Kong municipality is seeking self-empowerment within the existing web of state authority. However, the horizontal ties between Hong Kong and other Chinese cities are largely limited to their governments; the nongovernmental space in different municipalities has not been overly keen on building closer ties for climate change mitigation, which otherwise could have further facilitated municipal empowerment.

Nevertheless, low carbon transitions need to be localised, which in turn empowers municipal governments. There are functional distinctions between the state and subnational actors, and low carbon development falls within the policy portfolio (i.e. the building and transport sectors) of the latter. Low carbon initiatives are localised by the committed leadership and policy entrepreneurship of these cities. Such a localisation process involves accommodating the low carbon initiatives with local governing conditions by adjusting the implementation capacity and assessing the institutional feasibility. Committed leadership in localities politically drives policy innovations and steers them through administrative impediments. This 'committed leadership' is not held by the mayors, the Guangdong provincial governor or the chief executive of Hong Kong; instead, it is spread through the *intra*governmental networks of the municipal governments.

Without an instituted electoral system at the city level, city mayors in China are selected by upper-level government authorities based on candidates' past performance as managers of state-owned enterprises, as administrative leaders in district or county governments or as deputy heads of provincial governments. The major concern of mayors is how to stay on the promotion track by accommodating the preferences of upper-level government officials, as they cannot gain popularity through a democratic electoral system. Mayors are not expected to showcase policy entrepreneurship in city governance but merely to serve as the appointed leader of the party and administration system in which the municipal governing machinery operates. This also explains why transnational municipal networks (such as C40 and ICLEI, which have been successful in European and US cities) have been of no consequence in China. The members of these transnational municipal networks are mayors of participating cities, yet practical leadership of low carbon transitions in Chinese cities does not come from mayors but from the leaders of the different mitigation sectors.

Likewise, in Hong Kong, the leadership for low carbon transition does not come from the chief executive. In his exchanges with the national government in Beijing, the chief executive, as Hong Kong's head of government, often needs to concentrate his efforts on maintaining the city's relationship with the nation on matters related to political reform and economic cooperation. The chief executive therefore relies heavily on his political appointees (heads of policy bureaus) to lead the governing of the city's climate change mitigation issues.

The actual leading role in decarbonising cities is played by development planners. These are the local development and reform commissions in Guangzhou and Shenzhen and the Development and Environment Bureaus in Hong Kong that are directly linked with the national government, in particular the National Development and Reform Commission. Interagency leadership is fragmented across policy sectors. Some sectoral coordinating agencies, such as the Shenzhen Housing and Construction Department (SZ-HCD), are proactively taking the lead in the decarbonising of functional sectors under their administrative jurisdiction, especially when there are opportunities to bid for low carbon pilot initiatives in their own sectors. Other sectoral coordinating agencies have become more hands off within their sectoral jurisdictions in dealing with governing issues of low carbon transitions. This has resulted in an unbalanced development across sectors and in fragmented leadership in *intra*governmental networks.

Our study also found that policy entrepreneurship has emerged in sectoral departments and in the nongovernmental network organisations engaged in collaboration. Sectoral departments (buildings, transport, planning, etc.) are responsible for devising effective governing strategies to integrate sectoral innovations, taking on tasks such as resource mobilisation. This all comes down to the same basic quest for incentives – the positioning of political incentives for local governing elites in low carbon transitions, that is, what they can possibly gain or lose from taking governing risks with sectoral innovations and experiments. To facilitate cross-sector interaction, the *intra*governmental networks are then extended to the nongovernmental actors. Nongovernmental organisations that have successfully entered the core circle of municipal networks for low carbon transitions share certain characteristics, including potential access to abundant resources, strong connections with the state system (sometimes the networks are chaired by retired government officials) and good opportunities to campaign for external support and endorsement.

Reconsidering the network governance paradigm

We conceptualised climate municipal networks as *the complex process of interaction among a plurality of state and nonstate actors, with the objective of achieving climate mitigation goals through functions such as knowledge transfer and resource exchange, while maintaining municipal autonomy in vertical or horizontal coordination and enhancing municipal capacity in horizontal collaboration. These interactions are manifested in collective decision making, which includes consensus building and joint rule making.* The theorisation and empirical findings in this study have helped us to specify and refine the initially proposed conceptualisation of CMNs.

First, municipal autonomy is constrained in its *inter*governmental context. Such constraints have motivated state and nonstate actors to seek political and economic incentives in the low carbon transition process. Second, the functions of CMNs are diverse. Beyond knowledge transfer and resource exchange, networking functions involve *intra*governmental and *inter*governmental communication, nongovernmental engagement, the effective positioning of incentives to mobilise resources and protection of and experimentation with sectoral innovations prior to broader integration. Third, the process of collective decision making happens in both *intra*governmental coordination and cross-sectoral collaboration, combined with the legitimation of governing sectoral innovations through building consensus and aligning the expectations of different network participants.

Governance practices in the CMNs are embedded in different network forms and carried out through the dynamics of coordination, collaboration and legitimation. The empirical realities and theoretical interpretations of such new forms of governance form the basis of the proposed theoretical framework in Chapter 2.

Networks

The CMNs that have emerged in cities experimenting with low carbon initiatives and sectoral innovations are embedded in the crisscross of vertical and horizontal *inter*governmental relations. Refining the theoretical insights rendered by Torfing *et al.* (2012) in their conceptual development of 'interactive governance', our study identified diverse forms of interactions in the CMNs connecting city-level network actors in low carbon transitions: i) networks mandated from above through the entrusted mode of governing low carbon experimental projects (Chapter 7); ii) networks self-initiated from below through the proposing and applying of modes of governing low carbon pilot initiatives (Chapter 7); iii) formal and relatively exclusive networks embedded in the horizontal *inter*governmental initiatives; iv) informal and relatively open (inclusive) networks that have emerged in the expanding nongovernmental space to broaden the support base of sectoral integration (Chapters 5 and 6); and v) *intra*governmental networks established between municipal government agencies to coordinate the increased administrative tasks (Chapter 4).

Networking capacity and intensity varies across different forms of network interaction. Networks that are self-initiated from below are often more motivated and committed than networks that are mandated from above. The weakest network interactions are found in the formal and relatively close (exclusive) networks, whereas the informal and relatively open (inclusive) networks provide a legitimate networking environment in which to protect and nurture the market niche of sectoral innovation. The *intra*governmental networks cater for the changing administrative needs and demands of the complex governing issue of climate change mitigation, which have developed into steady network interactions with relatively neutral networking intensity.

The theoretical framework of CMNs (Figure 2.2) distinguishes the external process of network development from the internal networking dynamics, with the former denoting the institutional design during the governance networks' entire life cycle. Examined through the empirical cases in this study, the indicators of institutional design are specified for each stage of the external process and summarised in Table 9.1. Each network stage is reflected in the different empirical realities of the emerging governance networks connected within and across the municipalities.

This external process of network development is extrapolated from and manifested by the governing interactions for low carbon transitions in the three cities under study. The active and inactive networks of the three cities are illustrated in Figure 9.1, which brings together the empirical and theoretical findings of the book.

In Figure 9.1, each rounded square in dashed line represents the political space of a municipality, which can be disaggregated into governmental and nongovernmental space. The governmental space for low carbon transitions in the cities we studied is occupied by the *intra*governmental networks. In Guangzhou, the leading actor is the city's Development and Reform Commission (GZ-DRC), and stronger governing focus in low carbon transitions is placed on the building sector, coordinated by the city's Urban–Rural Construction Commission (GZ-URCC).

Table 9.1 External processes of the collaborative municipal networks

External process	Institutional design indicators	Empirical reality in china
Network Prototype	**Infrastructure:** • Coordination office • Resource mobilisation mechanism **Norms:** • Initial contacts and principled engagement • Shared vision **Rules and Regulations:** • Basic guidelines and action protocols	– Established external linkages of *intra*governmental coordination mechanisms – Initiated informal networks of core and resourceful nongovernmental actors
Network Formation	**Infrastructure:** • Incorporation of staff or liaison office • Enhanced communication channels **Norms:** Formal endorsement/mandates for policy actions • Consensus building among core network actors • Role settings among participants (to build discursive legitimacy) • Expanded mobilisation across sectors **Rules and Regulations:** • Baseline policy and rule setting with reduction targets	– Expanded nongovernmental space reaching out to connect with the *intra*governmental networks
Network in Action	**Infrastructure:** • Formation of cross-sectoral and transboundary taskforces on the implementation of low carbon initiatives • Platform for carrying out governing activities of sectoral integration **Norms:** • Engagement with peripheral network actors to provide broadened support base **Rules and Regulations:** • Detailed implementation procedures and division of labour	– ***Weak*** horizontal *inter*governmental networks enabling different municipalities to collaborate with ***minimal effectiveness*** – Strengthened ties between the *intra*governmental networks and the cross-sectoral networking in the nongovernmental space
Network Outcomes	**Infrastructure:** • Feedback mechanisms • Regeneration channels for accumulating experience and resources **Norms:** • Societal awareness and customised actions **Rules and Regulations:** • Evaluation and assessment methods • Revised policies with broader jurisdictional applicability	– ***Weak*** horizontal *inter*governmental networks for knowledge and experience transfer – ***Strong*** vertical *inter*governmental interactions for resources and endorsements

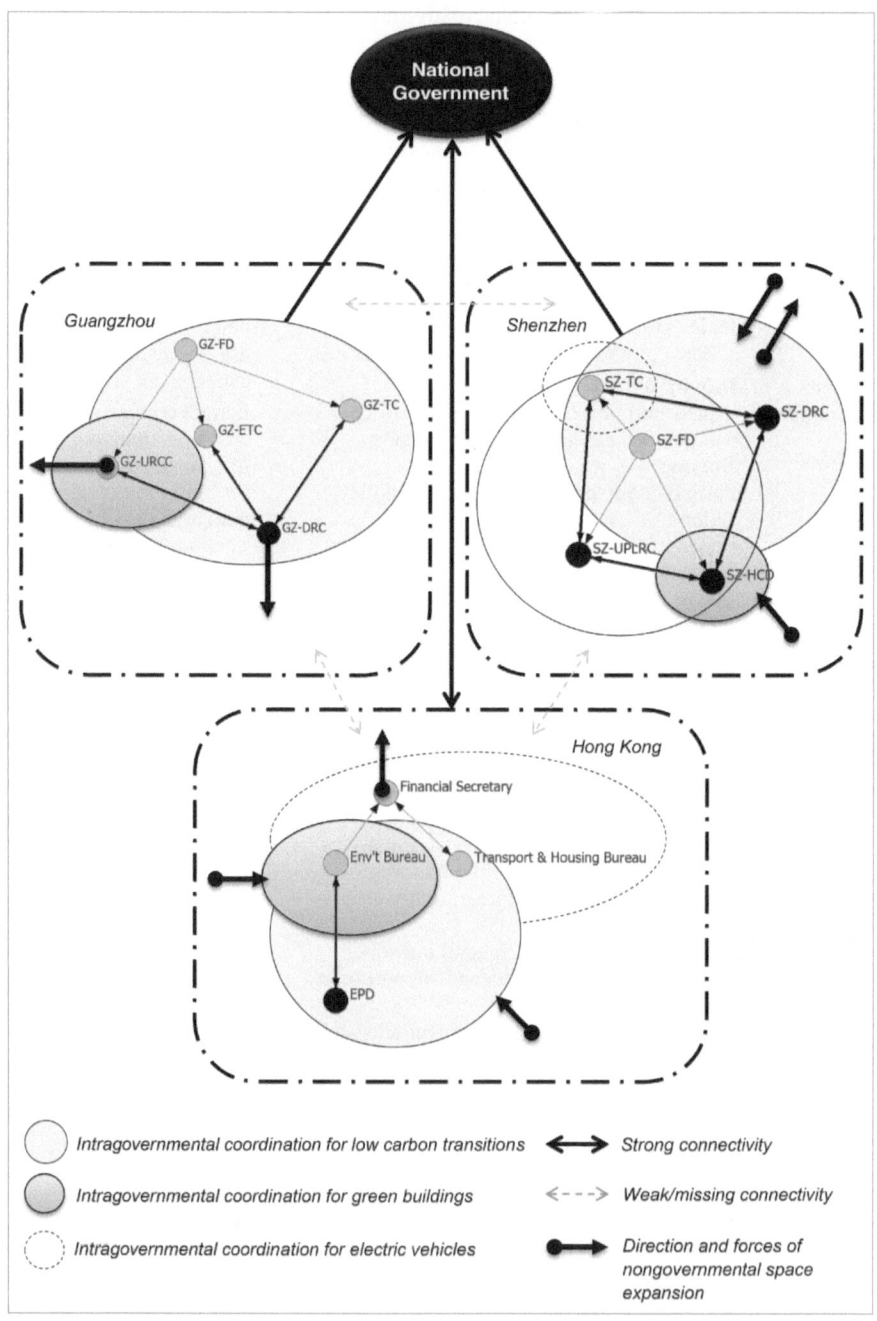

Figure 9.1 Collaborative municipal networks in low carbon transitions of Guangzhou, Shenzhen and Hong Kong

Source: Compiled by authors

In Shenzhen, there are multiple leading actors – the city's Development and Reform Commission (SZ-DRC), the Urban Planning and Land Resources Commission (SZ-UPLRC) and the Housing and Construction Department (SZ-HCD). The SZ-UPLRC has sought to compete with the SZ-DRC for the leading role in low carbon transitions, with the former convening a 'shadow' *intra*governmental coordination structure in parallel to the one established by the latter. In the meantime, coordination, both *intra*governmental and *inter*governmental, to decarbonise the building and transport sectors has been respectively initiated by the SZ-HCD and the Transport Commission (SZ-TC). The competitive departmentalism in Shenzhen has clearly overstretched the municipality's governing attention toward climate change mitigation.

In Hong Kong, the sectoral *intra*governmental coordination structures for green buildings and electric vehicles are initiated by a high-level agency of the administrative hierarchy (the Financial Secretary and two policy bureaus), whereas the coordination on climate change is convened by a low-ranking agency with few resources (the Environmental Protection Department, or EPD). These three structures display fragmented *intra*governmental capacity in low carbon transitions, mutually excluding the core network participants in the coordination process. For instance, the EPD – the leading actor in climate actions – is not included in the *intra*governmental coordination of the two mitigation sectors.

The arrows emerging from the black dots (Figure 9.1) signify the direction and forces of nongovernmental space expansion in each municipality, i.e., how the links between the *intra*governmental networks and the cross-sectoral networking in the nongovernmental space are gradually being strengthened in the governing process of low carbon transitions. In Guangzhou, such expansion of the nongovernmental space is typically government led, charged with beefing up the input legitimacy of low carbon actions. Associations for low carbon development, energy efficiency and building technology have been established in Guangzhou, but they were all instigated and supported by the municipal government. Shenzhen's expansion of nongovernmental space involves a combination of government-led and professional-led forces. Whereas the associations for energy efficiency were established in line with a municipal directive by people with strong governmental affiliation, the relatively independent organisation for green buildings (i.e. the Shenzhen Green Building Council, SGBA) was initiated by architects and engineers. The nongovernmental space in Hong Kong has been powerful and continues to grow. The forces that led to its expansion were diverse, involving both business, the professions and governing elites. Both of the leading nongovernmental organisations advocating for climate change issues (i.e. Climate Change Business Forum) and the green-building accreditor (i.e. the BEAM Society) were initiated by the private sector. Meanwhile, local academics formed a nongovernmental group – the China Green Building (Hong Kong) Council – to support an alternative green building assessment. To promote electric vehicles, governing elites have attempted to seek support from academia, the private sector and local environmental nongovernmental organisations.

The horizontal *inter*governmental networks (grey dashed arrows in Figure 9.1) are weak across the municipalities. Intermunicipal interactions have occurred only in the form of formal and exclusive activities – international events such as conferences, diplomatic site visits and official exchanges – and cooperation framework agreements such as the MOU on Hong Kong–Shenzhen cooperation on the promotion of electric vehicles, the Qianhai codevelopment plan and the Framework Agreement on the Hong Kong-Guangdong Cooperation to create a 'Quality Living Area'. Masked by these formal endorsements, there is neither real discussion nor practical collaboration on climate change mitigation across municipalities. These formal and close network ties, which potentially exclude resourceful nongovernmental actors, are also diminishing any pluralistic participation in low carbon transitions.

There are merely symbolic connections among the nongovernmental organisations in the three different municipalities, with no real mutual involvement despite their geographical proximity. Both the Shenzhen Green Building Association (SGBA) and the Hong Kong Green Building Council (HKGBC) are part of an umbrella organisation, the World Green Building Council. However, the two barely knew about each other's existence, let alone their ongoing activities. An architectural engineer affiliated to the Shenzhen Institute of Building Research – the initiator of the SGBA – was once a registered BEAM professional listed in the 2012 HKGBC directory, but he was the only member from a counterpart organisation outside Hong Kong and was removed from the list in 2013. There were a few intersecretariat meetings and visits between the SGBA and the Guangdong Building Energy Efficiency Association in Guangzhou, but collaboration between them has remained minimal. Furthermore, the existing transnational municipal networking opportunities provided by the C40 or the ICLEI are not effective in connecting or bringing in representatives of Chinese cities. They are unable to mobilise the weak horizontal *inter*governmental networks.

On the contrary, the interactions along the vertical *inter*governmental relations (double-headed black arrows in Figure 9.1) are robust and active. Such intense connectivity is driven by the competition among local leaders for additional financial resources (economic incentives), excellence in their jobs (political incentives) and stronger policy support, that is, devolved from the national government. In a semi-authoritarian state context such as China's, national mandates for low carbon action are still important sources of power, despite the pioneering role of cities in climate change mitigation.

Network dynamics: coordination–collaboration nexus

The various forms of network interactions listed in the previous section are further elucidated by the nexus of coordination and collaboration in the legitimation process. Our case studies show that both coordination and collaboration have their own limitations in networking interactions.

*Intra*governmental coordination mechanisms serve as institutionalised channels of *intra*governmental communication, sectoral negotiations and resource distribution. To cope with the new governing demands of climate change mitigation, the three cities have instituted various *intra*governmental coordination structures:

leadership groups and joint-session systems on low carbon development and climate change; the promotion of building energy efficiency, green buildings and electric vehicles; and emissions reductions in the transportation industry (see Table 4.2).

The performance and capacity of these *intra*governmental coordination structures are positively determined by the political significance attached to their respective agendas, the authority and political mandates granted in *intra*governmental interactions and the internal resources (manpower) to support their routine operations. Thus, we can conclude that higher political significance, mandated authority and internal manpower result in stronger coordination, with less sectoral fragmentation than in the *intra*governmental networks.

Yet it is worth noting that effective *intra*governmental coordination does not necessarily lead to more intensive interactions in the CMNs for climate change mitigation. For example, *intra*governmental coordination in Shenzhen suffers from interdepartmental power conflicts that result in fragmented governing authority in the city, and in Hong Kong there is weak mandated authority in the *intra*governmental coordination for climate change. The *intra*governmental coordination for low carbon transitions appears to be more effective in Guangzhou than in Shenzhen or Hong Kong. However, the external support from the nongovernmental space in the governance of climate change mitigation in Guangzhou is generally weaker than in Shenzhen or Hong Kong.

Governance through CMNs is deployed not only through effective coordination but also through cross-sectoral collaboration. In addition to municipal coordination, the collaborative functions and legitimation process in the CMNs shape the interactions and add to the transformation of the environmental state in response to climate change.

Mitigating climate change requires renovating sectoral practices. Networking to enable the integration of sectoral innovations in the broader environmental state involves enhancing input legitimacy, which has shifted the governing focus from the exclusive *intra*governmental and hierarchical coordination structure towards a more pluralistic and inclusive cross-sectoral collaboration that mobilises the nongovernmental space. The collaborative functions of effective communication through information disclosure, together with resource exchanges through incentive mechanisms, are mutually reinforcing the legitimation outcomes of sectoral innovations. Input legitimacy is bolstered by widening the participation base across sectors. This relies on enhanced engagement and strategic boundary-spanning activities, not only with the core but also with the peripheral organisational networks in the nongovernmental space for sectoral innovation. Through cross-sector collaboration, input legitimacy is being transformed into output legitimacy by aligning the expected environmental and development outcomes across the public sector, the market sphere (supply/demand of sectoral innovations) and the nongovernmental space, which increases citizens' awareness by providing a bridge between the market and public policies.

In examining empirical cases of institutionalising low carbon initiatives in Chinese cities, our study also considered how output legitimacy – with regards to the expectations of environmental outcomes – is generated and stabilised to transform

the environmental state. Different actors in the CMNs may have divergent perceptions and expectations in the implementation of low carbon pilot initiatives. Prior
to the actual launch of these initiatives, consensus building takes place mainly
among the core actors to align the differences in the output legitimacy of the proposed initiatives. Given the experimental nature of low carbon pilot initiatives,
the implementers encounter uncertainties in the implementation of initiatives.
A broader range of network actors is brought into the implementation stage. To
establish an equivalent level of output legitimacy, the leading actors of the initiative use mobilisation strategies with the peripheral actors through face-to-face
meetings that engage local industry, potential sponsors and communication agencies (media and nongovernmental organisations). The final stage of institutionalisation involves multilevel legitimation with actors at the subnational, national and
transnational levels to fit the experimental practices in a consensual and politically savvy manner for future replication.

The dynamic nexus between coordination and collaboration is a defining feature
of CMNs for low carbon transitions in a semi-authoritarian regime. Collaboration
is nested within democratic values, while coordination tends to be underlined
by authoritarian leadership. With the expanded nongovernmental space, but still
under the tight grip of state control, local actors in the municipal governing of
low carbon transitions in China have developed a new governance strategy that
combines coordination and collaboration, which is reconfiguring the traditional
state-based political authority. Thus, the reinforced collaborative functions in a
semi-authoritarian regime are gradually transforming the Chinese environmental
state into a more pluralistic and legitimate structure. The emerging municipal networks empower local actors in the low carbon transition process – the uncertainties and complexities of incorporating global practices give rise to smaller-scale
experiments in localities that are inevitably managed by local actors. These local
pilot initiatives should feed back to the national authorities to devise new functions for the environmental state. However, we suggest that the existing institutional design is still inadequate to accommodate such major transitions. Enhanced
channels of communication need to be instituted to facilitate the institutionalisation of sociotechnical innovations for low carbon transitions. Despite the release
of low carbon policies and plans, the administrative structure and procedures have
to be streamlined to remove the constant barriers to institutional innovations.

Collaboration happens in various forms. As O'Leary and Vij (2012: 518) note,
'we need more precise theoretical models of behaviour' to advance the study and
practice of collaborations in public management and governance and 'to understand how collaboration is itself interpreted and practiced differently by public
administrators in many countries around the world' (O'Leary and Vij, 2012: 518).
Existing theories and research on network governance, collaboration and collaborative governance are established under the assumption of a liberal democratic
state. Our study has not ignored these assumptions but has instead attempted to
demonstrate the applicability of these theories to an authoritarian context and how
they can be modified and incorporated into a new theoretical framework that fits
the Chinese state. This study has argued that democratic values are not a necessary
precondition of collaborative network arrangements. Collaborative governance

in different political contexts encounters different challenges and thus requires tailor-made institutional designs. The internal process of interaction dynamics proposed in the original theoretical framework of CMNs (see Figure 2.2) is thus refined in Figure 9.2, which specifies the networking purposes and respective institutional requirements.

In Figure 9.2, the internal networking in coordination has two foci – process and outcome. The coordination in the governing *process* seeks to achieve dynamic

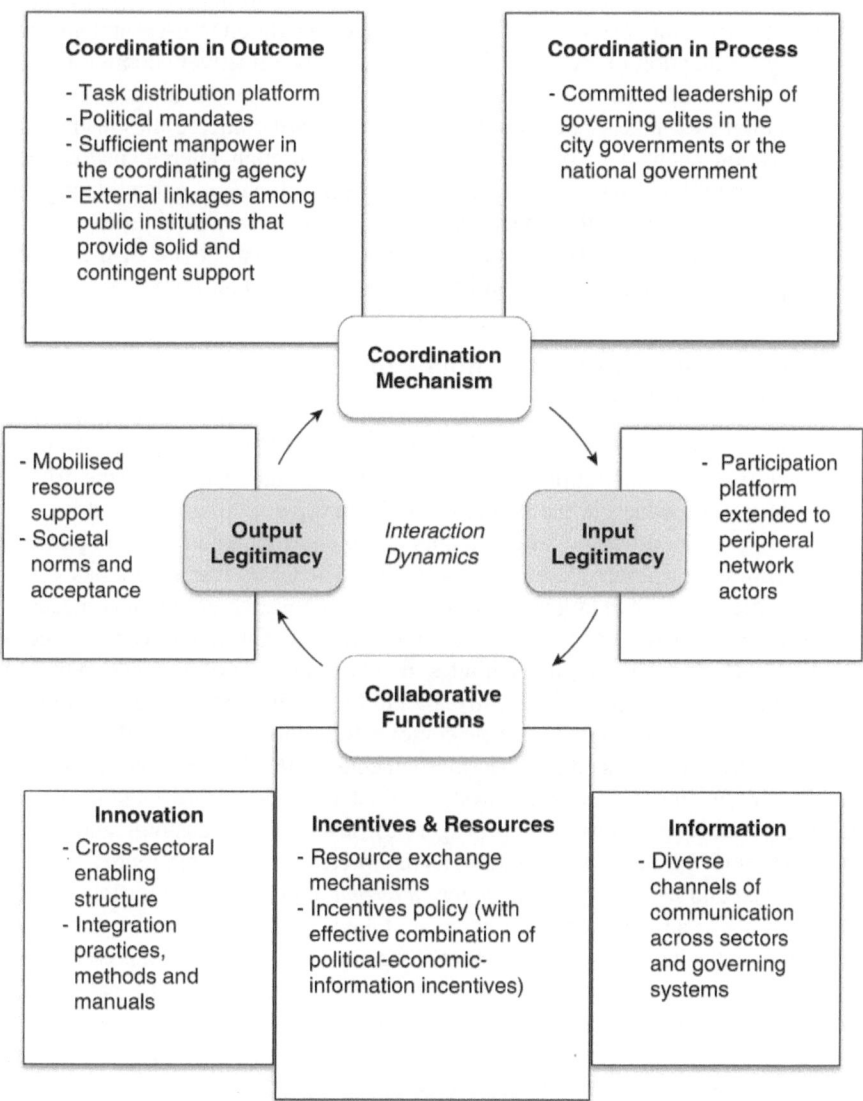

Figure 9.2 Internal process of interaction in collaborative municipal networks
Source: Compiled by authors

stability with *inter*organisational equilibrium – through negotiation, conflict resolution and balancing between coercion and consent – wherein the leadership of national and subnational governing elites is critical in streamlining transitions in sectoral practices and state functions. *Outcome* coordination is concerned with the capacity of integrating segregated policy sectors, which requires political, financial and human resources. To strengthen input legitimacy, it is necessary to institute a participation platform that enhances engagement and renders a venue for boundary-spanning activities with both core and peripheral network actors. The three collaborative functions – innovation, incentive and information – are mutually reinforcing. While the incentive policies encourage innovation transfer and integration, any promulgation of incentive policies needs to be effectively communicated across sectors to take effect. Effective communication relies on diverse channels of information exchange – at the moment, China's bureaucratic politics restrict the communication channel to official correspondence through internal instructions (*pishi*). In output legitimacy, the governance networks focus on consolidating the mobilised resources from the nongovernmental space and on establishing norms and acceptance across societal sectors regarding the critical nature of the policy problem *per se* (i.e., climate change mitigation) and the need for immediate actions.

Extrapolating the findings to China's environmental state: the governing norms

In relation to state transformations, this study examined 1) Chinese cities' capability for reconfiguring the traditional state-based political authority and 2) the drivers and barriers – embedded in the process of collaborative municipal networks – to transforming the institutional arrangements of the environmental state, in particular in a semi-authoritarian context such as China. The CMN framework has stronger explanatory power than general models of collaborative governance because it incorporates the essential aspect of coordination in the Chinese governance context. The CMN framework thus facilitates the understanding of how collaboration is 'interpreted and practiced differently' by public administrators and governance network actors in their regime-specific contexts (O'Leary and Vij 2012).

Conceptually, what constitutes the environmental state? The environmental state is essentially an 'institution' – the 'system of established and prevalent rules that structure social interactions', comprising rules, laws, norms, customs and habits (Hodgson 2006: 2). Such an institution, as interpreted in this book, represents customary social and legal rules that *structure interaction* in the CMNs for low carbon transitions. We examined how such interaction feeds back to *reconfigure* and *channel* changes in the institutional component of the environmental state. It is commonly agreed that networking organisations are not institutions, though the networking *interactions* among organisations are (Bromley 1982; Ménard 2005; North 1990; Williamson 1996). According to Sørensen and Torfing (2005: 197), such networking 'take[s] place within a relatively institutionalised framework of contingently articulated rules, norms, knowledge, and social imaginaries, that is self-regulating within limits set by external agencies; it ultimately 'contribute[s] to the production of public purpose in the broad sense of visions, ideas, plans, and regulations' (see Chapter 2 for full discussion). In investigating the interactions in CMNs, we distinguished the *norms* for collaboration, accountability, values for legitimacy and committed leadership from the *process* of resource exchange and knowledge transfer.

Several indicators were identified for measuring the transformation of the environmental state as a matter of institutional change. Because such transformation is manifested in changes in institutional capacity and institutional elements, it can be measured by the analytical aspects of *institutional feasibility*, that is, the adaptability of the mitigation actions, which covers political reality, stakeholder acceptance, social norms, legal legitimacy, bureaucratic structures, human capital, infrastructure and political will and leadership. As mentioned, the implications of our study are concerned with the social norms in relation to the restructuring of China's environmental state.

Local governments have tried to promote low carbon awareness among citizens. Government-sponsored advertisements advocating 'low carbon living' are seen everywhere, in subways, bus stations, stadia and other public venues, but these messages are often overridden by other environmental issues that are more obviously detrimental to people's daily quality of life, such as food safety, the siting of nuclear plants and air pollution. The climate change issue itself is often misinterpreted as simply the problem of air pollution, although solutions to the two environmental problems indeed have cobenefits.

The question is clearly not whether China has the 'social infrastructure for addressing climate change at the local level' to build trust, consensus and to organise collective action (Li 2011) but rather whether the climate change problem itself has been institutionalised in social norms as a 'real' problem. We conclude that this is not the case. There *are*, however, social infrastructures for organising collective action on other environmental problems, such as protests against the uranium plant in Jiangmen and against the building of an incinerator in Panyu, Guangzhou (Chan and He 2013; Watts 2009). There are also established environmental nongovernmental organisations (ENGOs) that are either government organised or independently funded.

The crux of the problem thus lies in the lack of customised social norms, that is, the limited public understanding of climate change and its potential consequences. Most independent grassroots ENGOs in Chinese cities (and there are very few) are not at all concerned with climate change mitigation issues, mainly because they lack the resources to address this complex problem. State-funded public information programmes are clearly needed. The climate change problem itself is necessarily involved with the state, whether national or subnational. The question is to what extent the state is involved, how much of an effect local governments can have and how many resources they can mobilise within the dynamic interactions of *inter*governmental relations, *intra*governmental management, sectoral integration and the collaborative functions of the CMNs.

A significant notion emphasised by institutional theorists is that if a regulation or policy is not binding, not known or not practiced by the people, it does not suffice to be an institution (Bromley 2004, 2006; Hodgson 2006). Only 'working rules' can serve as 'institutional foundations' for 'new technical opportunities' and environmental quality (Bromley 2004: 74). For new laws to become institutional rules, 'they have to be enforced to the point that the avoidance or performance of the behaviour in question becomes customary and acquires a *normative status*' (Hodgson 2006: 6).

Although there are many low carbon plans and policies promulgated by the national and subnational governments in China, this does not mean that they have actually been institutionalised in China's environmental state. For instance, a mandatory green building requirement was imposed on all new buildings in Guangzhou, but whether this requirement has been institutionalised depends on whether it is effectively enforced and has a binding effect on people. The environmental state as a system of institutional rules is composed of established environmental laws, regulations and policy instruments that have been turned into 'instances of institutional rules', that is, these laws and regulations need to become customary, and widely ignored laws or obsolete regulations are not institutional rules (Hodgson 2006).

How, then, are newly promulgated policies enabled or constrained by the interactions through the CMNs? The customary nature of institutional rules prescribes them to be 'socially transmitted', with established interaction structures (Hodgson 2006: 3), which highlights the indispensable value of the nongovernmental space in the CMNs. One of the essential findings of this book is that the nongovernmental space provides a legitimation environment for innovation transfer. As discussed in Chapter 5, in Guangzhou, the policy action regulating the mainstreaming of green buildings in the city has fallen back on the newly established GONGOs to generate discursive and input legitimacy from citizens and the professional community. In Hong Kong, the nongovernmental space renders a diverse representation of organised interests in the building sector and among business stakeholders, and this representation maintains the relatively high input legitimacy of practices and regulations on green buildings, which are continually evolving. In Shenzhen, the epistemic community has been proactive and closely connected with the local government, which has ensured the rapid development of green buildings. To sustain the initial change, the city's nongovernmental space is continually being expanded to strengthen the discursive and input legitimacy of the sectoral innovation by enabling relatively independent nongovernmental actors. This diversity in the nongovernmental space fosters interactions that integrate different segments of the CMNs, connecting administrators in the *intra*governmental mechanism with policy entrepreneurs in the nongovernmental space while at the same time tying up the technical sector and awareness advocates within the general public. Whereas an increase in the aggregate publicness of organisations (i.e. government involvement) in the nongovernmental space raises the significance of sectoral integration, an expanded scope of engagement is needed to sustain the output legitimacy of new institutions.

Although Chinese cities are not located in a liberal democratic state, it is pertinent to ask how much their nongovernmental space (as an important component of the political space in municipalities) has expanded the extent of democratic traits. If we view low carbon transitions as involving climate advocates and social movements[1] campaigning for behavioural change at multiple levels of government and in different aspects of social life – which is accommodated within the municipal political space and enabled by the CMNs – then studying the institutionalisation of low carbon innovations in the environmental state involves examining the fluid changes that occur in the governing and interaction process to implement the transition.

Magnusson (1996: 68) summarised the traits of social movements that engage people democratically in municipalities, which include 1) pluralistic – many simultaneous movements happening in one place; 2) impermanent – living on enthusiasm, prior to bureaucratisation or institutionalisation; 3) inchoate – 'no definite membership, authority relations, purposes, or programs'; 4) inclusive – open for anyone to join, with 'no definite means of excluding a person'; and 5) unbounded – 'tak[ing] in as many people, as much territory, and as many issues as seems appropriate to the people involved'. These democratic traits were found to be partially represented by the organisational interactions in the CMNs of the three Chinese cities under study: the sectoral integration has been pluralistic, involving simultaneous actions carried out by different actors with the same ultimate goal of mitigating climate change; the nongovernmental space supporting such transition has been inclusive; and the advocated norms for climate change mitigation are unbounded across territories and communities.

Magnusson argued that 'democracy is a movement within the movements', for instance, of environmentalists' advocacy for climate change awareness and mitigation practices or of feminists' campaigning for gender to become mainstreamed in decisions (1996: 69). If a democratic pursuit is a necessary yet implicit foundation for all of these movements and social transition processes, collaboration (or theories of collaborative governance) should not be grounded in any ideological claims or in the nature of the current regime, regardless of whether it is a liberal democracy or otherwise. Answering the basic question that motivated our study, *whether CMNs have reconfigured China's semi-authoritarian environmental state*, requires a conceptual debate on whether the network interactions have elicited the aforementioned democratic traits, which may gradually transform the Chinese state system. The democratic standard of political pluralism should be extended to cover pluralism among the governed, in the mobilised activities and in the governing authorities. Pluralism among the governed should focus on network participants in the expanding nongovernmental space and on their diverse representations, interests and campaign approaches. Cities are often not governed from a single focus. Therefore, especially for localities involved with world cities' connectivity, analysing pluralism in governing authorities should include not only the nation-state but also multinational corporations, global media and international authorities and rule makers such as the UN and the World Bank.

Limitations and future research

Our goal in writing this book was to advance our understanding of the governance dynamics and challenges of urban climate mitigation in China. Although we have made significant inroads into understanding how climate governance is deployed, there are still theoretical, empirical and methodological challenges in applying our theory of CMNs to multiple and comparative studies across other Chinese cities and policy sectors. Thus, there are limitations to our study that will require further research in the years ahead.

First, the implications of the governing norms and the enhanced role of collaboration in emerging governance networks are applicable to the cities we have studied and to China's other economically developed and internationally connected cities, for example, Shanghai, Beijing and Tianjin and possibly also Qingdao, Nanjing, Hangzhou and Chengdu. China exhibits large regional disparities in economic and sociotechnical development. However, the governance and trajectory of low carbon transitions in China's 'third- and fourth-tier' cities, which cover a substantial area of the country and contain large additional potential for carbon reduction, remain largely unknown. In addition, the network dynamics and interaction characteristics that we have debated in this book may not be found in other Chinese cities, given the unique sociopolitical statuses of the three cities we studied. Future research can focus on the governance of mitigation transitions in less-developed cities in China to examine how they may resemble developed, coastal cities.

Second, we focused on analysing the network interactions at the level of organisations and sectors – a relatively aggregate level of analysis. Further studies may apply a more disaggregate analytical approach, such as applying survey instruments among citizens of vulnerable areas to obtain a more in depth and complete view of the multifaceted responses – societal, technical, sectoral and geographical – to climate change in China.

Note

1 The 'social movement' perspective has been adopted to analyse the low carbon transitions in some localities of the UK (Bulkeley *et al.* 2011; Pickerill 2011; Smith 2011).

References

Bromley, D. W. 1982. Land and Water Problems: An Institutional Perspective. *American Journal of Agricultural Economics,* 64, 834–844

Bromley, D. W. 2004. Reconsidering Environmental Policy: Prescriptive Consequentialism and Volitional Pragmatism. *Environmental and Resource Economics,* 28, 73–99

Bromley, D. W. 2006. *Sufficient Reason: Volitional Pragmatism and the Meaning of Economic Institutions,* Princeton, NJ, Princeton University Press.

Bulkeley, H., Broto, V. C., Hodson, M. & Marvin, S. 2011. Introduction, in Bulkeley, H., Broto, V. C., Hodson, M. & Marvin, S. (eds.) *Cities and Low Carbon Transitions,* New York, NY, Routledge.

Chan, M. & He, H. 2013. Jiangmen Uranium Plant is Scrapped after Thousands Take Part in Protests. *South China Morning Post,* 13 July, 2013 [Online]. Available: www.scmp.com/news/china/article/1281748/jiangmen-uranium-plant-scrapped-after-protest [Accessed 3 March 2014]

Geels, F. W. 2004. From Sectoral Systems of Innovation to Socio-Technical Systems: Insights about Dynamics and Change from Sociology and Institutional Theory. *Research Policy,* 33, 897–920

Hodgson, G. M. 2006. What Are Institutions? *Journal of Economic Issues,* XL, 1–25

Li, W. 2011. Engaging with the Climate Change Regime: China's Challenges and Activities. *The China Monitor,* October (66) [Online]. Available: www.ccs.org.za/wp-content/uploads/2011/10/China_Monitor_OCT_2011_final.pdf [Accessed 7 August 2014]

Magnusson, W. 1996. *The Search for Political Space,* Cambridge, Cambridge University Press.

Ménard, C. 2005. The New Institutional Approach to Organization, in Menard, C. & Shirley, M. M. (eds.) *Handbook of New Institutional Economics,* Dordrecht: Springer.

North, D. C. 1990. *Institutions, Institutional Change and Economic Performance,* Cambridge, Cambridge University Press.

O'Leary, R. & Vij, N. 2012. Collaborative Public Management: Where Have We Been and Where Are We Going? *The American Review of Public Administration,* 42, 507–522

Pickerill, J. 2011. Building Liveable Cities: Urban Low Impact Developments as Low Carbon Solutions?, in Bulkeley, H., Broto, V. C., Hodson, M. & Marvin, S. (eds.) *Cities and Low Carbon Transitions,* New York, NY, Routledge.

Smith, A. 2011. Community-Led Urban Transitions and Resilience: Performing Transition Towns in a City, in Bulkeley, H., Broto, V. C., Hodson, M. & Marvin, S. (eds.) *Cities and Low Carbon Transitions,* New York, NY, Routledge.

Sørensen, E. & Torfing, J. 2005. The Democratic Anchorage of Governance Networks. *Scandinavian Political Studies,* 28, 195–218

Torfing, J., Peters, B. G., Pierre, J. & Sørensen, E. 2012. *Interactive Governance: Advancing the Paradigm,* Oxford, Oxford University Press.

Watts, J. 2009. Chinese Protesters Confront Police over Incinerator Plans in Guangzhou: Residents Say Government Is Lying over Health Dangers as Chinese Protesters Gain Confidence and Support. *The Guardian,* 23 November 2009 [Online]. Available: www.theguardian.com/environment/2009/nov/23/china-protest-incinerator-guangzhou [Accessed 3 March 2014]

Williamson, O. E. 1996. *The Mechanisms of Governance,* Oxford, Oxford University Press.

Appendix I

Interviews logbook

Code	Organisations	Informants
	Interview	
1/GZ/16-08-2011	Guangdong Energy Conservation Association	Vice Director
2/GZ/17-08-2011	Department of Electricity and Energy Conservation in Guangzhou Economy and Trade Commission	Deputy Director of Department
3/GZ/26-12-2011	Guangdong Low Carbon Association	General Secretary
4/GZ/26-12-2011	Guangzhou Energy Saving Academy	General Secretary and Membership Manager
5/GZ/02-02-20125/ GZ/17-02-2012	Guangdong Development and Reform Commission	Director
6/GZ/06-02-2012	SGS Group (Guangzhou) Limited	Marketing Manager in Carbon Programme
7/GZ/07-02-2012	South China Climate Change Network	Initiator
8/GZ/07-02-2012	Institute of Sustainable Communities, Guangzhou	Programme Manager
9/GZ/08-02-2012	Department of Resource and Environment in Guangzhou Development and Reform Commission	Deputy Director of Department
10/GZ/16-02-2012	Expert Committee of Low Carbon Economy for Guangdong Government	Expert Member
11/GZ/17-02-2012	Department of Vehicle Technology Management in Guangzhou Transport Commission	Director of Department
12/GZ/21-02-2012	Guangzhou Building Energy Efficiency and Wall Materials Innovation Office subordinated to Guangzhou Urban–Rural Construction Commission	Chief and Vice Chief of Section
13/GZ/22-02-2012	Guangzhou Environmental Protection Industrial Association	General Secretary

(*Continued*)

Code	Organisations	Informants
14/GZ/22-02-2012	Guangdong Building Energy Efficiency Association	Secretariat Director
15/HK/30-04-201215/ HK/18-08-201215/ HK/14-11-2012	Climate Change Policy Coordination Section, Environmental Protection Department, HKSAR Government	Environmental Protection Officer;Senior Environmental Protection Officer;
	Environment Bureau	Assistant to the Secretary for the Environment
16/SZ/07-05-2012	Department of Energy and Circular Economy in Shenzhen Development and Reform Commission	Chief of Division
17/SZ/11-05-2012	Scientific Information, Survey and Mapping Department in Shenzhen Urban Planning, Land and Resources Commission	Chief of Division
	Shenzhen Planning and Land Development Research Center	Researcher
18/SZ/14-05-2012	Shenzhen Human Habitat and Environment Commission	Environmental Protection Officer
19/SZ/14-05-2012	Shenzhen Human Habitat and Environment Commission	Manager of Greenway Development
20/SZ/23-05-2012	Office of Building Energy Efficiency in Department of Shenzhen Housing and Construction Bureau	Engineer and Chief of Division
21/SZ/24-05-2012	Department of Safety Management in Shenzhen Transport Commission	Chief of Division
22/SZ/28-05-2012	European Union Chamber of Commerce in China – PRD Chapter Shenzhen Liaison Office	Manager of Business and Government Affairs
23/SZ/29-05-2012	Shenzhen Energy Conservation Expert Alliance Committee	General Secretary
24/SZ/04-06-2012	Shenzhen Energy Conservation Association	Manager
25/SZ/18-06-2012	Shenzhen Green Building Association	General Secretary
26/HK/29-08-2012	Climate Change Business Forum, Business Environment Council, Hong Kong	Director and Research Manager
27/HK/20-08-2012	Climate Change Business Forum, Hong Kong	Initiator (former director)
28/HK/27-08-2012	China Green Building (Hong Kong) Council	Council Director
28/HK/19-11-2013	Ibid.	Ibid.
29/HK/22-10-2012	BEAM Society, Hong Kong	Initiator

(*Continued*)

Code	Organisations	Informants
30/HK/16-11-2012	Hong Kong Green Building Council	Chairman and Research Manager
30/HK/19-11-2013	Ibid.	Chairman
31/HK/25-02-2013	Hysan Development Company Limited	Property Manager
32/HK/04-02-2013	Business Environmental Council	Programme Manager
33/BE/28-05-2013	National Climate Center, China Meteorological Administration	Chief Scientist
34/BJ/06-08-2013	National Climate Center, China Meteorological Administration	Research Team on Climate Change Impact Assessment
35/BJ/08-08-2013	Division of Climate Change Economics, Institute for Urban and Environmental Studies, Chinese Academy of Social Sciences	Institute Director, Division Chief, and Researches,
36/HK/16-12-2013	Environment Bureau (Steering Committee on the Promotion of Green Buildings	Senior Environment Officer
37/HK/07-01-2014	Steering Committee on Promotion of Electric Vehicles	Member
Focus group		
1FG/HK/28-05-2012	Business Environment Council	Director, Climate Change Business Forum
	School of Energy and Environment, City University of Hong Kong	Adjunct Professor
	HSBC	Senior Corporate Sustainability Manager
	CLP Holdings Ltd.	Director, Group Environmental Affairs
	Swire Properties	Head of Technical Services and Sustainability
2FG/HK/29-05-2012	Greenpeace China	Senior Campaigner
	WWF-Hong Kong	Business Engagement Leader – Climate
		Business Engagement Manager – Climate

Note: All interviews and focus groups were conducted in Guangzhou (GZ), Shenzhen (SZ), Hong Kong (HK) and Beijing (BJ), with the exception of 33/BE/28-05-2013 (Berlin, BE). The interviews are coded according to a numeral identifier, the city where the interview took place and the date of interview.

Appendix II

Policies, laws and regulations, and government documents

Basic Law Promotion Steering Committee. 1990. The Basic Law of the Hong Kong Special Administrative Region of the People's Republic of China.

Buildings Department. 2011. *Practice Note for Authorized Persons, Registered Structural Engineers and Registered Geotechnical Engineers: Building Design to Foster a Quality and Sustainable Built Environment.* Hong Kong [Online]. Available: www.bd.gov.hk/english/documents/pnap/APP/APP151.pdf [Accessed 26 November 2013] Code: PNAP APP-151.

Buildings Department, Land Department & Planning Department. 2001. *Joint Practice Note No.1: Green and Innovative Buildings.* Hong Kong [Online]. Available: www.bd.gov.hk/english/documents/joint/JPN01.pdf [Accessed 20 March 2013] Code: JPN01.

Buildings Department, Land Department & Planning Department. 2002. *Joint Practice Note No.2: Second Package of Incentives to Promote Green and Innovative Buildings.* Hong Kong [Online]. Available: www.bd.gov.hk/english/documents/joint/JPN02.pdf [Accessed 20 March 2013] Code: JPN02.

CASS. 2003–2004. *Chinese Urban Development Report 2003–2004.* Chinese Academy of Social Sciences. Beijing, China. [Accessed 25 June 2013] (中国城市发展报告)

Census and Statistics Department. 2010a. *Hong Kong Energy Statistics Annual Report.* Hong Kong [Online]. Available: www.statistics.gov.hk/pub/B11000022010AN10B0100.pdf [Accessed 25 October 2011]

Census and Statistics Department. 2010b. *Hong Kong Statistics on Key Economic and Social Indicators.* Hong Kong [Online]. Available: www.censtatd.gov.hk/hkstat/sub/bbs.jsp [Accessed 25 October 2011]

Census and Statistics Department. 2010c. *Population and Household Statistics Analysed by District Council District.* Hong Kong [Online]. Available: www.statistics.gov.hk/pub/B11303012010AN10B0100.pdf [Accessed 25 October 2011]

Central Government, P. R. C. 2009. *State Council Executive Meeting Decided on Target of Controlling the Greenhouse Gas Emission in China.* Beijing, China [Online]. Available: www.gov.cn/ldhd/2009–11/26/content_1474016.htm [Accessed 15 June 2013] (国务院常务会研究决定我国控制温室气体排放目标)

Chief Executive. 1997–2013. *Policy Address.* Hong Kong. [Accessed 25 Feburary 2013]

CPC Central Committee. 2013. Decision on Major Issues Concerning Comprehensively Deepening Reforms.

Delegation of the European Union in China. 2012. *China's 12th Five-Year Plan 2011–2015 (English Translation).* British Chamber of Commerce in China Beijing [Online]. Available: www.britishchamber.cn/content/chinas-twelfth-five-year-plan-2011–2015-full-english-version [Accessed 12 May 2014]

Development Bureau. 2010. *Legislative Council Brief: Measures to Foster a Quality and Sustainable Built Environment.* Hong Kong [Online]. Available: www.susdev.org.hk/ susdevorg/archive2009/download/legco_brief_eng.pdf [Accessed 20 March 2013] Code: DEVB(PL-CR) 12/2010.

EMSD. 2012. *Hong Kong Energy End-Use Data 2012.* The Energy Efficiency Office of Electrical & Mechanical Services Department. Hong Kong [Online]. Available: www. emsd.gov.hk/emsd/e_download/pee/HKEEUD2012.pdf [Accessed 12 May 2013]

Environment Bureau, H. K. 2008. *Progress Report – Motion Debate on 'Responding to the Problem of Climate Change' Legislative Council Meeting on 28 November 2007.* Legislative Council. Hong Kong [Online]. Available: www.legco.gov.hk/yr07–08/english/ counmtg/motion/cm1128-m2-prpt-e.pdf [Accessed 20 April 2012]

Environment Bureau, H. K. 2009. *Administration's Paper on Promoting the Use of Electric Vehicles.* Legislative Council Panel on Environmental Affairs. Hong Kong [Online]. Available: www.legco.gov.hk/yr08–09/english/panels/ea/papers/ea0622cb1–1945–4-e. pdf [Accessed 25 March 2013]

Environment Bureau, H. K. 2010a. *Hong Kong's Climate Change Strategy and Action Agenda – Consultation Document.* Environment Bureau and Government Logistics Department, HKSAR. Hong Kong [Online]. Available: www.epd.gov.hk/epd/english/ climate_change/files/Climate_Change_Booklet_E.pdf [Accessed 20 April 2012]

Environment Bureau, H. K. 2010b. *Proposal to Increase the Share of Nuclear Power in the Fuel Mix.* Legislative Council – Panel on Environmental Affairs. Hong Kong [Online]. Available: www.epd.gov.hk/epd/english/news_events/legco/files/EA_Panel_101022b_ eng.pdf [Accessed 20 April 2012]

Environment Bureau, H. K. 2011a. *Administration's Paper on Consultancy Report – a Study of Climate Change in Hong Kong.* Legislative Council, Panel on Environmental Affairs. Hong Kong [Online]. Available: www.legco.gov.hk/yr10–11/english/panels/ea/papers/ ea0228cb1–1370–5-e.pdf [Accessed 20 April 2012] Code: CB(1)1370/10–11(05).

Environment Bureau, H. K. 2011b. *The Impacts of the Development of Nuclear Energy for Local Power Generation on Hong Kong.* Legislative Council – Panel on Environmental Affairs. Hong Kong [Online]. Available: www.epd.gov.hk/epd/english/news_events/legco/ files/EA_Panel_110429a_eng.pdf [Accessed 20 April 2012] Code: CB(1) 2022/10–11(14).

Environment Bureau, H. K. 2013. *2013 Policy Address: Policy Initiatives of Environment Bureau – Environmental Protection.* Legislative Council – Panel on Environmental Affairs. Hong Kong [Online]. Available: www.epd.gov.hk/epd/english/news_events/ legco/files/EA_Panel_20130128a_eng.pdf [Accessed 20 November 2013]

Financial Secretary. 2007–2012. *Budgetary Reports.* Hong Kong.

Gazette. 1994. *Moter Vehicles (First Registration Tax) Ordinance 1994 Amendment.* Legislative Council. Hong Kong [Online]. Available: www.legislation.gov.hk/blis_ind.nsf/ curallengdoc/4DF141105CB2FCD4482579D7000FA331?OpenDocument [Accessed 11 Dec 2013] Code: Cap.330 Sec.5 (35 of 1994 s.2).

GD-DRC. 2008. *People's Action Plan for Guangdong Low Carbon Economy.* Guangzhou, Guangdong, Provincial Government of Guangdong Province.

GD-DRC. 2011. *Annual Report on the Key Working Areas of Low Carbon Pilot Province (2011).* Guangzhou, Guangdong, Provincial Government of Guangdong Province.

GD-DRC. 2012. *Annual Report on the Key Working Areas of Low Carbon Pilot Province (2012).* Guangzhou, Guangdong, Provincial Government of Guangdong Province.

Guangdong Government. 2007. *Energy Conservation and Emission Reduction Comprehensive Strategy.* Guangzhou, Guangdong. (广东省节能减排综合行动工作方案)

Guangdong Government. 2011a. *Guangdong's 12th Five-Year Plan.* Guangzhou, Guangdong. (广东省"十二五"规划纲要)

Guangdong Government. 2011b. *Guangdong Climate Change Programme*. Guangzhou, Guangdong. (广东省应对气候变化方案)

Guangdong Government. 2012. *The Implementation Strategy of Low Carbon Pilot Province*. Guangzhou, Guangdong.

Guangdong Provincial Government. 2013. *Guangdong Action Plan on Green Building*. Guangdong [Online]. Available: zwgk.gd.gov.cn/006939748/201311/t20131115_452660.html [Accessed 26 November 2013] (广东省绿色建筑行动实施方案) Code: 粤府办 (2013)49号.

Guangzhou Municipal Government. 2010. *Guidelines on Rapid Development of Low Carbon Economy*. Guangzhou Municipal Government. Guangzhou [Online]. Available: www.gz.gov.cn/publicfiles/business/htmlfiles/gzgov/s2811/201010/694228.html [Accessed 25 March 2013] (关于大力发展低碳经济的指导意见)

Guangzhou Municipal Government. 2011a. *Guangzhou 12th Five-Year Plan*. Guangzhou. (广州市国民经济和社会发展第十二个五年规划)

Guangzhou Municipal Government. 2011b. *Guangzhou 12th Five-Year Plan*. Guangzhou. (广州市国民经济和社会发展第十二个五年规划纲要) Code: 穗府(2011)4号.

Guangzhou Municipal Government. 2012. *Guangzhou Municipal Government Notice on Accelerating Development of Green Building*. Guangzhou [Online]. Available: http://sfzb.gzlo.gov.cn/sfzb/file.do?fileId = FF8080813658A88201365C161C62000C [Accessed 20 April 2012] (广州市人民政府关于加快发展绿色建筑的通告) Code: 穗府 (2012)1号.

Guangzhou Municipal Government. 2013. *Guangzhou Management Regulations on Green Building and Building Energy Efficiency*. Mayor Office [Online]. Available: zwgk.gd.gov.cn/007482532/201304/t20130412_372145.html [Accessed 26 November 2013] (广州市绿色建筑和建筑节能管理规定) Code: 广州市人民政府令(第92号).

Guangzhou Municipal Party Committee. 2010a. *Guangzhou's Implementation Details of P.R.D. Outline Plan*. Available: www.gzplan.gov.cn/rdzt/zsjfzgh/zywj/201002/t20100206_9245.htm [Accessed 6 Febuary 2010] (广州市贯彻落实《珠江三角洲地区改革发展规划纲要 (2008–2020年)》实施细则) Code: 穗字 (2010)5号.

Guangzhou Municipal Party Committee. 2010b. *Guangzhou's Work Programme of Implementing the P.R.D. Outline Plan and Realizing the 'Four-Years Grand Development'*. Guangzhou [Online]. Available: www.gz.gov.cn/publicfiles/business/htmlfiles/gzgov/s2810/201101/749008.html (广州市实施〈珠江三角洲地区改革发展规划纲要(2008–2020)〉实现"四年大发展"工作方案) Code: 穗字(2010)17号.

Guangzhou Statistics Bureau. 2007–2012. *Guangzhou Statistical Yearbook*. Guangzhou Statistics Bureau. Guangzhou.

Guangzhou Statistics Bureau. 2007–2012. Expenditure of Local Government (2006–2011), in *Guangzhou Statistical Yearbooks 2007–2012*. Guangzhou Statistics Bureau. Guangzhou.

GZ-DRC. 2011a. *Guangzhou Work Plan of Implementing the <Pearl River Delta Outline Development Plan> to Realize 'Four-Year Grand Development'* Guangzhou [Online]. Available: www.gzplan.gov.cn/rdzt/zsjfzgh/gyjd/201101/t20110121_12365.htm (广州市实施《珠江三角洲地区改革发展规划纲要》实现"四年大发展"工作方案有关情况)

GZ-DRC. 2011b. *Management Methods of Guangzhou Energy Efficiency Special Fund* Guangzhou [Online]. Available: http://sfzb.gzlo.gov.cn/sfzb/file.do?fileId = 71EA51E50D1047CC9C893E3D7FC84D51 [Accessed 20 April 2012] (关于印发广州市节能专项资金管理办法的通知) Code: 穗发改资环(2011)11号.

GZ-EPD. 2011. *2011 Year-End Departmental Financial Settlement of Guangzhou Environmental Protection*. Guangzhou Environmental Protection Department Guangzhou Municipal Government. Guangzhou [Online]. Available: www.gzepb.gov.cn/zwgk/gs/cgyzb/201211/P020121122641497161308.pdf [Accessed 15 June 2013] (广州市环境保护局 (2011)年部门决算)

GZ-URCC. 2010. *Guidelines of Green Building Development in Guangzhou.* Guangzhou Urban–Rural Construction Commission. Guangzhou [Online]. Available: www.gzcc.gov. cn/zwgk/zhzxinfo.aspx?id = 39149&name = %E5%9F%8E%E5%BB%BA%E5%8A %A8%E6%80%81 [Accessed 18 November 2013] (*广州市发展绿色建筑指导意见*) Code: 穗建技函(2010)2047号.

GZ-URCC. 2011. *Guangzhou Building Energy Efficiency Special Plan in 12th Five-Year Plan Period.* Guangzhou Building Energy Efficiency Leadership Group. Guangzhou [Online]. Available: www.gzcc.gov.cn/back/edit/uploadfile/20111014165411511.doc [Accessed 20 April 2012] (*广州市建筑节能"十二五"专项规划*)

HK-EPD. 2008. *Estimates of Government Resources Devoted to Environmental Protection and Conservation Work in 2007.* Environmental Protection Department. Hong Kong [Online]. Available: www.epd.gov.hk/epd/english/resources_pub/spending/archives_2007. html [Accessed 15 June 2013]

HK-EPD. 2009. *Estimates of Government Resources Devoted to Environmental Protection and Conservation Work in 2008.* Environmental Protection Department. Hong Kong [Online]. Available: www.epd.gov.hk/epd/english/resources_pub/spending/archives_2008. html [Accessed 15 June 2013]

HK-EPD. 2010. *Estimates of Government Resources Devoted to Environmental Protection and Conservation Work in 2009.* Environmental Protection Department. Hong Kong [Online]. Available: www.epd.gov.hk/epd/english/resources_pub/spending/archives_2009. html [Accessed 15 June 2013]

HK-EPD. 2011. *Estimates of Government Resources Devoted to Environmental Protection and Conservation Work in 2010.* Environmental Protection Department. Hong Kong [Online]. Available: www.epd.gov.hk/epd/english/resources_pub/spending/archives_2010. html [Accessed 15 June 2013]

HK-EPD. 2012. *Estimates of Government Resources Devoted to Environmental Protection and Conservation Work in 2011.* Environmental Protection Department. Hong Kong [Online]. Available: www.epd.gov.hk/epd/english/resources_pub/spending/spending_govt.html [Accessed 15 June 2013]

HK-EPD. 2013a. *Greenhouse Gas Emissions and Carbon Intensity in Hong Kong.* Hong Kong [Online]. Available: www.epd.gov.hk/epd/english/climate_change/files/HKGHG_ CarbonIntensity_201304.pdf [Accessed 1 May 2013]

HK-EPD. 2013b. *Greenhouse Gas Emissions in Hong Kong by Sector.* Hong Kong [Online]. Available: www.epd.gov.hk/epd/english/climate_change/files/HKGHG_Sectors_ 201304.pdf [Accessed 12 May 2013]

Hong Kong Housing Authority. 2010. *Housing in Figures.* Hong Kong.

Legislative Council. 2013. *Item for Finance Committee: New Item* "Capital Injection into the Environment and Conservation Fund" *(Approved 14 June 2013).* Finance Committee, Legislative Council. Hong Kong [Online]. Available: www.legco.gov.hk/yr12–13/ english/fc/fc/papers/f13–15e.pdf [Accessed 15 June 2013] Code: FCR(2013–14)15.

MOHURD. 2006. *Evaluation Standard for Green Buildings.* Ministry of Housing and Urban–Rural Development. Beijing, China. (*绿色建筑评价标准*) Code: GB/T 50378–2006.

MOHURD & MOF. 2012. *Implementation Opinions on Accelerating Green Building Development in China.* Ministry of Housing and Urban–Rural Development, Ministry of Finance. Beijing, China [Online]. Available: jjs.mof.gov.cn/zhengwuxinxi/tongzhigonggao/ 201205/t20120507_648962.html [Accessed 26 November 2013] (*关于加快推动我国 绿色建筑发展的实施意见*) Code: 财建(2012)167号.

MOT. 2011. *Annex of the Notice on Carrying Out Low Carbon Transportation System Pilot Cities.* Ministry of Transport. Beijing, China.

MOT. 2012. *Notice on Carrying out the Low Carbon Transportation System Pilot Cities (Second Batch).* Ministry of Transport. Beijing, China [Online]. Available: www.moc.gov.cn/zhuzhan/zhengwugonggao/jiaotongbu/qita/201202/t20120207_1191546.html [Accessed 17 March 2013] (关于开展低碳交通运输体系建设第二批城市试点工作的通知)

NDRC. 2005. *Initial Communications on Climate Change.* National Development and Reform Commission. Beijing, China.

NDRC. 2007. *China's National Climate Change Programme.* National Development and Reform Commission. Beijing, China.

NDRC. 2008. *The Outline of the Plan for the Reform and Development of the Pearl River Delta (2008–2020)* Beijing, China.

NDRC. 2008–2012. *Annual Report on China's Climate Policies and Actions.* Beijing, China.

NDRC. 2009. *China's Policies and Actions on Climate Change: Annual Progress Report 2009.* National Development and Reform Commission. Beijing, China. (中国应对气候变化的政策与行动 – 2009 年度报告)

NDRC. 2010. *NDRC Notice on Implementing Public Programmes for Low Carbon Provinces and Cities.* Beijing, China [Online]. Available: www.sdpc.gov.cn/zcfb/zcfbtz/2010tz/t20100810_365264.htm [Accessed 25 October 2013] (国家发展改革委关于开展省区和低碳城市试点工作的通知) Code: 发改气候(2010)1587 号.

NDRC & DECC. 2011. *Memorandum of Understanding between the National Development and Reform Commission of the People's Republic of China and the Department of Energy and Climate Change of the United Kingdom of Great Britain and Northern Ireland Concerning Co-Operation on Low-Carbon.* Climate Change and Energy Section, British Consulate General, Guangzhou. China–UK [Online]. Available: http://ukinchina.fco.gov.uk/resources/zh/pdf/21023937/534335182 [Accessed 20 April 2012]

Patten, C. 1993–1996. *Address by the Governor.* Hong Kong.

Planning Department. 2012. *Broad Land Usage Distribution.* Hong Kong.

SDC. 2010. *Council for Sustainable Development: Report on the Public Engagement Process on Building Design to Foster a Quality and Sustainable Built Environment* [Online]. Available: www.susdev.org.hk/susdevorg/archive2009/download/councilreport_june2010_eng.pdf [Accessed 20 March 2013]

Shenzhen Construction Department. 1995. *Shenzhen Management Measure of Wall Reform Bail and Wall Reform Foundation* [Online]. Available: http://203.91.55.40:9000/b5/www.sz.gov.cn/zfgb/2008/gb622/200810/t20081030_231448.htm [Accessed 24 March 2013] (深圳市墙改保证金墙改基金管理办法) Code: 深建字(1995)206 号.

Shenzhen Department of Quality and Technical Supervision. 2009. *Evaluation Specification of Green Buildings.* Shenzhen. (绿色建筑评价规范) Code: SZJG 30–2009.

Shenzhen Finance Bureau. 2011. *Approval of the Revenue & Expenditure Plan About 2011 New Wall Materials Special Fund* Shenzhen. (关于2011年度新型墙体材料专项基金收支计划的批复) Code: 深财居(2011)57号.

Shenzhen Human Habit and Environment Commission. 2011. *Announcemet of Projects Supported by 2011 Environmental Protection Special Fund.* Shenzhen [Online]. Available: http://203.91.55.40:9000/b5/www.sz.gov.cn/cn/xxgk/zjxx/gbmzxjj/201112/t20111231_1795856.htm [Accessed 15 June 2013] (2011年度环境保护专项资金资助项目公告)

Shenzhen Municipal Government. 2008a. *Action Plan of Constructing Green Building Metropolis.* Shenzhen Housing and Construction Department. Shenzhen [Online]. Available: www.sz.gov.cn/zfgb/2008/gb590/200810/t20081019_93517.htm [Accessed 25 May 2012] (关于打造绿色建筑之都的行动方案) Code: 深府(2008)42 号.

Shenzhen Municipal Government. 2008b. *Constructing Joint Session System on Promoting Building Energy Efficiency and Developing Green Building in Shenzhen.* Shenzhen Housing and Construction Department. Shenzhen [Online]. Available: www.sz.gov. cn/zfgb/2008/gb601/200810/t20081019_93400.htm [Accessed 31 May 2012] (建立深圳市推行建筑节能和发展绿色建筑联席会议制度) Code: 深府办(2008)60 号.

Shenzhen Municipal Government. 2009a. *Shenzhen New Energy Industry Revitalization and Development Plan (2009–2015)* [Online]. Available: www.sz.gov.cn/zfgb/2010/gb681/201001/t20100112_1424158.htm [Accessed 5 Feburary 2014] (深圳新能源产业振兴发展规划(2009–2015年)) Code: 深府(2009)239 号.

Shenzhen Municipal Government. 2009b. *Shenzhen New Energy Industry Revitalization and Development Policy.* Guangdong Provincial Government. (深圳新能源产业振兴发展政策) Code: 深府(2009)240号.

Shenzhen Municipal Government. 2010a. *Establishing Leadership Group on Responding to Climate Change, Energy Efficiency, and Emission Reduction Works in Shenzhen.* Shenzhen Municipal Government Secretariat. Shenzhen [Online]. Available: www.sz.gov. cn/zfgb/2011/gb728/201101/t20110125_1632591.htm [Accessed 20 April 2012] (成立深圳市应对气候变化及节能减排工作领导小组) Code: 深府办(2010)112.

Shenzhen Municipal Government. 2010b. *Overall Development Plan on Shenzhen-Hong Kong Cooperation on Modern Service Industries in Qianhai Area.* Shenzhen [Online]. Available: www.sz.gov.cn/cn/xxgk/xwfyr/wqhg/20101220/ [Accessed 28 January 2014]

Shenzhen Municipal Government. 2011a. *Policy Programme of Co-Building Low-Carbon Ecological City.* Shenzhen Urban Planning, Land and Resources Commission (SZ-UPLRC). Shenzhen [Online]. Available: www.suprc.org/items/items_686.aspx?typeid=5 [Accessed 31 May 2012] (国家住房和城乡建设部与深圳市人民政府共建国家低碳生态示范市工作方案)

Shenzhen Municipal Government. 2011b. *The Shenzhen 12th Five-Year Plan.* Shenzhen.

Shenzhen Municipal Government. 2012. *Shenzhen 12th Five-Year Plan on Energy Efficiency.* Shenzhen.

Shenzhen Statistics Bureau. 2007, 2008, 2010, 2012. *Shenzhen Statistical Yearbook.* Shenzhen Statistics Bureau. Shenzhen.

State Council, P. R. C. 2009. *Automotive Industry Readjustment and Revitalization Plan.* State Council Office. Beijing, China [Online]. Available: www.gov.cn/zwgk/2009–03/20/content_1264324.htm [Accessed 3 May 2013] (汽车产业调整和振兴规划)

State Council, P. R. C. 2010. *The State Council's Approval Reply About the Overall Development Plan on Shenzhen-Hong Kong Cooperation on Modern Service Industries in Qianhai Area.* Beijing [Online]. Available: www.tid.gov.hk/english/aboutus/tradecircular/cic/asia/2010/files/ci2010616a.pdf [Accessed 28 January 2014] (国务院关于前海深港现代服务业合作区总体发展规划的批复)

State Council, P. R. C. 2011. *The 12th Five-Year Plan for National Economic and Social Development of the People's Republic of China.* Xinhua News Agency. Beijing, China [Online]. Available: www.cmab.gov.hk/doc/12th_5yrsplan_outline_full_text.pdf [Accessed 3 May 2013] (中华人民共和国国民经济和社会发展第十二个五年规划纲要)

SZ-DRC. 2012. *Medium and Long Term Plan of Shenzhen Low Carbon Development (2011–2020).* Shenzhen Development and Reform Commission. Shenzhen. (深圳市低碳发展中长期规划) Code: 深发改(2012)512.

SZ-HCD. 2010. *Notice on Applying Green-Building Standards to and Implementing Energy-Efficiency Measures in the Construction of Shenzhen's Indemnificatory Housing.* Shenzhen Housing and Construction Department. Shenzhen. (关于我市保障性住房应按照绿色建筑标准建设并落实节能减排措施的通知) Code: 深建节能(2010)131号.

SZ-TC. 2011. *Implementation Programme of Constructing Pilot Low Carbon Transportation System.* Shenzhen Transportation Commission. Shenzhen.

SZ-TC & MOT. 2011. *Shenzhen Implementation Program for Building Low Carbon Transportation System Pilot.* Shenzhen. (深圳市建设低碳交通运输体系试点实施方案)

SZ-UPLRC. 2011. *Shenzhen White Paper of Creating National Low Carbon Ecological Demonstration City.* Shenzhen Urban Planning, Land and Resources Commission. Shenzhen. (深圳创建国家低碳生态示范市白皮书(2010–2011))

The Editorial Committee for Shenzhen Real Estate Yearbook. 2010. *Shenzhen Real Estate Yearbook 2010.* Haitian Press. Shenzhen.

The Editorial Committee for Shenzhen Real Estate Yearbook. 2011. *Shenzhen Real Estate Yearbook 2011.* Haitian Press. Shenzhen.

Transport and Housing Bureau. 2012. *Legislative Council Brief on Road Traffic Ordinance (Chapter 374) Amendments.* Legislative Council. Hong Kong [Online]. Available: www.legco.gov.hk/yr11–12/english/subleg/brief/77–79_brf.pdf [Accessed 6 May 2013]

Index

Note: page numbers with *f* indicate figures; those with *t* indicate tables.